THE EFFECTIVE CRYSTAL FIELD POTENTIAL

THE EFFECTIVE CRYSTAL FIELD POTENTIAL

JACEK MULAK and **ZBIGNIEW GAJEK**
*W. Trzebiatowski Institute of Low Temperature and
Structure Research, Polish Academy of Sciences,
Wrocław, Poland*

2000

ELSEVIER
Amsterdam - Lausanne - New York - Oxford - Shannon - Singapore - Tokyo

ELSEVIER SCIENCE Ltd
The Boulevard, Langford Lane
Kidlington, Oxford OX5 1GB, UK

First edition 2000

Library of Congress Cataloging in Publication Data

British Library Cataloguing in Publication Data
A catalogue record from the British Library has been applied for.

Mulak, J.
    The effective crystal field potential / Jacek Mulak and Zbigniew Gajek.
      p. cm.
    Includes bibliographical references and index.
    ISBN 0-08-043608-0 (hc)
      1. Crystal field theory. I. Gajek, Zbigniew. II. Title.

    QD475 .M765 2000
    538'.43--dc21
                                                                    00-028030

ISBN: 0 08 043608 0

♾ The paper used in this publication meets the requirements of ANSI/NISO Z39.48-1992 (Permanence of Paper).

Printed and bound in the United Kingdom
Transfered to Digital Printing, 2011

PREFACE

As it results from the very nature of things, the spherical symmetry of the surroundings of a site in a crystal lattice or an atom in a molecule can never occur. Therefore, the eigenfunctions and eigenvalues of any bound ion or atom have to differ from those of spherically symmetric respective free ions. In this way, the most simplified concept of the crystal field effect or ligand field effect in the case of individual molecules can be introduced.

The conventional notion of the crystal field potential is narrowed to its non-spherical part only through ignoring the dominating spherical part which produces only a uniform energy shift of gravity centres of the free ion terms. It is well understood that the non-spherical part of the effective potential "seen" by open-shell electrons localized on a metal ion plays an essential role in most observed properties. Light absorption, electron paramagnetic resonance, inelastic neutron scattering and basic characteristics derived from magnetic and thermal measurements are only examples of a much wider class of experimental results dependent on it. The influence is discerned in all kinds of materials containing unpaired localized electrons: ionic crystals, semiconductors and metallic compounds including materials as intriguing as high-T_c superconductors or heavy fermion systems. It is evident from the above that we deal with a widespread effect relative to all free ion terms except those which can stand the lowered symmetry, e.g. S-terms.

Despite the universality of the phenomenon, the available handbooks on solid state physics pay only marginal attention to it, merely making mention of its occurrence. This comes as no surprise, since this is a pure local effect relative to the point symmetry of ionic core and considering fundamental solid state characteristics, i.e. its translational symmetry and band structure, it is in fact a problem of secondary importance. On the other hand the crystal field effect is somehow involved in the majority of studies on the properties of solid state. This glaring disproportion between the "supply and demand" prompted us to take the writing of this monograph in hand.

During the period of over seventy years, which have passed since the crystal field concept was introduced, the crystal field potential (and effect) itself has not been a primary problem in the majority of studies on solid state physics and chemistry. The best works, based on experimental data, have been confined to phenomenological parameterization of crystal field potential that, exploiting only the point symmetry of the paramagnetic ion, quantitatively reproduced its energy level spectrum. Unfortunately, no clear physical meaning could be attributed to parameters themselves. Moreover, there is a menace, commonly met, of incorrect parameterization resulting from the diversity of formally equivalent sets of the parameters, i.e. from co-existence of several local minima of fitting procedure. Verification of such uncertain parameterizations is, in principle, impossible without a quantitative or at least semi-quantitative estimation of the actual parameters. This is the main aspect of the monograph.

Present understanding of the origins of the crystal field potential differs essentially from the pioneering electrostatic picture postulated in the twenties. The considerable development of the theory that has been put forward since then can be traced in many regular articles scattered throughout the literature. The last two decades have left their impression as well but, to the authors' best knowledge, this period has not been closed with a more extended review. This has also motivated us to compile the main achievements in the field in the form of a book.

The state of knowledge on the crystal field effect is far from satisfactory. The quantitative inadequacy of the models used still remains a challenge to solid state physicists and chemists. Two basic approaches put into practice, one based on perturbation expansion of the effective Hamiltonian and the other developed in terms of the global charge density, are unfortunately stipulated by a number of rough approximations some of which are formulated on the ground of heuristic only arguments. Simultaneously, the approaches become more and more complex and cumbersome to be recommended to experimentalists for interpretation practice.

There is a common request for a well argued simplified model of crystal field, at least a semi-empirical one, which could be derived from the exact sophisticated models. To this end, again, the recipe for the ab initio calculation of the crystal field parameters is unavoidable. We hope that in this respect the book will be particularly helpful. All known basic contributions to the crystal field potential for both ionic and metallic systems are individually discussed. The outlined evolution of the approaches from the point charge model via several intermediate ones up to the more commonly, recently employed density functional theory is especially instructive.

All chapters are in principle autonomous although chapter four entitled: *"Ionic complex or quasi-molecular cluster. Generalised product function"* plays a key role. Nevertheless, the entirety of the problem may properly be seen only from the point of view of all nineteen chapters.

This book is, first of all, devoted to experimentalists: mainly spectrocopists, EPR men and people engaged in inelastic neutron scattering, magnetic measurements and Schottky effect in heat capacity. The main emphasis is directed toward a method of calculating values of the crystal field parameters, explaining their, often unexpected, variability and giving directions for acceptable choice of the parameters.

Some acquaintance with the basic solid state physics, elements of quantum chemistry, vector coupling theory and the group theory, including the three dimensional rotational group, is assumed.

Many people gave their contribution to this enterprise. It gives us great pleasure to thank Dr M. Mulak for his important contribution to writing this book. We are grateful to Prof. W. Suski who recommended us to Elsevier Science Ltd.

J. Mulak, Z. Gajek
Wroclaw, March, 2000

CONTENTS

1. Introduction   **1**

2. Parameterization of crystal field Hamiltonian   **11**
   2.1. Operators and parameters of the crystal field Hamiltonian 12
      2.1.1. Operators related to spherical harmonics 13
      2.1.2. Operators related to tesseral harmonics 13
   2.2. Basic parameterizations 14
      2.2.1. $C_q^{(k)}$ tensor operator parameterization 14
      2.2.2. O_k^q Stevens operator equivalent method 16
   2.3. Symmetry transformations of the operators 18
   2.4. The number of independent crystal field parameters 23
   2.5. Standardization of the crystal field Hamiltonian 26
   2.6. Final remark 29

3. The effective crystal field potential.
Chronological development of crystal field models   **31**

4. Ionic complex or quasi-molecular cluster.
Generalized product function   **41**
   4.1 Concept of the generalized product function 41
   4.2 The density functions and the transition density functions 43
   4.3 Model of the generalized product functions 44
   4.4 Crystal field effect in the product function model 49

5. Point charge model (PCM)   **53**
   5.1 PCM potential and its parameters 53
   5.2 Simple partial PCM potentials 56
   5.3 Extension of PCM – higher point multipole contribution 61

6. One-configurational model with neglecting
the non-orthogonality. The charge penetration
and exchange effects   **65**
   6.1 Classical electrostatic potential produced by the ligand
     charge distribution 65
   6.2 The charge penetration effect and the exchange interaction
     in the generalized product function model 68
   6.3 The weight of the penetration and exchange effects
     in the crystal field potential 71
   6.4 Calculation of the two-centre integrals 73
   6.5 Final remarks 74

7. **The exclusion model. One-configurational approach with regard to non-orthogonality of the wave functions**   **77**

7.1 Three types of the non-orthogonality 77

7.2 The renormalization of the open-shell Hamiltonian H_a owing to the non-orthogonality of the one-electron functions 79

7.3 The contact-covalency – the main component of the crystal 84 field potential ....

7.4 The contact-shielding 87

7.5 The contact-polarization 88

7.6 Mechanisms of the contact-shielding and contact-polarization in terms of the exchange charge notion 88

8. **Covalency contribution, i.e. the charge transfer effect**   **91**

8.1 The one-electron excitations. Group product function for the excited state 91

8.2 The renormalization of the open-shell Hamiltonian due to the covalency effect 94

8.3 Basic approximations 96

8.4 The one-electron covalency potential V^{cov} 97

8.5 The one-electron covalency potential V^{cov} in the molecular-orbital formalism 101

8.6 Remarks on the covalency mechanism 102

9. **Schielding and antishielding effect: contributions from closed electron shells**   **105**

9.1 Phenomenological quantification of the screening effect 106

9.2 Microscopic model of the screening effect 107

9.3 General expressions for the screening factors 109

9.4 The screening factors 116

10. **Electrostatic crystal field contributions with consistent multipolar effects. Polarization**   **119**

10.1 Expansion of the electrostatic potential of point charge system into the multipole series 119

10.2 Extended formula for the crystal field parameters including all multipole moments of the surroundings 121

10.3 The self-consistent system of permanent and induced multipole moments in crystal lattice 126

10.4 The off-axial polarization terms in local coordinate systems . 127

10.5 Typical examples of dipole and quadrupole polarization contributions to the crystal field potential 129

**11. Crystal field effect in the Stevens
    perturbation approach**   **131**
    11.1 The Wannier functions 132
    11.2 The perturbation scheme for degenerate systems
        employing projection operators 133
    11.3 The crystal field effect 136

**12. Specific mechanisms of metallic states
    contributing to the crystal field potential**   **143**

13. Screening the crystal field in metallic materials   **147**
    13.1 The Fourier form of the crystal lattice potential 149
    13.2 The dielectric static screening function $\varepsilon(\mathbf{q})$ 153
    13.3 The dynamic mechanism of the screening –
        zero-point plasmon 159

**14. Virtual bound state contribution to the
    crystal field potential**   **163**
    14.1 The resonance scattering of conduction
        electrons by a central potential 163
    14.2 The nature of the virtual bound state 166
    14.3 Spin-polarization of the virtual bound state 167
    14.4 Experimental manifestations of existing
        the virtual bound states and methods of
        estimating their localization degree 167
    14.5 The crystal field splitting of the virtual bound state 168
    14.6 The primary crystal field effect relative to
        the open-shell states $(4f)$ 169
    14.7 Corrections to the simple model of the
        virtual bound state mechanism 174

**15. Hybridization or covalent mixing between
    localized states and conduction band
    states in metallic crystals**   **177**
    15.1 The essence of the hybridization 177
    15.2 Hybridization contribution to the crystal
        field parameters 178
    15.3 The scale of the hybridization effect 182
    15.4 Contribution to the crystal field potential
        from a split-off state from the conduction band
        in impurity systems 184

16. Density functional theory approach   **185**
16.1 Electron density as a key variable 185
16.2 The Kohn–Sham equations 188
16.3 Local density approximation 190
16.4 Extensions 192
    16.4.1 Degenerate ground state and excited states 192
    16.4.2 Multicomponent system 196
    16.4.3 Local spin density approximation 197
    16.4.4 Relativistic effects 198
16.5 Exchange-correlation energy 201
    16.5.1 Self-interaction correction 201
    16.5.2 Generalized gradient approximation 203
16.6 Mapping DFT on effective Hamiltonian 204
16.7 Applications 206
    16.7.1 Ionic compounds 206
    16.7.2 Intermetallic compounds 207
    16.7.3 Final remarks 208

17. Analysis of the experimental data. Interpretation of
crystal field parameters with additive models   **211**
17.1 Phenomenological Hamiltonian 212
17.2 Simplified crystal field models 215
    17.2.1 Decomposition of the CF potential. Virtual ligands ... 216
    17.2.2 Superposition model and angular overlap model 218
    17.2.3 Limitations 221
    17.2.4 Non-equivalent ligands 223
17.3 Towards applications 224

18. Lattice dynamics contribution   **229**
18.1 Adiabatic and harmonic approximations 230
18.2 Collective (normal) coordinates and
    the "quasi-molecular" model 233
18.3 The Jahn-Teller effect 235
18.4 Lattice dynamics and the crystal field effect 241

19. Extension of the crystal field potential
beyond the one-electron model   **247**
19.1 Two-electron correlation effect in the crystal
    field model 247
19.2 Parameterization of the two-electron potential 248
19.3 The term dependent crystal field 250
19.4 Spin correlated crystal field (SCCF) 252
19.5 Many-electron approach to the crystal field effect 254

20. Appendices   **257**
A. Transformation from local to the global coordinate system 257
B. 3n-j symbols 259
C. Methods of orthogonalization of functions 261

21. References   **263**
22. Author index   **287**
23. Subject index   **293**

CHAPTER 1

Introduction

Using the notion of the crystal–field potential in the interpretation practice we do not, in general, realize the fact that this notion is based on the crucial assumption – any bound ion, i.e. that being inherent in crystal lattice or in molecule of a chemical compound feels the global interaction of its surroundings in the form of one–electron potential. Then, the effective Hamiltonian of the open–shell electrons of metal ion (in atomic units) reads

$$\mathcal{H}_{\text{eff}} = \sum_{i=1}^{N_a} \left[-\frac{\nabla_i^2}{2} + V(r_i) \right] + \sum_{i>j}^{N_a} \frac{1}{r_{ij}} + \sum_{i=1}^{N_a} \zeta_{na} \mathbf{l}_i \cdot \mathbf{s}_i + \sum_{i=1}^{N_a} V_{\text{CF}}(\mathbf{r}_i) \quad (1.1)$$

where the summation is limited to the open–shell electrons only. In general, any open–shell considered below will be denoted by a. These can be $3d$, $4d$, $5d$, $4f$ and $5f$ shells.

The Hamiltonian (Eq.1.1) consists of the free–ion part being composed of, in sequence of appearance, the kinetic energy of the electrons, their spherically symmetric potential energy, the inter–electron direct and exchange Coulomb interaction and the spin–orbit coupling energy, as well as of the ion surroundings potential, the non–spherical part of which is called the crystal–field potential $V_{\text{CF}}(\mathbf{r}_i)$. The spherically symmetric part of the surroundings potential which is dominating as for its magnitude but causes no energy splittings of electronic levels is usually absorbed in the central potential of the ion $V(r_i)$. This leads to somewhat modified radial distributions of individual spinorbitals which differentiates the bound ion from its free counterpart and changes both the Slater integrals (or the Racah coefficients) and the spin–orbit coupling constant ζ_{na}.

This assumption is not a priori obvious particularly considering various complications arising from non–orthogonality of interacting states and possibility of different electronic excitations admixturing the initial states. Fortunately, it has turned out, that the phenomenological parameterization of crystal–field effect by means of the one–electron potential expanded in

1

the series of spherical tensor operators works exceptionally well. This observation has been many times effectively confirmed by a vast amount of experimental material, mainly by data of electron spectroscopy [1, 2, 3, 4, 5, 6, 7, 8], magnetic measurements [9, 10, 11], EPR [12, 13], inelastic neutron scattering [14, 15, 16, 17, 18], heat capacity [19], electrical resistivity [20, 21] and others [22, 23, 24, 25] from the whole seventy years of the crystal–field theory use. Later on, it has appeared, by way of ab initio calculations, that the assumption, i.e. the adequacy of Eq.1.1, has also its theoretical justification.

In order to understand the idea of the effective potential let us come back to the initial state of affairs, i.e. that from before introducing the notion of the crystal–field potential. Then, the eigenstates of the bound ion and their eigenvalues could be found by way of solution of appropriately developed Schrödinger equation Hamiltonian of which should include, apart of all terms occurring in the component free-ion Hamiltonians, every possible interaction of the electrons with extraneous nuclei and all interelectron interactions. Consequently, the solutions in request would belong to a set of multicore many–electron functions.

In practice, to solve this many–body problem several simplifying assumptions has to be introduced. The two basic of them are

(I) Partition of the whole surroundings of a chosen ion, the so–called central–ion, into the closer neighbourhood confined to the closest ligands, the so–called ionic complex or cluster, and further surroundings. Within the ionic complex the important electronic states are taken into account in the form of their full wave–functions and quantum chemistry methods are employed in the calculations. Within the further neighbourhood the ions are treated as the electrostatic point charges or at most as dipoles and quadrupoles.

(II) The many–electron functions are expressed in the form of the so–called generalized product functions built up in a natural way of mutually orthogonalized groups of spinorbitals belonging to various shells and different ions, respectively, and with regard to, if need be, different types of admixtures being introduced through different excitation mechanisms.

In the solution obtained in this way the so–called crystal–field effect is included automatically and manifests itself first of all in the form of splitting of electron terms originating from the open–shell of the central ion.

However, the problem can be solved in another way. One may return, through some formal conversions, from the basis of the generalized product functions to the basis of the pure free–ion functions of the open–shell and in such a way at that in order to keep all values of the matrix elements constant. This aim is attained at the cost of suitable extending or otherwise renormalization of the Hamiltonian. This new Hamiltonian is called the effective Hamiltonian [26]. The rest after subtracting from it the spherically symmetric part resolves itself by means of appropriate averagings into a certain one–electron Hamiltonian which is just the effective crystal–field potential.

The effective crystal–field potential should be always related to the model on the ground of which it is defined. The number of the additional terms, i.e. the contributions to the effective potential introduced during the renormalization process, depends obviously on degree of extending the model.

Analysing thoroughly the form of the effective potential it is easy to notice that the global crystal–field effect can be with a satisfactory accuracy expressed in the form of sum of contributions arising from individual ions. These contributions can be calculated within the local coordinate systems in principle of axial symmetry with the z-axis joining the central ion and the chosen neighbour. Next, the intrinsic contributions can be transformed to the global system with appropriate rotation of the local coordinate system defined for each the lattice site or the ion separately. This assumption makes the ground for the so–called superposition model (SM) as well as the angular overlapping model (AOM). There are however contributions, rather of secondary meaning, which do not obey the additive rule. They will be pointed out at the right places.

Knowledge of the effective potential makes allowance for quantitative calculation of the crystal field effect, which in the parameteric quantification corresponds with a recipe for calculation of the crystal–field parameters. Three types of the parameters can be employed for description of the crystal–field effect

(I) the e_μ parameters used in the AOM which represent changes in energy of individual spinorbitals $|l\mu\rangle$ of the open–shell in the local field of the particular ligand. Values of the e_μ parameters are clearly correlated with the chemical characteristic of the ligand and vice versa. The e_σ parameter correspond to the effect of σ bond, e_π to that of π-bond etc.

(II) the intrinsic parameters b_k (or $\bar{B}_{k0}$) relating also to a local binary central ion–ligand system but connected with different quantification of the energy changes in the local crystal field. In contrast with the e_μ parameters the intrinsic parameters do not concern the changes in energy of individual spinorbitals but the changes in energy of the whole open–shell produced by the k-th order multipole components of the local potential. Within the local system of axial symmetry only the axial component, i.e. with $q = 0$, of any multipole moment is effective. This change is an appropriately weighted sum of the energy changes of all the occupied spinorbitals. Therefore no wonder these parameters are not so readable as the previous ones, especially for chemists.

(III) The conventional global parameters B_{kq} (and all their variations) describe the change in energy of the whole open–shell of the central ion induced by the q-component of the 2^k-multipole moment of the whole surroundings of the central ion. The global parameter B_{kq} is approximately a superposition of all the intrinsic parameters by means of suitable coordination factors. Thus, these parameters are coded doubly.

These three types of the parameters are mutually related. The global B_{kq} parameters with the intrinsic b_k ones by the relation

$$B_{kq} = \sum_t C_q^{(k)*}(\theta_t, \phi_t) b_k^t \tag{1.2}$$

where $C_q^{(k)}(\theta, \phi)$ is the normalized spherical harmonics, θ_t, ϕ_t are the angles between the z-axis of the local system t and the z-axis of the main coordinate system, and b_k^t is the intrinsic parameter of k-degree for ligand t. In turn, the intrinsic parameters b_k^t can be expressed with the e_μ^t parameters by the formula

$$b_k^t = \frac{2k+1}{2l+1} \left[\begin{pmatrix} l & k & l \\ 0 & 0 & 0 \end{pmatrix} \right]^{-1} \sum_{\mu=0}^{l-1} (-1)^\mu (2-\delta_{\mu 0}) \begin{pmatrix} l & k & l \\ \mu & 0 & -\mu \end{pmatrix} e_\mu^t \tag{1.3}$$

where the $3 - j$ symbols and Kronecker delta are used.

So, knowing one of the three sets of the parameters and the crystallographic geometry of the system the two other sets can be calculated. In the case of f-electrons ($l = 3$) only three parameters: e_σ, e_π and e_δ corresponding to $\mu = 0, 1$, and 2 are effective. The fourth parameter is inessential from the point of view of energy differences between the spinorbitals. Correspondingly, for d-electrons ($l = 2$) only the e_σ and e_π are effective.

The more extended model is considered the more laborious calculations are required giving, in general, an incommensurably weak improvement of the results. A characteristic evolution of the models from the simplest PCM to the advanced models based on the generalized product function formalism of the ionic complex with regard to the non–orthogonalities of component wave functions and effects of covalency, screening and polarization arising from different kinds of electronic excitations sketched in chapter 3 is very instructive indeed. Whereas the PCM does not give, in principle, any reliable results at all, the last generation model allows the parameters to be calculated with the accuracy not worse than $20 - 30\%$.

As evidently seen the nature of the crystal–field interaction is enough intricate and goes far beyond the simple electrostatic model. The above observations are well illustrative of scale of difficulties connected with the problem in this approach.

Within the frame of the analytical approach to the crystal–field mechanism, i.e. for separate treating all component contributions in the linear approximation, the impact of the three main non–orthogonalities in the ionic complex, and the most important electron excitations on the crystal–field effect is considered by turns. In this perturbational model the functions are expressed with accuracy to the first order with respect to such quantities as the overlap integrals or admixture amplitudes whereas the energies up to the second order with respect to the same quantities. Such a level of accuracy allows us for independent and additive treatment of the individual mechanisms or contributions. Here it is their list

- the point charge contribution,

- the contribution arising from space distribution of electron density on ligands or otherwise the penetration effect,

- the exchange effect being derived from the Coulomb exchange interaction between the central–ion open–shell electrons and ligand outer electrons.

The next three contributions result from the non–orthogonality of one–electron wave–functions entering into the composition of various groups of the generalized product function

- the contact covalency issuing from the non–orthogonality of the central–ion open–shell wave–functions (a) and the ligand outer spinorbitals (χ),

- the contact screening as the result of the non–orthogonality of the central–ion spinorbitals from its outer closed shells (ξ) and those of ligand outer shells (χ),

- the contact polarization arising from the non–orthogonality of spinorbitals belonging to different ligands $(\chi_t, \chi_{t'})$.

Further improvement of the model eigenfunctions of lowered energy can be gained by making allowance for admixtures of excited states. This leads to

- the covalency contribution related to interionic excitations of charge transfer type: ligand $\rightarrow$ metal ion, i.e. the effect connected with transition of electron from an outer closed shell of a ligand to the open–shell of the central–ion. One ought to realize that such an one–electron approach to the interatomic charge transfer effect is only a rough approximation of this many–electron process.

In turn, the intraionic excitations are the source of

- the screening effects resulting from specific distortions of the outer closed shells of the central ion which come of admixturing excited states to them via the crystal–field interaction, and

- the polarization as the result of excitations in outer closed shells of ligands and further ions.

In metals, intermetallic compounds and alloys the only important source of the crystal–field potential, with neglecting the presence of the conduction electrons, would remain the electrostatic interaction of positively charged and compact atomic cores for which, for evident reasons, the effects of charge penetration, exchange and polarization as well as that of the orthogonalizing potential are considerably weaker compared with those in ionic compounds. However, new specific mechanisms contributing to the crystal field effect are related to the conduction electrons present in the considered solid. These are

- the screening with the conduction electrons,

- the virtual bound states contribution which can form on the central ion as a consequence of the so–called Friedel condensation of conduction electrons in the resonance states,

- the hybridization, i.e. the effect of covalency mixing the localized and conduction band states.

This is the complete list of the basic contributions into the effective crystal–field potential in both localized and metallic systems which are considered in the monograph.

The analytical construction of the effective crystal–field potential taking up the best part of the monograph keeps still, apart from the historical aspect, its cognitive and instructive values. There is no other way to understand the mechanisms of the effect. Nevertheless, one should not forget that the fundamental assumptions of the construction concerning the mutual independence and linearity of the particular contributions are rather strong ones and not fully justified. Besides, this approach is characterized by two inconveniences:

- it does not lead to satisfactory results, i.e. the correct values of the parameters,

- it is not a practical method for experimentalists for each separate case requires a series of complex and cumbersome calculations.

The evident shortcomings have caused some time ago a decrease of interest in the approach. There are some promising attempts to rationalize and simplify the method through reducing the number of the independent parameters which in the complete expansion may seem to be excessive. Such a reduction can be gained without loosing the physico–chemical meaning of the parameters by exploiting the additivity of the global potential, i.e. the superposition model and particularly the AOM in which the crystal–field effect is attributed to energy effects of the chemical bonds between the central ion and ligands. First of all, the reduced parameterizations lead to simplified though correct phenomenological schemes. From the ab initio quantitative point of view they allow us not exactly to improve the results but to verify effectively any well–fitting parameterizations and consequently the splittings diagrams and reject those of them which are unrealistic being the effect of accidental fittings (chapter 17).

From about a decade the development of the crystal–field methods has concerned another, non–analytical model. Leaving the detailed microscopic quantification of the effect aside the fundamental assumption that the crystal field potential is explicitly determined by the density charge distribution about the central ion nucleous has been postulated.

To find this distribution the density functional theory (DFT) is used limited however by the local density approximation (LDA). Consequently, according to this approach it is enough to know the charge density distribution only but with sufficient accuracy in order to find the crystal field parameters. Unfortunately, this method has also its immanent drawbacks.

To pass over the problem of validity of the fundamental assumption about the direct correspondence between the local charge density and the crystal–field potential and an inadequacy of the LDA there is an additional inborn weakness of this approach, viz. a minimal weight of the crystal–field potential, in the strict sense, in relation to the total potential. The quantitative precision of the method resembles that of the story about weighing a captain of a ship from difference in draught of the ship with and without the captain. This disproportion may be a source of an additional error. Nevertheless, this method is nowadays more handy in use and of accuracy not worse than that offered in the most advanced conventional methods.

It is not out of place to realize a typical magnitude of the crystal field effect against a background of other intraatomic energy effects. In general, it is an intermediary effect. Its value, measured e.g. by the total splitting of the ground electron term, is differentiated according to the open–shell type (d or f), its main quantum number n, as well as the surroundings of the central ion. Roughly, this effect may be ordered as follows

$$\text{open} - \text{shell} \qquad 5d > 4d > 3d > 5f > 4f$$
$$\text{crystal} - \text{field effect}, [\text{cm}^{-1}] \quad 10^5 - 10^4 - 10^3 - 10^2 - 10^1$$

According to position of V_{CF} (Eq.1.1) on the energy scale in relation to the Coulomb repulsion and the spin–orbit coupling three cases: the strong, intermediate and weak ones are possible, respectively. In the strong–field case this is the crystal field which governs the distribution of the open–shell electrons. In turn, in the weak–field case the Hund's rules dominate and J may be considered as the good quantum number.

In ionic compounds this effect strongly rises with the valence of the central ion. The series of ions (ligands) ordered in accordance with the strength of crystal–field produced is called the spectrochemical series. The cyanide ion is known as a strong one, the halogenide ions are rather weak ligands, whereas the water molecule is an intermediate one in the series. The crystal-field effect in metals is usually weaker than that observed for the same ion in ionic compounds.

The crystal–field, irrespective of its symmetry, is unable to lift the degeneracy of Kramers doublets. Their twofold degeneracy resulting from the time–reversal symmetry can not be removed in any field of electrostatic origin.

The universality and fundamental significance of the crystal–field effect results from the fact that this effect determines the partition function of the system, especially in the range being comparable with kT, and in consequence its whole thermodynamics.

This monograph is aimed at the readers who intend to get to know both the general formalism leading to the effective crystal–field Hamiltonian and particular mechanisms forming the potential. The numerical data given in many places are to convey a proper opinion of the weight of the considered mechanisms.

The short sketch presenting the evolution of the conventional crystal field theory (chapter 3) has an introductory character. In further chapters from 5 to 10 the basic contributions to the effective potential in the case, first of all, of localized electron systems are considered. Next, in chapters from 13 to 15 the mechanisms specific for the metallic state are presented. They are preceded by chapter 11 devoted generally to the problem of crystal–field potential in highly degenerate periodic systems given by Stevens and chapter 12 on general characteristics of crystal–field effect in metals. Chapter 16 is devoted to the alternative approach based on the density functional theory and the local density approximation. The last three chapters of the main body of the book refer to the practical consequences of the ab initio data (chapter 17), the lattice dynamics contribution (chapter 18) and finally the extended approach going beyond the fundamental assumption of one–electron character of the potential (chapter 19).

Additionally, three short appendices referring to the problem of rotational transformations of coordinate systems, the methods of orthogonalization of wave functions and the $3n - j$ symbols are enclosed.

Particular chapters are to some extent autonomous and may be read, in principle, in any order. Undoubtedly, the list of quoted references does not comprise the whole enormous bibliography of the field. In many cases the original old papers are replaced by those published later, especially the review articles. However, based on the presented reference list each reader can easily come to the deepest roots of the problem.

CHAPTER 2

Parameterization of crystal field Hamiltonian

As is well known the essence of the crystal field effect lies in the aspherical part of environment interaction on localized states of unpaired electrons from an unfilled shell of a paramagnetic ion or atomic core in metals.
The nature of this interaction is multi-sourced and highly sophisticated
far exceeding that of the simple electrostatic field [27]. It is unquestionable that operators chosen to parameterization of the effect should be one
way or another related to the states on which they act. The commonly
used initial states of zero order approximation are the hydrogenlike one-
electron Hartree-Fock spinorbitals obtained in the central self-consistent
field. Their angle distributions are described by appropriate spherical harmonics $Y_l^m(v, \phi)$, where l and m are the quantum numbers of total angular
momentum and its projection on a chosen direction (z −axis), respectively.
Therefore it is advisable to expand the environment interaction Hamiltonian, $\mathcal{H}_{CF}$, about the central ion position into the series of spherical tensor
operators of the same type transformation properties as the electronic states.

The irreducible spherical tensor operator $\hat{\mathbf{T}}_k$ is defined as a set of $2k+1$
operators $\hat{T}_{kq}$ ($q = -k, -k+1, \ldots, k-1, k$) which transform under rotations
of the frame of coordinates like the components of the spherical tensor $\hat{Y}_k^q$
[28], namely as

$$\hat{D}(\omega)\hat{T}_{kq}\hat{D}^{-1}(\omega) = \sum_{q'=-k}^{k} \hat{T}_{kq'}\mathcal{D}_{q'q}^{(k)}(\omega) \qquad (2.1)$$

where $\omega \equiv (\alpha, \beta, \gamma)$ denotes the three Euler angles, $\hat{D}(\omega)$ is the rotation
operator and $D^{(k)}(\omega)$ is the rotation matrix of $2k+1$ order. The irreducible
spherical tensor operator may be equivalently defined as an operator components of which obey the following commutation rules with total angular

11

momentum operators [28, 29]

$$\left[\hat{J}_{\pm}, \hat{T}_{kq}\right] = \hat{T}_{kq+1}\hbar[(k \mp q)(k \pm q + 1)]^{1/2}$$

$$\left[\hat{J}_0, \hat{T}_{kq}\right] = \hat{T}_{kq}\hbar q \tag{2.2}$$

It results from the fact that the total angular momentum operators are the infinitesimal operators (generators) of the rotation group $SO(3)$ [28, 30]. Such an expansion of the crystal field potential reduces otherwise laborious calculations of its matrix elements to a simple finding the tabulated data. A typical integral in the crystal field theory has the form

$$\int_0^\pi dv \sin v \int_0^{2\pi} d\varphi Y_{l'}^{m'*}(v,\varphi)\hat{Y}_k^q(v,\varphi)Y_l^m(v,\varphi) =$$

$$= \langle l'm'|\hat{Y}_k^q|lm\rangle = (-1)^{l'-m'}\langle l'||\hat{Y}_k||l\rangle \begin{pmatrix} l' & k & l \\ -m' & q & m \end{pmatrix} \tag{2.3}$$

where

$$\langle l'||\hat{Y}_k||l\rangle = (-1)^{l'}\left[\frac{(2l'+1)(2k+1)(2l+1)}{4\pi}\right]^{1/2} \begin{pmatrix} l' & k & l \\ 0 & 0 & 0 \end{pmatrix} \tag{2.4}$$

is known as the reduced (double bar) matrix element and the factors within the round parentheses are 3-j symbols [28, 31]. The expansion of $\hat{\mathcal{H}}_{\mathrm{CF}}$ into the series of spherical tensor operators is, in spite of appearances, a natural one turning back to the classical electrostatic model of the potential. Then, the $\mathcal{H}_{\mathrm{CF}}$ obeys from the nature of things the Laplace equation $\Delta\mathcal{H}_{\mathrm{CF}} = 0$, that is it has to be expressible just as a combination (conditioned by the symmetry) of the spherical harmonics. In spite of the complex and not till the end comprehensible nature of the crystal field Hamiltonian its expansion into the series of spherical tensor operators treated phenomenologically together with assumption on its one-electron character have turned out to be surprisingly adequate in most of the parameterizations.

2.1 Operators and parameters of the crystal field Hamiltonian

There are several variants of the parameterization differing in normalization of the tensor operators or their combinations applied. In general, they may be divided into two classes – those related to spherical harmonics: $\hat{Y}_k^q$, $\hat{C}_q^{(k)}$ and $\hat{O}_q^{(k)}$ – the Buckmaster-Smith-Thornley (BST) operators [33, 34] differing only in constant factors from the original Buckmaster operators introduced earlier [32, 34], and those related to tesseral harmonics (real operators defined for $q > 0$: $\hat{Z}_{kq}^c = \frac{1}{\sqrt{2}}\left[\hat{Y}_k^{-q} + (-1)^q\hat{Y}_k^q\right]$,

$\hat{Z}_{kq}^s = \frac{i}{\sqrt{2}} \left[\hat{Y}_k^{-q} - (-1)^q \hat{Y}_k^q \right]$): $\hat{O}_k^q$ – the Stevens operator equivalents [35], and $\hat{O}'^q_k$ – the normalized Stevens operator equivalents [36, 37].

2.1.1 Operators related to spherical harmonics

As for the first class of the operators then except the oldest works [9, 38, 39, 61] in which $\hat{Y}_k^q$ have directly been used, the parameterization by Wybourne [3] is nowadays dominating

$$\hat{\mathcal{H}}_{\text{CF}} = \sum_i \sum_{k,q} B_{kq} \hat{C}_q^{(k)}(v_i, \varphi_i) \tag{2.5}$$

where

$$\hat{C}_q^{(k)} = \left(\frac{4\pi}{2k+1} \right)^{1/2} \hat{Y}_k^{(q)} \tag{2.6}$$

where i runs over all unpaired electrons of the unfilled shell of the metal ion, and k and q over all effective q components of the spherical tensor operator of rank k. More rarely, and that in the spin Hamiltonian theory rather, the BST operators are employed. This parameterization takes use of operators having on the one hand the transformation properties of the spherical harmonics but on the other character of the Stevens operator equivalents. Symbolically, in order to differentiate notations corresponding to various parameterizations and to avoid misunderstandings the following notation is recommended for the BST parameterization [34]

$$\hat{\mathcal{H}}_{\text{CF}} = \sum_{k,q} B_q^k \hat{O}_q^{(k)}(\hat{J}_z, \hat{J}_\pm) \tag{2.7}$$

where $\hat{J}_z$, $\hat{J}_\pm$ are the spherical components of total angular momentum.

The main difference between both the above parameterizations lies in the fact that in the former (Eq.2.5) the operators act on one-electron states whereas in the latter (Eq.2.7) on resultant many-electron states which are products of definite couplings. Using the first class operators enables straightforward application of the Wigner rotation matrices (see eg. [63]) for transformation of $\hat{\mathcal{H}}_{\text{CF}}$ expressed in terms of these operators.

2.1.2 Operators related to tesseral harmonics

To the second class belong widely used the Stevens operator equivalents or so-called extended (i.e. non-normalized) Stevens operators defined by

$$\hat{\mathcal{H}}_{\text{CF}} = \sum_{k,q} B_k^q \hat{O}_k^{(q)}(\hat{J}_z, \hat{J}_\pm) \tag{2.8}$$

and the normalized Stevens operators

$$\hat{\mathcal{H}}_{CF} = \sum_{n,m} B'^m_n O'^m_n(\hat{J}_z, \hat{J}_\pm) \tag{2.9}$$

where the norm of the parameters, i.e. $\sum_m \left(B'^m_n\right)^2$ is invariant under rotations of the frame of the coordinates. Notice the intentionally differentiated notations in Eq.2.7, Eq.2.8 and Eq.2.9. Both the extended and normalized Stevens operators are directly related (see below) [34, 37]. In the spin Hamiltonian theory for describing so-called zero-field splitting (ZFS) term [34] several modifications of the above parameterizations are occasionally employed which is rather confusing. Chronologically first (1952) and elegant method of the Stevens operator equivalents [35] particularly effective for individual electronic terms and systems of higher symmetry had been in course of time gradually replaced by the tensor parameterization by Wybourne. The reason of it was on the one hand steeped in troubles with rotational transformation of the equivalent operators $\hat{O}^q_k$ and in underdevelopment of their theory for $q < 0$ and on the other hand in progress of vector-coupling theory and tensor formalism [28, 29, 31, 3, 40, 44, 62].

2.2 Basic parameterizations

Nowadays the hindrances no longer come into consideration. The transformation properties of the Stevens operators are not much more complicated than those of $\hat{C}^{(k)}_q$ [37, 41] and whole range of $\hat{O}^q_k$ operators including those of $q < 0$ are at our disposal.

Two of the above mentioned parameterizations are undoubtedly the prevailing ones. These are the parameterizations by Wybourne, Eq.2.5, and by Stevens, Eq.2.8. They and relations between them will be considered in detail.

2.2.1 $\hat{C}^{(k)}_q$ tensor operator parameterization

Let us start from the Wybourne parameterization (Eq.2.5). From hermiticity of Hamiltonian (Eq.2.5) (its eigenvalues have to be real), and according to the Condon-Shortley convention [28]

$$\hat{C}^{(k)*}_q = (-1)^q \hat{C}^k_q \tag{2.10}$$

Hence, it results that

$$B^*_{kq} = (-1)^q B_{k-q} \tag{2.11}$$

In other words, the Hamiltonian is hermitean and even real but the crystal field parameters can be complex provided they are complex-conjugate in couples, respectively. The expansion (Eq.2.5) is described in the compact form. Separating explicitly the real ($\mathrm{Re}B_{kq}$) and imaginary ($\mathrm{Im}B_{kq}$) parts of B_{kq}, i.e. writing $B_{kq} = \mathrm{Re}B_{kq} + \mathrm{i}\mathrm{Im}B_{kq}$, expansion (Eq.2.5) may be presented in the expanded form

$$
\mathcal{H}_{\mathrm{CF}} = \sum_{i} \left\{ B_{k0}\hat{C}_0^{(k)}(i) + \sum_{q=1}^{k} \left[\mathrm{Re}B_{kq} \left(\hat{C}_q^{(k)}(i) + (-1)^q \hat{C}_{-q}^{(k)}(i) \right) + \right. \right.
$$
$$
\left. \left. + \quad \mathrm{Im}B_{kq}\mathrm{i} \left(\hat{C}_q^{(k)}(i) - (-1)^q \hat{C}_{-q}^{(k)}(i) \right) \right] \right\} \tag{2.12}
$$

from where evidently

$$
\mathrm{Re}B_{kq} = \tfrac{1}{2} \left(B_{kq} + (-1)^q B_{-q}^k \right)
$$
$$
\mathrm{Im}B_{kq} = \tfrac{1}{2\mathrm{i}} \left(B_{kq} - (-1)^q B_{-q}^k \right) \tag{2.13}
$$

and as a consequence of that

$$
\mathrm{Re}B_{k-q} = (-1)^q \mathrm{Re}B_{kq}
$$
$$
\mathrm{Im}B_{k-q} = (-1)^{1+q} \mathrm{Im}B_{kq} \tag{2.14}
$$

Both the parameters $\mathrm{Re}B_{kq}$ and $\mathrm{Im}B_{kq}$ as well as the combinations of operators (including the imaginary unit i, if any) associated with them are all real. So, one should not attach the literal meaning to the term "imaginary" in the case of $\mathrm{Im}B_{kq}$. Nevertheless, the operators $\mathrm{i}\left(\hat{C}_q^{(k)}(i) - (-1)^q \hat{C}_{-q}^{(k)}(i) \right)$ which are equivalent to the $\hat{O}_k^{-q}$ operators yield imaginary matrix elements indeed according to the relation [37, 42]

$$
\langle M'|\hat{O}_k^{-q}|M \rangle = \pm \mathrm{i} \langle M'|\hat{O}_k^q|M \rangle \tag{2.15}
$$

where the sign $+$ is for $M' < M$ and $-$ for $M' < M$, and complex (hermitean) matrix $\mathcal{H}_{\mathrm{CF}}$. Diagonalization of a complex matrix can be replaced by diagonalization of a real matrix of double dimension [43]. The combinations of operators within the round brackets of Eq.2.12 correspond exact to a constant factor to the tesseral operators, that is to the Stevens operators too.

Crystal field Hamiltonian in the Wybourne formulation acts on each i-th one-electron wave function separately. On the other hand its matrix elements are to be calculated in most cases within bases of many-electron functions specified by resultant quantum numbers S, L, M_s, M_L or J, M_J. This complication is effectively overcome owing to use of the transformation properties of the tensor operators involved. Consider the most general case of the final coupling (through the Russel-Saunders coupling) to $|JM_J\rangle$ states.

Then

$$\langle l^N SLJM_J|\hat{\mathcal{H}}_{\text{CF}}|l^N SL'J'M_J'\rangle =$$

$$\sum_{k,q} B_{kq}\langle l^N SLJM_J|\hat{U}_q^{(k)}|l^N SL'J'M_J'\rangle\langle l||\hat{C}^{(k)}||l\rangle \qquad (2.16)$$

where according to custom instead of $\hat{C}_q^{(k)}$ operator its normalized equivalent $\hat{U}_q^{(k)}$ for which $\langle l||\hat{U}^{(k)}||l\rangle = 1$ is used. Since $\hat{\mathcal{H}}_{\text{CF}}$ does not act on spin states the requirement $S = S'$ has to be fulfilled. Continued using the tensor formalism yields [3, 40, 44, 2]

$$\langle l^N SLJM_J|\hat{U}_q^{(k)}|l^N SL'J'M_J'\rangle =$$

$$= (-1)^{J-M_J}\begin{pmatrix} J & k & J' \\ -M_J & q & M_J' \end{pmatrix}\langle l^N SLJ||\hat{U}_q^{(k)}||l^N SL'J'\rangle \qquad (2.17)$$

and finally

$$\langle l^N SLJ||\hat{U}^{(k)}||l^N SL'J'\rangle = (-1)^{S+L'+J+K}\left[(2J+1)(2J'+1)\right]^{1/2} \times$$

$$\times \begin{Bmatrix} J & J' & K \\ L' & L & S \end{Bmatrix}\langle l^N SL||\hat{U}^{(k)}||l^N SL'\rangle \qquad (2.18)$$

where the factor within the braces is a 6-j symbol. The reduced matrix elements of $\hat{U}^{(k)}$ operator have been compiled by Nielson and Koster [46] whereas the 3-j and 6-j symbols can be found in tables by Rotenberg et. al. [31]

2.2.2 $\hat{O}_k^q$ Stevens operator equivalent method

Contrary to the simple tensor operators $\hat{C}_q^{(k)}$ the equivalent Stevens operators act directly on many-electron states being resultants of various coupling mechanisms in the atoms under consideration. To each free-ion electron term being characterized by the total angular momentum quantum number J (or L) such proportionality factors (one for each k in $\hat{\mathcal{H}}_{\text{CF}}$) can be assigned that the matrix elements of the $\mathcal{H}_{\text{CF}}$ being expressed in one-electron space coordinates $(\hat{C}_q^{(k)}(i))$ are equal to those of the $\mathcal{H}_{\text{CF}}$ being expressed in components of the total angular momentum $(\hat{O}_k^q(J_z, J_\pm))$ multiplied by the factor. This may be done on the strength of the Wigner-Eckart theorem [28, 30, 44], i.e. from identity of the transformation properties of x, y, z coordinates on the one hand and l_x, l_y, l_z and J_x, J_y, J_z (or L_x, L_y, L_z) components on the other with regard however their commutation rules. This is the concept of the Stevens operator equivalents.

These are the optimum operators from the point of view of electronic states on which they act and their use allows cumbersome calculations to replace into a simple arithmetic. The whole coupling effect $\sum_i (l_{ix}, l_{iy}, l_{iz}) \rightarrow J_x, J_y, J_z$ is inherent in the proportionality coefficient (Stevens multiplication factor) κ_k. This conversion is thoroughly presented in the original paper by Stevens [35] and the κ_k factors are tabulated in a number of papers and textbooks [35, 39, 42, 46]. The most direct definition of the Stevens equivalent operators is based on the Buckmaster operators [32, 47]

$$O_q^0 = O_l^0$$

$$O_k^q = \tfrac{1}{2} \left(O_l^{+m} + O_l^{-m} \right) \qquad \text{for} \quad q > 0 \text{ and } m > 0 \tag{2.19}$$

$$O_k^q = \tfrac{1}{2i} \left(O_l^{+m} - O_l^{-m} \right) \qquad \text{for} \quad q < 0 \text{ and } m > 0$$

The Stevens operator are on the left side, the Buckmaster ones on the right. Using the tables of matrix elements of the Stevens operator equivalents one should keep in mind that [42]

$$\langle M|\hat{O}_k^q|M - q\rangle = (-1)^{k+q}\langle -M|\hat{O}_k^q| - M + q\rangle = \langle M - q|\hat{O}_k^q|M\rangle$$

$$\langle M|\hat{O}_k^{-q}|M - q\rangle = (-1)^{k+q+1}\langle -M|\hat{O}_k^{-q}| - M + q\rangle =$$

$$-\langle M - q|\hat{O}_k^{-q}|M\rangle \tag{2.20}$$

$$\langle M|\hat{O}_k^{-q}|M - q\rangle = -i\langle M|\hat{O}_k^q|M - q\rangle$$

One reason for the prevailing use of the Stevens operators is the availability of tables of their matrix elements and computer programs for fitting procedures [39, 42, 12, 57].

Another starting-point of working out the equivalent operators $\hat{O}_k^q$ (for fixed k) is the complete separation of the Maclaurine expansion of the k-th term of $\hat{\mathcal{H}}_{CF}$ into uniform polynomials of k-th degree of course which, exact to constant factors depending on q, are equal to appropriate Legendre polynomials. This is always feasible due to the Laplace equation validity. However, a choice of the factors, from which the $\hat{B}_q^k$ values are automatically determined, is, in principle, arbitrary. These polynomials in their most convenient form have been defined by Stevens [35] and to this option the crystal field parameters are related. This aspect of the equivalent operator origin manifests itself in older way of notation of the associated parameters in the form of $A_k^q \langle r^k \rangle$ where $\langle r^k \rangle$ are the mean radii powers of the unpaired electrons

$$\langle r^k \rangle = \int_0^\infty dr P_{nl}^2(r) r^k \tag{2.21}$$

As the radial part of the wave function $P_{nl}(r)$ is never accurately known [48, 49] the radial integral is often taken as a parameter (together with the geometric factor A_k^q).

In Table 2.1 presented below relations between the respective $\hat{\mathcal{H}}_{CF}$ parameters originating from the two basic parameterizations are given [3, 1]

Table 2.1 The ratio $B_{kq}/(B_k^q = A_k^q \langle r^k \rangle)$ of crystal field parameters defined by Wybourne (Eq.2.5) and Stevens (Eq.2.8) parameterizations, respectively

k	q	$B_k^q / A_k^q \langle r^k \rangle$
2	0	$2 = 2.00$
2	1	$-\sqrt{6}/6 = -0.408$
2	2	$\sqrt{6}/3 = 0.816$
4	0	$8 = 8.00$
4	1	$-2\sqrt{5}/5 = -0.894$
4	2	$2\sqrt{10}/5 = 1.265$
4	3	$-2\sqrt{35}/35 = -0.338$
4	4	$4\sqrt{70}/35 = 0.956$
6	0	$16 = 16.00$
6	1	$-4\sqrt{42}/21 = -1.234$
6	2	$16\sqrt{105}/105 = 1.561$
6	3	$-8\sqrt{105}/105 = -0.781$
6	4	$8\sqrt{14}/21 = 1.425$
6	5	$-8\sqrt{77}/231 = -0.304$
6	6	$16\sqrt{231}/231 = 1.053$

In [3] the ratios for $q = 2$ and 4 are incorrect and in [1] the signs of those for $q = 3$ are wrong.

Since in general case B_{kq} are complex whereas B_k^q real the given ratios are *de facto* equal to $\mathrm{Re}B_{kq}/B_k^q$. The identical relations are valid for $\mathrm{Im}B_{kq}/B_k^{-q}$ ratios. Apart from differences in absolute values of the parameters particularly large for the axial ones the differences in their signs for odd q is worthy of noticing.

2.3 Symmetry transformations of the operators

This is a good moment to pay attention to essential differences between these two parameterizations and to their consequences. Postponing consideration of the whole point symmetry group to which $\hat{\mathcal{H}}_{CF}$ has to be invariant for later on let us take into account two fundamental symmetries – parity (under spatial inversion $\mathbf{r} \rightarrow -\mathbf{r}$) and time inversion symmetry (under time inversion $t \rightarrow -t$). A vector which under spatial inversion changes sign is a polar vector, otherwise an axial vector or pseudovector, e.g. each cross product $\mathbf{a} \times \mathbf{b}$ is an axial vector. Hence, tensor operators constructed from polar vector operators are polar for odd ranks and axial for even ranks. On the other hand tensor operators constructed from axial vector operators retain axiality regardless of their rank. Operators which under time inversion change sign are called time-odd operators, otherwise time-even operators. The electric nature of the crystal field implies $\hat{C}_q^{(k)}$ are polar operators for odd k and axial for even k and the requirement of the time inversal invariance of $\hat{\mathcal{H}}_{CF}$, as constructed from time-even vector operators, is always fulfilled.

On the other hand, $\hat{O}_k^q$ operators constructed from components of angular momentum $\hat{\mathbf{J}}$ which is axial and time-odd operator, are always axial and time-even (odd) for even (odd) k. In consequence $\hat{\mathcal{H}}_{CF}$ constructed from the $\hat{O}_k^q$ operators restricts k to even values only. This important distinction between both the parameterizations is not always explicitly realized which may lead to illusive misunderstandings.

Finding out the number of independent parameters in the operator expansion of $\hat{\mathcal{H}}_{CF}$ is the next crucial problem. Let us start from calculation procedure of the crystal field potential in a model approach. The problem of fitting a model Hamiltonian to experimental data will be considered afterwards. In order to do it in the frame of whichever model a reference coordinate system has to be fixed unambiguously. Such a system may be chosen arbitrarily and all quantities defined in it can be transformed to each other system exploiting their characteristic transformation properties under rotations. The form of $\hat{\mathcal{H}}_{CF}$ expansion in an arbitrarily chosen coordinate system is adequately complicated, i.e. respectively general although the scalar quantities including energies of states, i.e. roots of the characteristic equation of the $\hat{\mathcal{H}}_{CF}$ are invariant. However, for each environment symmetry except those of triclinic system there is always such a particular coordinate system (or systems!) in which the $\hat{\mathcal{H}}_{CF}$ expansion assumes its simplest possible form distinguishing itself by the minimum number of independent parameters. This is the so-called symmetry adapted system. One may call here a geometrical analogy, e.g. the transformation (canonical) of the conic curves to the main axes.

Each point symmetry group of $\hat{\mathcal{H}}_{CF}$ is characterized by its own minimum number of the independent crystal field parameters. Before we review all the 32 crystallographic point groups from this point of view it is advisable to remind the transformation properties of the operators occurring in the crystal field Hamiltonian with full particulars.

A fundamental feature of the tensor operators used in the crystal field theory is their transformation behavior with respect to rotation of the frame of reference about its initial point. This transformational characteristic is directly exploited in the superposition model of the crystal field potential

[27, 50, 65] and in the angular overlap model (AOM) [51, 26]. The systematization and standardization [52, 53, 54, 55] of results presented in different coordinate systems are also based on this property. In particular, in the case of various orientation of the magnetic field intensity **H** with respect to the crystallographic axes (magnetic problems, EPR) the appropriate formulation of the Hamiltonian requires the transformations to be known. As an example the transformation by Jones et al [56] may serve which is the result of two successive rotations: $(\phi_1 = 0, \theta_1 = \pi/2)$ and $(\phi_2 = \pi/2, \theta_2 = \varphi/2)$ where ϕ is the azimuthal angle of the rotation about the initial z-axis and θ – the polar angle of rotation about the new y-axis. Since the spherical harmonics $Y_q^k(v, \varphi)$, and the normalized spherical harmonics $C_q^{(k)}(v, \varphi) = \left(\frac{4\pi}{2k+1}\right)^{1/2} Y_q^k(v, \varphi)$ and other relative quantities are the basis functions of $2k + 1$ dimensional irreducible representation of the three-dimensional rotation group (the unimodular orthogonal group $SO(3)$), the components of the aforementioned tensor operators transform under rotations according to the relation

$$\hat{D}(\alpha, \beta, \gamma)\hat{Y}_q^k(v, \varphi)\hat{D}^{-1}(\alpha, \beta, \gamma) = \sum_{q'=k}^{k} \hat{Y}_{q'}^k(v, \varphi)\mathcal{D}_{q'q}^{(k)}(\alpha, \beta, \gamma) \qquad (2.22)$$

where, as in the specification of Eq.2.1, $\hat{Y}_q^k(v, \varphi)$ is the q-component of the operator of rank k, $\hat{D}(\alpha, \beta, \gamma)$ is the operator of rotation by three Euler angles α, β, γ and $D^{(k)}(\alpha, \beta, \gamma)$ is the rotation matrix elements of which are equal to [28]:

$$\mathcal{D}_{q'q}^{(k)}(\alpha, \beta, \gamma) = \langle Y_k^{q'}|\hat{D}(\alpha, \beta, \gamma)|Y_k^q\rangle = \exp(iq'\alpha)d_{q'q}^{(k)}(\beta)\exp(iq'\gamma) \qquad (2.23)$$

and

$$d_{q'q}^{(k)}(\beta) = \left[\frac{(k+q')!(k-q')!}{(k+q)!(k-q)!}\right]^{1/2} \sum_{\sigma} \binom{k+q}{k-q'-\sigma}\binom{k-q}{\sigma} \times$$

$$\times (-1)^{k-q'-\sigma}(\cos\beta/2)^{2\sigma+q'+q}(\sin\beta/2)^{2k-2\sigma-q'-q} \qquad (2.24)$$

where the sum is over all positive values of σ such that the Newton's symbols have the sense. The above relations refer to a general case if all three Euler angles are needed to define the position of the system transformed. However, if knowledge of the complete orientation of the initial system is not necessary, e.g. in the case of axial symmetry, but only its z-axis orientation with respect to the final system, two angles: $\alpha = \phi$ and $\beta = \theta$ are sufficient. They are nothing else than the spherical coordinates of the initial z-axis in the final coordinate system. It corresponds to the three Euler angles according to the relation $(\alpha, \beta, \gamma) = (\phi, \theta, 0)$ and the geometrical sense of the angles has been defined previously at description of the Jones transformation. Then, Eq.2.22 takes the simpler form

$$\hat{D}(\phi, \theta, 0)\hat{Y}_k^q(v, \varphi)\hat{D}^{-1}(\phi, \theta, 0) = \sum_{q'=k}^{k} \hat{Y}_k^{q'}(v, \varphi)\mathcal{D}_{q'0}^{(k)}(\phi, \theta, 0) \qquad (2.25)$$

where

$$\mathcal{D}_{q'0}^{(k)}(\phi, \theta, 0) = (-1)^{q'} \left(\frac{4\pi}{2k+1}\right)^{1/2} Y_k^{q'}(\theta, \phi) \qquad (2.26)$$

What is needed now is knowledge of the elements in the middle column of the $\mathbf{D}^{(k)}(\phi, \theta, 0)$ matrix only which are easily available from tables of the spherical harmonics [28, 39]. $\mathbf{D}^{(k)}$ is the unitary matrix , i.e. $\left(\mathbf{D}^{(k)}\right)^{-1} = \left(\mathbf{D}^{(k)}\right)^{\dagger}$, the inverse of $\mathbf{D}^{(k)}$ is equal to the hermitean conjugate of $\mathbf{D}^{(k)}$. It is attended by invariance of $\sum_q |B_{kq}|^2$ with respect to rotations for each k separately, where the B_{kq} are parameters associated with the corresponding operators.

The so-called coordination factor K_k^q [27, 41] – the key-notion in the superposition approach defined by

$$B_{kq} = \sum_i K_k^q(\theta_i, \phi_i)\overline{B_k}(R_i) \qquad (2.27)$$

describes contribution of the k-degree intrinsic parameter of ligand i, $B_k(R_i)$, to the global parameter B_{kq}. The sum in Eq.2.27 runs over all ligands and θ_i, ϕ_i, R_i are the spherical coordinates of ligand i in the global coordinate frame. In general, the coordination factors are the elements of the middle row ($q' = 0$) of the inverse transformation matrix (from a local (i) to the global system) $\left[D^{(k)}(\theta_i, \phi_i, 0)\right]^{-1}$. From the unitarity of the rotation matrix and Eq.2.26 it results that the coordination factors are equal to

$$K_k^q(\theta_i, \phi_i) = \mathcal{D}_{0q}^{(k)^*}(\theta_i, \phi_i, 0) = C_q^{(k)^*}(\theta_i, \phi_i) \qquad (2.28)$$

As for the tesseral type of operators their transformation properties under rotations are a little more complicated [37]. The lack of their exact recognition has hindered in the past more extensive employing the Stevens operator equivalents. Now, let us focus our attention on these operators. The transformation properties of the Stevens operators $\hat{O}_k^q$ are given by the matrices $\mathbf{S}_k(\phi, \theta)$ defined by [37, 41]

$$\begin{Bmatrix} \hat{O}_k^{+k} \\ \vdots \\ \hat{O}_k^0 \\ \vdots \\ \hat{O}_k^{-k} \end{Bmatrix} = \left(S_k(\phi, \theta)_{ij} \right) \begin{bmatrix} \hat{O}_k^{+k} \\ \vdots \\ \hat{O}_k^0 \\ \vdots \\ \hat{O}_k^{-k} \end{bmatrix} \qquad (2.29)$$

i.e.

$$\{\mathbf{O}_k\} = \mathbf{S}_k(\phi, \theta)[\mathbf{O}_k]$$

where the curly brackets denote a column matrix of the Stevens operators $\hat{O}_k^q$ in the original axis system, whereas the square brackets denote the operator in the transformed axis system which is rotated by an azimuthal angle ϕ about the original z-axis and a polar angle θ about the new y-axis. The $\mathbf{S}_k$ matrices have been derived [37] by means of operators introduced by Buckmaster et al [47] which are operator equivalents to the individual spherical harmonics. For instance, the transformation (Eq.2.29) for the three second order Stevens operators are given below [37]

$$\{\hat{O}_2^0\} = \frac{3}{2}\sin^2 2\theta[\hat{O}_2^2] - 3\sin 2\theta[\hat{O}_2^1] + \frac{1}{2}(3\cos^2 2\theta - 1)[\hat{O}_2^0]$$

$$\begin{aligned}\{\hat{O}_2^1\} =\ & -\frac{1}{4}\sin 2\theta\cos\phi[\hat{O}_2^2] + \cos 2\theta\cos\phi[\hat{O}_2^1] \\ & + \frac{1}{4}\sin 2\theta\cos\phi[\hat{O}_2^0] - \cos\theta\sin\phi[\hat{O}_2^{-1}] + \frac{1}{2}\sin\theta\sin\phi[\hat{O}_2^{-2}]\end{aligned}$$

$$\begin{aligned}\{\hat{O}_2^2\} =\ & \frac{1}{2}(\cos^2 2\theta + 1)\cos 2\phi[\hat{O}_2^2] + \sin 2\theta\cos 2\phi[\hat{O}_2^1] \\ & + \frac{1}{2}\sin^2 2\theta\cos 2\phi[\hat{O}_2^0] - 2\sin\theta\sin 2\phi[\hat{O}_2^{-1}] - \cos\theta\sin 2\phi[\hat{O}_2^{-2}]\end{aligned}$$

Certain rules can be noticed in the matrices $\mathbf{S}_k$:

- the expressions for $\{\hat{O}_k^{-q}\}$ are obtained from the $\{\hat{O}_k^q\}$ expansions on replacement everywhere in the first form $\cos q\phi$ by $\sin q\phi$ and $\sin q\phi$ by $-\cos q\phi$, therefore $\hat{O}_k^q$ and $\hat{O}_k^{-q}$ transform in a different way,

- in the $\hat{O}_k^0$ expansions there are no $[\hat{O}_k^{-q}]$ operators and these expansions do not depend on ϕ.

The matrices $\mathbf{S}_k(\phi, \theta)$ are real but unfortunately they are not orthogonal. Hence, their inverse matrices are not equal to the transposed ones. Nevertheless, the inverse matrix $\mathbf{S}_k^{-1}$ can be found from the $\mathbf{S}_k$ matrix according to the relation

$$\left(S_k^{-1}\right)_{ij} = \left(\frac{c_j^k}{c_i^k}\right)^2 (S_k)_{ij} \tag{2.30}$$

where the coefficients c_q^k for relating extended $\hat{O}_k^q$ and normalized $\hat{O'}_k^q$ Stevens operators ($\hat{O'}_k^q = c_q^k\hat{O}_k^q$ and correspondingly $\hat{B'}_k^q = (1/c_q^k)B_k^q$) are listed by Rudowicz [34, 37]. Consequently, the coordination factor amounts to

$$K_k^q = (c_q^k)^2(S_k)_{q0} \tag{2.31}$$

2.4 The number of independent crystal field parameters

Now, it is timely to answer the basic question – how does the number of independent crystal field parameters depend on the surroundings symmetry. The expansion of $\hat{\mathcal{H}}_{CF}$ into the spherical tensor operator series is not subjected, a priori, to any limitation as for the range of variability of k and q indices. Nevertheless, from among the infinite number of the terms not many of them are effective only. This restriction comes down from two sources:

- symmetry properties of electronic states on which the $\hat{\mathcal{H}}_{CF}$ acts, and

- symmetry of the surroundings.

In any case, from the triangle rule results that $k \leq l + l'$, where l and l' are the quantum numbers of angular momentum of the *bra* and *ket* one-electron states in the matrix element. It denotes that k can at most be equal to 6 if $l = l' = 3$, i.e. for f electrons, and to 4 for d-electrons, respectively. Such a truncation of the expansion has nothing to do with an approximation – all remaining terms of higher orders give exactly no contribution to $\hat{\mathcal{H}}_{CF}$. In turn, since the matrix elements are scalars, the parity of the states interacting via the $\hat{\mathcal{H}}_{CF}$ reduces the number of nonzero matrix elements because $k + l + l'$ has to be even. For the intraconfiguration interactions, i.e. for $l = l'$, k has to be even. Thus, it can be equal to 2, 4 and 6 only. The term with $k = 0$ of spherical symmetry produces only an uniform shift of all levels and from the point of view of spectroscopic or magnetic properties may be neglected without loosing generality. However, this is the largest term in the expansion responsible for a major part of lattice energy. Further reducing the number of parameters results from details of the central ion symmetry. The presence of inversion center eliminates the terms with odd k which at confining to electronic states of a single configuration does not change the effective form of $\hat{\mathcal{H}}_{CF}$. Nevertheless, e.g. the crystal field potential of tetrahedral symmetry (c.n.= 4) differs from that of cubic symmetry (c.n. = 8) not only by factor 1/2 but also by effective terms with odd k. Only in the case of non-existence of any element of symmetry except C_1 (triclinic system) the number of parameters of $\hat{\mathcal{H}}_{CF}$ with $k = 2, 4$ and 6 amounts to 27 whereas for all k from 1 to 6 as many as 48. For all the remaining cases the terms with $q = 1$ and 5 do not occur and the minimum number of the parameters reduces to 15. If there is a two-fold axis along the quantization axis or a symmetry plane perpendicular to it only the terms with even q are effective, i.e. there are, as above, 15 effective parameters. If there is a three-fold axis only q values divisible by 3 are permissible. Generally, the presence of a p-fold axis imposes the condition $q = np$, where n is a natural number. It automatically reduces the number of the parameters to $15, 9, 7$ and 5 for the two-, three-, four- and six-fold axis, respectively. The invariance of $\hat{\mathcal{H}}_{CF}$ with respect to transformation $\phi \rightarrow -\phi$ is necessary and sufficient condition of presence in the expansion of $\hat{\mathcal{H}}_{CF}$ of the real parameters only, i.e. in order to $\mathrm{Im}B_{kq} = 0$ or $B_k^{-q} = 0$ for all admissible k and q. What is enough for it is existing a symmetry plane or a two-fold axis (for $\hat{\mathcal{H}}_{CF}$ with even k only) in the central ion point symmetry group. However, in both the cases the appropriate choice of the coordinate system is required. For the symmetry plane the z-axis of the system has to lie in the plane. In the second case the z-axis must be perpendicular to the two-fold axis (even powers of z and $-z$ give identical contributions). However, not always these systems are the symmetry adapted systems of minimum numbers of the parameters.

From among 32 point symmetry groups of $\hat{\mathcal{H}}_{CF}$ (32 three-dimensional crystallographic classes) with $k = 2, 4$ and 6, seven classes absolutely (for any coordinate system) do not fulfill the above conditions and the corresponding Hamiltonians contain terms with complex parameters. These are: $C_1(12)$, $C_i(12)$, $C_3(3)$, $C_{3i} = S_6(3)$, $C_4(2)$, $S_4(2)$ and $C_6(1)$ where in the parentheses after the Schoenflies symbols of the point groups the numbers of the complex parameters (the pairs of $\mathrm{Re}B_{kq}$ and $\mathrm{Im}B_{kq}$) in the symmetry adapted systems are given, respectively.

If the condition is satisfied , i.e. only the real parameters are present, the number of independent parameters for central ion symmetry with two-, three-, four- and six-fold axis decreases to 9, 6, 5 and 4 respectively. For each further six point groups a coordinate system may be chosen in such a way that the $\hat{\mathcal{H}}_{CF}$ expansion is expressible with the real parameters only but these are not the symmetry adapted systems. Their conventional parameterizations of minimum number of parameters are composed of both real and complex parameters. These are: $C_2(6)$, $C_s(6)$, $C_{2h}(6)$, $C_{4h}(2)$, $C_{3h}(1)$ and $C_{6h}(1)$. For the odd ($k = 1, 3$ and 5) terms of $\hat{\mathcal{H}}_{CF}$ the complex parameters have to be taken into account for the following eleven point symmetry groups: $C_1(9)$, $C_2(3)$, $C_s(3)$, $D_2(3)$, $C_3(2)$, $D_3(2)$, $C_4(1)$, $S_4(1)$, $D_4(1)$, $T(1)$, $O(1)$, where as previously, the number of the complex parameters is given in the parentheses.

When in the $\hat{\mathcal{H}}_{CF}$ expansion the complex parameters occur there is always and independently at that on other symmetry elements present in the group, a possibility of eliminating one of them (it concerns each of them) by means of a certain rotation of the coordinate system about the z-axis by an angle χ_{kq} given by

$$\tan q\chi_{kq} = \frac{\mathrm{Im}B_{kq}}{\mathrm{Re}B_{kq}} = \frac{\mathrm{i}(B_{kq} - B_{k-q})}{(B_{kq} + B_{k-q})} = \frac{B_k^{-q}}{B_k^q} \qquad (2.32)$$

Then, the $\mathrm{Im}B_{kq}$ parameter vanishes and the new parameter $[B_{kq}]$ becomes real and equal to the modulus

$$[B_{kq}] = [(\mathrm{Re}B_{kq})^2 + (\mathrm{Im}B_{kq})^2]^{1/2} \qquad (2.33)$$

In this way, e.g. the number of parameters of the C_{3h} group may be reduced from 5 to 4 (B_{20}, B_{40}, B_{60}, $|B_{60}|$) and is identical with that of the D_{3h} group for which the $\mathrm{Im}B_{66}$ parameter is absent due to the higher symmetry. It is obvious that each such a rotation implicitly orientates the coordinate system.

Generally, in the case of low symmetry crystal fields the real $\mathrm{Re}B_{kq}$ and imaginary $\mathrm{Im}B_{kq}$ parts of their complex parameters have, in principle, a comparable weight. However, if we deal with a slight distortion of a system being parametrizable with the real parameters only, e.g. $D_{2d} \rightarrow S_4$ (rhombic system $\longrightarrow$ monoclinic system) the imaginary parameters can be a measure of the distortion.

The existence of additional elements of symmetry causes a further reducing the number of independent parameters and so, e.g. from 5 parameters for the tetragonal group C_{4v} to 2 only in the case of cubic symmetries O_h or T_d. All the cubic potentials generated by tetrahedron (c.n.= 4), cube (c.n.= 8), octahedron (c.n.= 6), cubooctahedron – $AuCu_3$ (c.n.= 12) and others can be described with two parameters, one of the fourth order ($k = 4$) and one of the six order ($k = 6$) because mutual relations between the parameters of the same k, i.e. B_{44}/B_{40} and B_{64}/B_{60} are fixed and amounts to $\sqrt{5/14}$ (5 in the Stevens parameterization) and $-\sqrt{7/2}$ (-21 in the Stevens parameterization), respectively, if the z-axis is in line with the four-fold axis. Then

$$\hat{\mathcal{H}}_{\mathrm{CF}}(\text{cubic}, C_4) = \sum_i \left\{ B_{40} \left(\hat{C}_0^{(4)}(i) + \left(\frac{5}{14}\right)^{1/2} \hat{C}_4^{(4)}(i) \right) + \right.$$
$$\left. + B_{60} \left(\hat{C}_0^{(6)}(i) - \left(\frac{7}{2}\right)^{1/2} \hat{C}_4^{(6)}(i) \right) \right\} \qquad (2.34)$$

If the quantization axis is colinear with the three-fold axis of the cubic system the same potential has the form

$$\hat{\mathcal{H}}_{\mathrm{CF}}(\text{cubic}, C_3) = \sum_i \left\{ B_{40} \left(\hat{C}_0^{(4)}(i) - \left(\frac{10}{7}\right)^{1/2} \hat{C}_3^{(4)}(i) \right) + \right.$$
$$+ B_{60} \left(\hat{C}_0^{(6)}(i) + \frac{1}{4}\left(\frac{35}{6}\right)^{1/2} \hat{C}_3^{(6)}(i) + \right.$$
$$\left. \left. + \frac{1}{8}\left(\frac{77}{3}\right)^{1/2} \hat{C}_6^{(6)}(i) \right) \right\} \qquad (2.35)$$

In turn, for the C_{6h} point symmetry group of hexagonal system of $c/a = 2\sqrt{2}/3$ [59]

$$\hat{\mathcal{H}}_{CF}(\text{hex.}, C_6) = \sum_i \left\{ B_{40}\hat{C}_0^{(4)}(i) + \right.$$
$$\left. + B_0 \left(\hat{C}_0^{(6)}(i) + \frac{1}{8}\left(\frac{77}{3}\right)^{1/2}\hat{C}_6^{(6)}(i) \right) \right\} \quad (2.36)$$

which likewise the previous ones are the two-parameter Hamiltonians. In such cases a set of two other independent parameters defined with the B_{40}/B_{60} ratio is often used. For the cubic system it is the popular parameterization and all possible splitting diagrams of free ion electronic terms has been given by Lea, Leask and Wolf [46] and for the hexagonal system those by Segal and Wallace [58].

The potential of the Archimedean antiprism symmetry $\mathcal{D}_{4d}$ [59] distinguishes itself with the exceptionally simple form

$$\hat{\mathcal{H}}_{CF}(D_{4d}, \bar{8}) = \sum_i \left\{ B_{40}\hat{C}_0^{(4)}(i) + B_{60}\hat{C}_0^{(6)}(i) \right\} \quad (2.37)$$

This is the purely axial field which does not mix states of different m but only splits them into the doublets $|\pm m\rangle$ and singlet $|0\rangle$ for integer l, and into the Kramers doublets for half-integer l. Generally, each potential with the symmetry axis of multiplicity greater than 6 for f-electrons and 4 for d-electrons is felt as the pure axial potential of $C_{\infty v}$ symmetry.

The effective expansions of some high-symmetric potentials are composed of the sixth order terms only, i.e. they are at most the one-parameter Hamiltonians. What does belong to them is e.g. the potential of icosahedron symmetry [42, 67] (c.n.= 12, twelve atoms at the corners of three crossing mutually perpendicular rectangles with sides of ratio 1.618)

$$\hat{\mathcal{H}}_{CF}(\text{icos.}, C_2) = \sum_i B_{60} \left\{ \hat{C}_0^{(6)}(i) - 1/2\left[(21)^{1/2}\hat{C}_2^{(6)}(i) + \right.\right.$$
$$\left.\left. + (14)^{1/2}\hat{C}_4^{(6)}(i) - (105/11)^{1/2}\hat{C}_6^{(6)}(i)\right] \right\} \quad (2.38)$$

where the two-fold axis of the icosahedron plays a role of the z-axis. For the f-electron ions this is an one-parameter potential whereas the d-electron ions see it as the spherical potential ineffective as for the splitting.

2.5 Standardization of the crystal field Hamiltonian

As was emphasized earlier calculation of crystal field potential and the crystal field parameters was unavoidable connected with the choice of a particular coordinate system. That is not the case in fitting procedure of a parameterized form of $\hat{\mathcal{H}}_{CF}$ to the spectrum of energy levels available from experiments, mainly spectroscopy. In the procedure we start from the canonical form of $\hat{\mathcal{H}}_{CF}$ for a given (known) point symmetry group attributed to the symmetry adapted system with the minimum number of independent parameters, without however specifying the coordinate system itself.

However, two problems arise. They concern

- the ambiguity of the symmetry adapted system and in consequence the equivocal character of the parameters fitted, and

- the available set of independent parameters.

For most of the systems, particularly for those of higher symmetry there is one to one correspondence between the symmetry adapted system and the set of crystal field parameters found in the fitting procedure. Obviously, we do not differentiate crystallographically equivalent systems which physically are identical, e.g. the x, y and z axes in the cubic system are equivalent and each of them may play the role of quantization axis (distinguished axis). However, there are surroundings symmetries which distinguish themselves by a finite or even infinite number of crystallographically inequivalent symmetry adapted reference systems. It leads to qualitatively identical form of $\hat{\mathcal{H}}_{CF}$ parameterization but to different although equivalent sets of parameters. One can fail to realize that outwardly different sets of parameters can be coordinated by means of appropriate symmetric transformations. This problem generally refers to all point symmetry groups $\hat{\mathcal{H}}_{CF}$ of which contains terms with complex parameters since its form is invariant with respect to any rotation about the z-axis.

Additionally, it concerns the point groups of rhombic system considering three possible orientations of the z-axis along the unit cell edges a, b and c, respectively, and of course those of triclinic system in which, owing to the lack of any distinguished direction, an arbitrary reference system can play the role of the symmetry adapted system. The crystal field potentials of monoclinic symmetry are free of this complication for owing to the favored direction (C_2 axis or that perpendicular to a plane) different forms of $\hat{\mathcal{H}}_{CF}$ correspond to various choices of the z-axis [55]. What remains is naturally the problem arising from occurring the complex parameters, if any. Arrangement such of the parameterizations, i.e. their reducing to a common coordinate system or so-called their standardization is the necessary operation previous to comparing and taking use of various data relating to different parameterizations of the $\hat{\mathcal{H}}_{CF}$. This is particularly important for many low symmetry systems frequently applied as laser materials [4]. In this case when the complex parameters are present the proper standardization relies on a strictly defined rotation of the coordinate system about its z-axis to eliminate one, a priori pointed out, $\mathrm{Im}B_{kq}$ parameter. It denotes that from the fitting procedure point of view the number of independent parameters is reduced by one, the parameterization of $\hat{\mathcal{H}}_{CF}$ is unambiguously defined but the orientation of the reference system related to our choice remains unknown. As a rule, the imaginary parameter $\text{Im}B_{kq}$ of the lowest (second) order, uncommonly that of the highest (sixth) order is eliminated, e.g. for triclinic system usually $\text{Im}B_{21}$, more rarely $\text{Im}B_{65}$ is eliminated and after the transformation the number of independent parameters amounts to 26. Deleting one after the other $\text{Im}B_{kq}$ one gets different, although equivalent sets of the parameters. It gives a certain insight into mutual correlations between the parameters being one of the central ideas of so-called multipole correlated fitting technique [55, 4].

A particular situation takes place in the case of rhombic system. According to the three possibilities of choice of the z-axis there are three apparently different parameterizations. The standardization in this case relies on reducing the ratio B_{22}/B_{20} into the range $(0,\ 1/\sqrt{6})$ or B_2^2/B_2^0 into the range $(0, 1)$, respectively. It may be achieved by means of one of the six transformations $(S_1 - S_6)$ of the reference system defined by Rudowicz and Bramley [54]. Only after such transformation we have at our disposal well-defined parameters being suitable for comparisons.

In the case of monoclinic system the transformation of both forms of $\hat{\mathcal{H}}_{CF}$ with and without complex parameters relies as above on reducing by one (to 14) the number of parameters with a suitable rotation. The standardization of the rhombic part of monoclinic $\hat{\mathcal{H}}_{CF}$ (similarly to that for rhombic system) can be routinely performed although it is not necessary on account of the differentiation of the $\hat{\mathcal{H}}_{CF}$ form [55]. At last, for triclinic system the reduction by one imaginary parameter from 27 to 26 is always feasible. Unfortunately, there are infinite number of such parameterizations which corresponds to infinite number of choices of the z-axis.

It is not true that in the fitting process magnitudes and signs of all the permissible parameters can be fully determined. This, how many magnitudes of the parameters are available and how many their signs have the absolute or relative character arises from the algebraic symmetry of the characteristic equation $|\hat{\mathcal{H}}_{CF} - E\mathbf{1}| = 0$ [55, 60, 66].

Knowledge of energy spectrum of $\hat{\mathcal{H}}_{CF}$ is equivalent to knowledge of coefficients of its characteristic equation which are invariants of the transformations of the coordinate system. In the case of coexisting various symmetry adapted systems the characteristic equation coefficients have the same structure, i.e. the $\hat{\mathcal{H}}_{CF}$ parameters enter them in the form of invariant combinations (invariant with respect to transformations converting one symmetry adapted system into another). This confines both the number and the signs of the parameters which can be determined by fitting. For instance, for the S_4 tetragonal symmetry the $\hat{\mathcal{H}}_{CF}$ expansion consists formally of seven parameters: B_{20}, B_{40}, B_{60}, $\text{Re}B_{44}$, $\text{Im}B_{44}$, $\text{Re}B_{64}$, $\text{Im}B_{64}$. However, by means of fitting only the first three (axial) parameters and the following three additional quantities (instead of four parameters) are available:

$(\mathrm{Re}B_{44})^2 + (\mathrm{Im}B_{44})^2$, $(\mathrm{Re}B_{64})^2 + (\mathrm{Im}B_{64})^2$ and $\mathrm{Re}B_{44}\,\mathrm{Re}B_{64} + \mathrm{Im}B_{44}$ $\mathrm{Im}B_{64}$. Deleting $\mathrm{Im}B_{44}$ parameter one gets the absolute magnitudes of $\mathrm{Re}B_{44}$, $\mathrm{Re}B_{64}$ and $\mathrm{Im}B_{64}$. The signs of $\mathrm{Re}B_{44}$ and $\mathrm{Re}B_{64}$ are mutually correlated and depend on the sign of $\mathrm{Im}B_{64}$ which may be assumed arbitrarily [60].

In turn, for monoclinic symmetry sites after eliminating $\mathrm{Im}B_{22}$ parameter fourteen independent parameters remain. In fitting the experimental data the following information can be determined [55]:

   (I) the magnitudes of all the fourteen parameters,

  (II) the absolute signs of $\mathrm{Re}B_{44}$ and $\mathrm{Re}B_{64}$,

 (III) the relative signs of $\mathrm{Re}B_{22}$, $\mathrm{Re}B_{42}$, $\mathrm{Re}B_{62}$, $\mathrm{Re}B_{66}$,

  (IV) the relative signs of $\mathrm{Im}B_{42}$ & $\mathrm{Im}B_{62}$ and $\mathrm{Im}B_{44}$ & $\mathrm{Im}B_{64}$

   (V) the relative signs of $\mathrm{Im}B_{42}$ & $\mathrm{Im}B_{44}$ and $\mathrm{Im}B_{44}$ & $\mathrm{Im}B_{66}$ with respect to the sign of $\mathrm{Re}B_{22}$.

In the light of these limitations a large part of crystal field interpretations met in the literature should be verified. And so, the results of fitting the complete set of seven parameters of S_4 point group (scheelite matrices) are probably incorrect for these parameters are not fully independent [55, 60].

2.6 Final remark

To the end with the section devoted to the crystal field potential parameterization a short comment should be given on the zero-field-splitting (ZFS) term in the spin Hamiltonian commonly used in EPR. Strictly, the term ZFS denotes a splitting no matter what origin in the magnetic field's absence. In the spin Hamiltonian this is a term of effective nature having often indirect only and secondary connection with the crystal field sensu stricto. Usually, it is a combined effect of various higher order interactions. Therefore, direct setting the crystal field interpretation to $\hat{\mathcal{H}}_{\mathrm{ZFS}}$ is incorrect and leads to misunderstandings.

CHAPTER 3

The effective crystal field potential. Chronological development of crystal field models

Conventionally narrowed notion of crystal field effect reduces itself to energy splitting of ionic, atomic or ionic core electronic terms in the field produced by their surroundings in the lattice or compound. Within the frame of the elementary approach this effect was identified with the Stark splitting in simple electrostatic field of definite symmetry [68, 69]. In reality the problem is far more complex. The crystal field effect from its very nature refers to open–shell systems, i.e. those with unfilled d or f shell being characterized by resultant vector terms with $L \neq 0$ or $J \neq 0$. Including the configuration mixing phenomenon via higher orders of perturbation approach can also lead to a weak but from theoretical viewpoint significant splitting of the S term as a part of so–called zero–field–splitting in the EPR domain. The splitting of the ^8S term of Gd^{3+} and Eu^{2+} ions [12, 70] may serve as an example.

Both the extensive experimental material and theoretical predictions show that the crystal field potential producing the splitting can be with satisfactory accuracy assumed as the one–electron potential seen by the open–shell electrons, each of them separately. The correlation effects for bound ions seem to be as in the free ions negligible. The ion parameters like the spin–orbit coupling and Slater integrals although somewhat different in solids and compounds than in the free ions are practically universal for all various crystal field states of the parent term [71].

The parameterization of the crystal field Hamiltonian $\mathcal{H}_{CF}(\mathbf{r})$ accepted in this book and commonly used elsewhere is derived from its expansion into a series of tensor operators, most often into the normalized spherical

31

harmonics $C_q^{(k)}$ after Wybourne [3]

$$\mathcal{H}_{CF}(\mathbf{r}) = \sum_k \sum_q \beta_{kq}(r) C_q^{(k)}(\mathbf{r}/r) \tag{3.1}$$

where $\beta_{kq}(r)$ is the radial coefficient.

In turn, since $\mathcal{H}_{CF}(\mathbf{r})$ acts only within the open–shell states which are characterized by the same radial distribution $P_{nl}(r)$, the Hamiltonian can be replaced by its modified form which depends on the angle coordinates only. Then, for i-th electron

$$h_{CF}(\mathbf{r}_i/r_i) = \sum_k \sum_q B_{kq} C_q^{(k)}(\mathbf{r}_i/r_i) \tag{3.2}$$

where B_{kq} parameter is the $\beta_{kq}(r)$ coefficient averaged over the radial distribution of the open–shell electrons which now is independent on i

$$B_{kq} = \langle\, \beta_{kq}(r)\, \rangle = \int P_{nl}^2(r)\beta_{kq}(r)dr \tag{3.3}$$

Sum of the one–electron Hamiltonians $h_{CF}(i)$ (Eq.3.2) over all open–shell electrons yields the parameteric form of the total crystal field Hamiltonian $\mathcal{H}_{CF}$

$$\mathcal{H}_{CF} = \sum_i h_{CF}(\mathbf{r}_i/r_i) = \sum_i \sum_k \sum_q B_{kq} C_q^{(k)}(\mathbf{r}_i/r_i) \tag{3.4}$$

where B_{kq} stands for the crystal field parameter in notation which is commonly used in the parameterization practice.

Several various but equivalent methods of the parameterization are met in the literature. Relations between them are considered in chapter 2.

The crystal field Hamiltonian can also be expressed within the second quantization formalism, i.e. in the occupation number representation as

$$(\mathcal{H}_{CF})_{s.q.} = \sum_m \sum_{m'} \sum_\sigma \langle lm|\mathcal{H}_{CF}|lm'\rangle a_{m\sigma}^+ a_{m'\sigma} \tag{3.5}$$

where $a_{m\sigma}^+$ and $a_{m'\sigma}$ are the creation and annihilation operators of electrons in states specified with the operator indices, $\langle lm|\mathcal{H}_{CF}|lm'\rangle$ is the corresponding matrix element of the $\mathcal{H}_{CF}$ operator (Eq.3.4) within the $|lm\rangle$ state basis.

Both expressions, Eq.3.4 and Eq.3.5, are entirely equivalent although the operators entering them act within quite different function spaces. This equivalence is proved in chapter 11. The problem of calculating the $C_q^{(k)}$ matrix elements is put aside in this book. It is fully solved and the results for all possible bases of states for various levels of vector coupling of the orbital and spin angular moments in d– and f– ions are tabulated [45, 1]. A reference summary is presented in chapter 2.

Two stages or two levels can be distinguished in interpretation of crystal field splittings. The first, of qualitative nature, refers to the point symmetry of the central ion, the number of levels arising from a free–ion term and their irreducible representations in the point group symmetry. This is entirely worked out problem [3, 2]. The second level concerns the splitting values or in other words the B_{kq} parameters and contrary to the former phenomenological representation needs a reliable microscopic model. The later is not effectively diagnosed so far and from the very beginning of the theory has been a challenge facing solid state physicists and chemists.

The space distributions of electron density in solid, their mutual penetration and overlap, mutual polarizability, potentiality of intra– and interatomic (charge transfer) excitations, existence of inter–electronic exchange interacting and standing the Pauli exclusion principle lead to several various mechanisms which combine to the effective crystal field potential and determine the crystal field parameters [27].

As the effective crystal field potential in a specified model we will understand such a potential matrix elements of which within the basis of initial (free–ion) wave–functions of the open–shell are identical to the corresponding matrix elements of the true initial Hamiltonian within the basis of appropriately renormalized states [26].

According to extent of the basis of electron states treated explicitly as well as assumptions relating to the interactions involved the crystal field theory possesses several levels of generality and each of them splits into different variants.

The primary conception – the point charge model (PCM) [72, 38, 39] was founded on point approximation of electrostatic ionic charges in crystals, and that for both the nearest neighbors, i.e. ligands and those from further coordination spheres. In the case of spherically symmetric ions and separateness of their charge clouds this assumption would be justified. Within the frame of this model no electronic states except the open–shell states of the central ion are treated explicitly. The charge ascribed to a lattice site (or ionic core) is the resultant charge of its electron shells and nucleus. An interaction of a neutral ligand can not be expressed in the model. Similarly, there is no possibility to take into account the exchange interaction between the open–shell central ion electrons and ligand electrons. Schematic outline of the model is sketched in Fig.3.1.

Unfortunately, the PCM has turned out to be clearly inadequate and useless for calculation of crystal field parameters. On the other hand in the case of further neighbours this approximation, if polarization effects are negligible, is correct. In reality a typical central ion–ligand distance is too short to treat the ligands as points. The first step in generalization of the model was taking into consideration a space distribution of ligand electron density and its penetration into the central ion electron density area [73, 74], Fig.3.2.

34

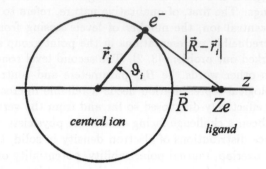

Fig. 3.1 Point charge model in the local central ion – ligand system, the ligand of Ze charge at R distance from the central ion

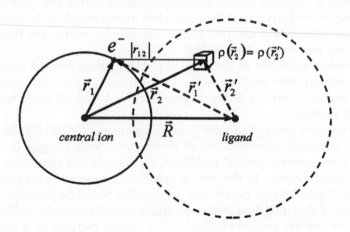

Fig. 3.2 Local model with space distribution of ligand electron density

A negative result of this kind extension of the model was a certain surprise. As often happens the first attempt to improve results of the model completely destroyed it.

Both above approaches are comprised within the limits of one–configurational approximation in which all the ions are assumed to be in their ground electronic configurations. Within the frame of this approximation another one step likening the model to reality can be done, viz. the non–orthogonality of the central ion wave functions, specifically its open–shell (a) and outer closed shells (ξ), in relation to ligand wave functions (χ) as well as that regarding the electron wave functions coming from various ligands (χ_{t_1}, χ_{t_2})

can be taken into account. These three types of nonorthogonality: $a - \chi$, $\xi - \chi$ and $\chi_{t_1} - \chi_{t_2}$ where t_1 and t_2 indices denote different ligands, yield the three so–called contact–renormalization corrections in the crystal field potential, respectively. Then the model is sometimes called the exclusion model [27]. Schematically, these non–orthogonality effects are shown in Fig.3.3, 3.4 and 3.5.

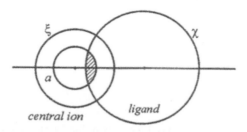

Fig. 3.3 Non-orthogonality of $a - \chi$ type responsible for the contact-covalency contribution

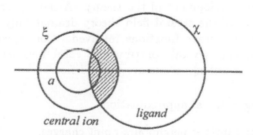

Fig. 3.4 Non-orthogonality of $\xi - \chi$ type responsible for the contact-screening contribution

The succeeding improvement of the model relies upon including electronic excitations and first of all the one–electron excitations leading to many–configurational approximation. These excitations can be either of intra–ionic or inter–ionic type. The former manifest themselves as screening and polarization mechanisms (this time the true polarization, not the contact one), the latter, in turn, manifests itself as a charge transfer from ligand to metal ion, or so–called covalency mechanism. In a very simplified and schematic way these mechanisms can be pictured as in Figs 3.6, 3.7 and 3.8.

The evolution of the crystal field models sketched shortly above has had its direct repercussion in history of employing the crystal field theory. Evidently, development of the crystal field theory distinguishes itself by

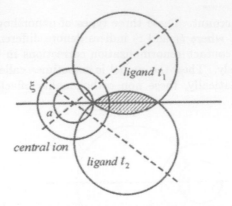

Fig. 3.5 Non-orthogonality of $\chi_{t_1} - \chi_{t_2}$ type responsible for the contact-polarization contribution

instructive fluctuations in its standing and popularity oscillating in time between periods of prosperity and fall [71, 75, 76]. Today we perceive it as natural phases of the development of the theory. A more precise analysis shows that correctness of the crystal field theory dramatically depends on appropriate choice of the wave functions involved. Consequently, in the localized electron systems several contributions combining to the crystal field potential can be specified. These are derived from:

within the one–configurational approximation:

1. ligands cores and further neighbours point charges,

2. space distribution of ligand electron density including $a - \chi$ exchange interaction,

3. $a - \chi$ non–orthogonality,

4. $\xi - \chi$ non–orthogonality,

5. $\chi_{t_1} - \chi_{t_2}$ non–orthogonality,

and within the many–configurational approach:

6. covalency,

7. screening,

8. polarization

In the metallic systems some specific mechanisms resulting from the presence of conduction electrons and broadening the energy levels ought to be additionally allowed for. The three fundamental contributions have been quantitatively (or rather semi–quantitatively) defined in the theory by Dixon and Wardlaw [77]. These are:

1. screening the ionic cores potentials with the conduction electrons,

2. contribution of so–called virtual bound states,

3. hybridization of localized and delocalized electron states.

In spite of the fact that the real crystal field potential is far from its electrostatic prototype its dominating part is still subordinated to the additivity law with respect to individual ions in the lattice (or compound) which means that the global crystal field potential is, with a good accuracy, the sum of all partial potentials coming from each of the ions. It makes allowance for partition of the whole system into simple binary subsystems of z-axis joining the central ion nucleous with other ion chosen. In such a local coordinate system, due to typical spherical symmetry of the ions the crystal field produced in the central ion area possesses an axial symmetry $C_{\infty v}$ and can be expressed with only three independent parameters B_{20}, B_{40}, B_{60} which in this particular case are called the intrinsic parameters [27] and are denoted by b_2, b_4 and b_6, respectively. Superposition of the partial potentials to obtain the global one and calculating conventional crystal field parameters B_{kq} from the intrinsic ones is performed by means of appropriate rotations of the local coordinate systems to adjust them with the main system and is the basis of the superposition model [27] and the angular overlap model (AOM) [78].

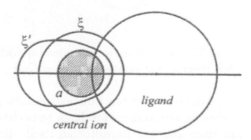

Fig. 3.6 Screening mechanism with the central ion outer electrons ξ. The ligand electronic wave functions are also subjected to polarization (deformation) which is not shown

However, there are some non–additive mechanisms, in general of secondary meaning, which break loose from the superposition approach and need non–local or global treatment. In the relevant points of the review this characteristic will be recalled.

The basic formalism being a framework of the considered approximate models consists in the conception of group product functions [79] (chapter 4) and perturbational expansion in the projection operator technics [26, 80]. Let us notice that already the zero–approximation Hamiltonian of open–shell electrons (Eq.33 in chapter 4) has a structure showing the natural partition of electrons into weakly interacting groups and is composed of several corresponding contributions.

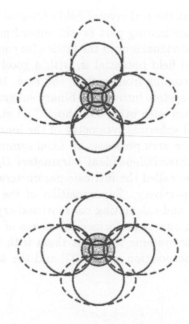

Fig. **3.7** Polarization of ligands, schematic examples: a) with preservation of symmetry (C_{4v}), b) with lowering the symmetry ($C_{4v} \rightarrow C_{2v}$)

Any improvement of the one–electron functions resulting from their orthogonalization or including admixtures produces respective complications both in the wave functions and the Hamiltonian itself. However, there is always a possibility to return at the cost of expansion of the Hamiltonian by appropriate corrections, to the initial basis of the open–shell states only. The Hamiltonian modified in this way becomes the effective Hamiltonian and its one–electron part of non–spherical symmetry – the effective crystal field potential.

In some domains of solid state physics and solid state chemistry a precise, quantitative knowledge of the electron eigenfunctions is not necessarily needed. Such a situation takes place in electron and neutron spectroscopy if we are interested in interpretation of energy level spectrum but not interested in intensities of the transitions or the cross–sections for inelastic neutron scattering. Then, the appropriate number of independent equations for energy level distances allows the sought for set of the parameters to be experimentally determined.

On the other hand to calculate these parameters ab initio the exact knowledge of the eigenfunctions is required. The lack of the knowledge is just the reason of general failure in theoretical calculations of the parameters. To understand the difficulties connected with the ab initio calculations of crystal field parameters one ought to realize that the crystal field potential is only a marginal part of otherwise large total potential.

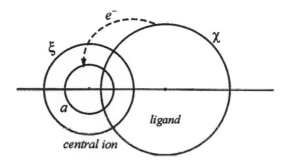

Fig. 3.8 Covalency mechanism – charge transfer $\chi \rightarrow a$ leading to the basic covalency admixture (a^{n+1}) of the ground configuration a^n.

In principle there are two approaches to the problem. The first relies on ab initio construction of the crystal field potential by making allowance for all specified mechanisms independently on a linear perturbational level. Unfortunately, in general we do not know if the requirements imposed by perturbation method regarding the linearity and independence of particular contributions are fulfilled. One may rather suspect that often they are not justified. This is a main weakness of the traditional approach. The second method consists in separation of the crystal field potential from the global potential on the ground of the total charge density distribution which however has to be known with respectively high accuracy. This method dominates in recent years but despite of its intensive development does not lead to fully satisfactory results. We are even not sure if the charge density distribution alone, irrespective of accuracy we know it, is the only factor determining the crystal field effect.

So, the conclusion arises that nowadays except the passive phenomenological description of the crystal field effect we have at our disposal no effective crystal field theory based on reliable microscopic model which would be enough rigorous and simultaneously useful and handy for experimentalists. Is our claim feasible? In the next chapters construction of the effective potentials for various models, their development with improvement of the models, the room for simplification and the physical meaning of the defined contributions are presented.

CHAPTER 4

Ionic complex or quasi–molecular cluster. Generalized product function

On account of the complex nature of the crystal field effect setting up its theoretical model will be preceded by suggesting itself and advisable partition of the whole surroundings of the chosen ion (the central ion) into two separate areas – the close vicinity of the central ion labeled as ionic complex or cluster and the outer surroundings. In the close vicinity the quantum–chemistry methods are rather necessary whereas the contribution from the outer space may be well enough described with classical electrostatics. The main part of the crystal field effect comes from the former area and it will be the subject of our considerations in this chapter.

Despite of the limitation introduced a system of electrons and nuclei forming the cluster remains in its nature a complex many–body problem. Unattainability of exact solution of the problem from the one hand and inadequacy of the simplified models from the other suggest an intermediate approach as the only practical resolution. In such an approach only a part of electrons of the considered cluster is treated explicitly whereas the remaining just as the nuclei are taken into account as point charges in the electrostatic outcome.

A natural partitioning the electrons to weakly interacting groups related to individual ions of the cluster allows the generalized product function formalism [79, 26] to be applied. This formalism enables us to reduce the initial N-electron system into several subsystems of lower dimensions.

4.1 Concept of the generalized product function

Let us suppose that the cluster contains N electrons which are divided into subsets labeled $A, B, \ldots, R, S, \ldots$ and $N_A + N_B + \ldots + N_R + N_S + \ldots = N$.

A trial initial wavefunction in the group product form is

$$
\begin{aligned}
\Phi_k(\mathbf{x}_1, \mathbf{x}_2, \ldots, \mathbf{x}_N,) \;=\;& M_k \hat{A}[\phi_{Aa}(\mathbf{x}_1, \ldots, \mathbf{x}_{N_A}) \times \\
&\times\; \phi_{Bb}(\mathbf{x}_{N_A+1}, \ldots, \mathbf{x}_{N_A+N_B}) \ldots] \\
(k \;=\;& Aa, Bb, \ldots, Rr, Ss, \ldots)
\end{aligned}
\tag{4.1}
$$

Here $\hat{A}$ is the antisymmetrizer making the product antisymmetric for all permutations of the N electrons, $\{\mathbf{x}_i\}$ are combined space $\{\mathbf{r}_i\}$ and spin $\{\mathbf{s}_i\}$ coordinates, the factor M_k simply normalizes the wave function to unity, ϕ_{Aa}, ϕ_{Bb}, ... are normalized antisymmetric wave functions for the groups of electrons A, B, ... with quantum numbers a, b, ..., respectively, and k is an index that picks out the combination $(a, b, \ldots, r, s, \ldots)$.

In the limit of zero interaction between the group functions (factors) in $\Phi_k(\mathbf{x}_1, \mathbf{x}_2, \ldots, \mathbf{x}_N,)$ (Eq.4.1) could in principle give an exact description of the state of the N electrons. Evidently, the antisymmetrized product of group functions may be regarded as a generalization of the antisymmetrized spin–orbital product, the individual factors now describing whole groups of electrons instead of a single electron. If the electron groups are in fact weakly interacting, e.g. as a result of their separation in space, a generalized product may be an accurate wave function because each factor may, in principle, describe correlation within the groups of electrons to which it refers. In one–electron self–consistent methods such as the molecular orbital (MO) method this effect is lost. A further refinement may be achieved by allowing the mixing of a number of generalized products functions, each product corresponding to a particular selection of states for the individual groups. This mixing is entirely analogous to configuration interaction.

Specification of the operator antisymmetrizing a group product function, $\hat{A}$, (Eq.4.1), under preservation of partitioning electrons into separate groups in not, in general case, a trivial task. The antisymmetry may generally be ensured by forming an appropriately developed superposition of the product functions. In the case of single product of determinants $|\ldots| \times |\ldots| \times |\ldots| \ldots$ the antisymmetry may be readily reached by erasing the middle $|x|$'s to get the one global determinant [76]. In general, if the group functions are individually antisymmetric the antisymmetrizer is a sum of transpositions which exchange variables between different groups only.

There is used a simplified approach in which instead of full group product function its one–determinant counterpart, so–called reference state function [27] comprising apart from the closed–shells spinorbitals only one chosen open–shell spin–orbital. This is directly harmonizing with the one–electron concept of the crystal field theory. A satisfying adequacy most of the interpretations within the frame of the model justifies such a choice of the trial function. The results are equivalent to those obtained in the MO model [81].

Generalized product functions can be handled with the same facility as spin–orbital products only if we impose certain orthogonality requirements. These are much stronger than those imposed on ordinary spin–orbitals and take the form

$$\int \phi_{Rr}^*(\mathbf{x}_1, \mathbf{x}_i, \mathbf{x}_j, \ldots)\phi_{Ss}(\mathbf{x}_1, \mathbf{x}_k, \mathbf{x}_l, \ldots)dx_1 \equiv 0, \qquad (R \neq S) \qquad (4.2)$$

The usual orthonormality requirement is

$$\int \phi_{Rr}^*(\mathbf{x}_1, \mathbf{x}_2, \ldots, \mathbf{x}_{N_R})\phi_{Rr'}(\mathbf{x}_1, \mathbf{x}_2, \ldots \mathbf{x}_{N_R})d\mathbf{x}_1\mathbf{x}_2 \ldots \mathbf{x}_{N_R} = \delta_{rr'}.$$

This means that the result of integrating over any one variable ($\mathbf{x}_1$) common to two different group functions must vanish identically for all values of the other variables $\mathbf{x}_i, \mathbf{x}_j, \ldots$. This is so–called strong orthogonality requirement.

In consequence, determination of matrix element expressions, the procedure and results run entirely parallel to those of the Slater method for Slater determinant functions. In the model of generalized product functions, especially in calculations of their matrix elements the idea of the density matrices or more generally the transition density matrices built up for both the whole products ($\Phi_k, \Phi_{k'}$) and group functions themselves (ϕ_{Rr}, ϕ_{Ss}) is very useful.

4.2 The density functions and the transition density functions

The one–electron density function $\varrho_1(\mathbf{x}_1)$, i.e. the probability of any electron being found at point $\mathbf{x}_1$ (spin included), or rather in the vicinity $d\mathbf{x}_1$, other electrons anywhere, is defined as [82]

$$\varrho_1(\mathbf{x}_1) = N \int \Psi(\mathbf{x}_1, \mathbf{x}_2, \ldots, \mathbf{x}_N)\Psi^*(\mathbf{x}_1, \mathbf{x}_2, \ldots \mathbf{x}_N)d\mathbf{x}_2 \ldots d\mathbf{x}_N \qquad (4.3)$$

where $\Psi(\mathbf{x}_1, \mathbf{x}_2, \ldots, \mathbf{x}_N)$ is the N-electron wave function. The indistinguishability of the electrons is accounted for the factor N. It should be noted that $\mathbf{x}_1$ on the left of Eq.4.3 refers to point $\mathbf{x}_1$ at which the density is evaluated rather than to the coordinates of electron "1". Similarly, the two–electron density function $\varrho_2(\mathbf{x}_1, \mathbf{x}_2)$, i.e. the probability of two electrons being found simultaneously at points $\mathbf{x}_1$ and $\mathbf{x}_2$ is

$$\varrho_2(\mathbf{x}_1, \mathbf{x}_2) = N(N-1) \int \Psi(\mathbf{x}_1, \mathbf{x}_2, \ldots, \mathbf{x}_N) \times$$
$$\times \Psi^*(\mathbf{x}_1, \mathbf{x}_2, \ldots \mathbf{x}_N)d\mathbf{x}_3 \ldots \mathbf{x}_N \qquad (4.4)$$

In order to express everything in terms of the density functions a very simple device is introduce. To make Ψ^* function (staying on the left in matrix element) immune from the effect of any operator we change the name of its variables from $\mathbf{x}$ to $\mathbf{x}'$. It means we denote $\Psi(\mathbf{x})\Psi^*(\mathbf{x}')$ by $\varrho(\mathbf{x};\mathbf{x}')$. Consequently, we have

$$\varrho_1(\mathbf{x}_1;\mathbf{x}_1') = N \int \Psi(\mathbf{x}_1,\mathbf{x}_2,\ldots,\mathbf{x}_N)\Psi^*(\mathbf{x}_1',\mathbf{x}_2,\ldots\mathbf{x}_N)d\mathbf{x}_2\ldots d\mathbf{x}_N \quad (4.5)$$

and

$$\begin{aligned}
\varrho_2(\mathbf{x}_1,\mathbf{x}_2;\mathbf{x}_1',\mathbf{x}_2') &= N(N-1)\int \Psi(\mathbf{x}_1,\mathbf{x}_2,\ldots,\mathbf{x}_N) \times \\
&\times \Psi^*(\mathbf{x}_1',\mathbf{x}_2',\ldots\mathbf{x}_N)d\mathbf{x}_3\ldots d\mathbf{x}_N \quad (4.6)
\end{aligned}$$

The two above density functions are used in calculating the diagonal elements of all operators (their expectation values).

To determine off–diagonal matrix elements the transition density functions are exploited.

$$\begin{aligned}
\varrho_1(kk'\mathbf{x}_1;\mathbf{x}_1') &= N \int \Psi_k(\mathbf{x}_1,\mathbf{x}_2,\ldots,\mathbf{x}_N) \times \\
&\times \Psi_{k'}^*(\mathbf{x}_1',\mathbf{x}_2,\ldots\mathbf{x}_N)d\mathbf{x}_2\ldots d\mathbf{x}_N \quad (4.7)
\end{aligned}$$

and

$$\begin{aligned}
\varrho_2(kk'|\mathbf{x}_1,\mathbf{x}_2;\mathbf{x}_1',\mathbf{x}_2') &= N(N-1)\int \Psi_k(\mathbf{x}_1,\mathbf{x}_2,\ldots,\mathbf{x}_N) \times \\
&\times \Psi_{k'}^*(\mathbf{x}_1',\mathbf{x}_2',\ldots\mathbf{x}_N)d\mathbf{x}_3\ldots d\mathbf{x}_N \quad (4.8)
\end{aligned}$$

Note, that for $k = k'$ the density function for a single state is obtained.

When we deal with generalized product functions (Eq.4.1) two kinds of density and transition density functions can be defined.

The first refer to the whole product functions Φ_k, $\Phi_{k'}$ and are denoted by $\varrho_1(kk'|\mathbf{x}_1;\mathbf{x}_1')$ and $\varrho_2(kk'|\mathbf{x}_1,\mathbf{x}_2;\mathbf{x}_1',\mathbf{x}_2')$, respectively.

The second kind refers to the group functions themselves, e.g. ϕ_{Rr}, $\phi_{Rr'}$ and they are $\varrho_1^R(rr'|\mathbf{x}_1;\mathbf{x}_1')$ and $\varrho_2(rr'|\mathbf{x}_1\mathbf{x}_2;\mathbf{x}_1'\mathbf{x}_2')$, respectively.

4.3 Model of the generalized product functions

From the strong orthogonality conditions follows that non–zero $\varrho_1(kk'|\mathbf{x}_1;\mathbf{x}_1')$ are only if k is different from k' at most in one group and $\varrho_2(kk'|\mathbf{x}_1\mathbf{x}_2;\mathbf{x}_1'\mathbf{x}_2')$ if k is different from k' at most in two groups. Analogously, $\varrho_1^R(rr'|\mathbf{x}_1;\mathbf{x}_1')$ and $\varrho_2^R(rr'|\mathbf{x}_1\mathbf{x}_2;\mathbf{x}_1'\mathbf{x}_2')$ are different from zero if r differs from r' at most in one and two spin–orbitals, respectively.

Summarizing, the following one– and two–electron density and transition density functions are effective for the generalized product functions.

Relations between the two kinds of density and transition density functions are revealed too.

$$(a) \quad \text{if} \quad k = k' = (a, b, \dots, r, s, \dots)$$

$$\varrho_1(kk|\mathbf{x}_1; \mathbf{x}_1') = \sum_R \varrho_1^R(rr|\mathbf{x}_1; \mathbf{x}_1')$$

$$\varrho_2(kk|\mathbf{x}_1, \mathbf{x}_2; \mathbf{x}_1', \mathbf{x}_2') = \sum_R \varrho_2^R(rr|\mathbf{x}_1, \mathbf{x}_2; \mathbf{x}_1', \mathbf{x}_2') +$$

$$+ \quad (1 - P_{12}) \sideset{}{'}\sum_{R,S} \varrho_1^R(rr|\mathbf{x}_1; \mathbf{x}_1') \varrho_1^S(ss|\mathbf{x}_2; \mathbf{x}_2')$$

$$(b) \quad \text{if} \quad k = (a, b, \dots, r, s, \dots)$$
$$\text{and} \quad k' = (a, b, \dots, r', s, \dots) - \text{one difference}$$
$$\varrho_1(kk'|\mathbf{x}_1; \mathbf{x}_1') = \varrho_1^R(rr'|\mathbf{x}_1; \mathbf{x}_1')$$
$$\varrho_2(kk'|\mathbf{x}_1, \mathbf{x}_2; \mathbf{x}_1', \mathbf{x}_2') = \varrho_2^R(rr'|\mathbf{x}_1, \mathbf{x}_2; \mathbf{x}_1', \mathbf{x}_2') +$$
$$+ \quad (1 - P_{12}) \sum_{S(\neq R)} \left[\varrho_1^R(rr'|\mathbf{x}_1; \mathbf{x}_1') \varrho_1^S(ss|\mathbf{x}_2; \mathbf{x}_2') + \right.$$
$$+ \quad \left. \varrho_1^S(ss|\mathbf{x}_1; \mathbf{x}_1') \varrho_1^R(rr'|\mathbf{x}_2; \mathbf{x}_2') \right] \tag{4.9}$$

$$(c) \quad \text{if} \quad k = (a, b, \dots, r, s, \dots)$$
$$\text{and} \quad k'' = (a, b, \dots, r', s', \dots) - \text{two differences}$$
$$\varrho_2(kk''|\mathbf{x}_1, \mathbf{x}_2; \mathbf{x}_1', \mathbf{x}_2') = (1 - P_{12}) \left[\varrho_1^R(rr'|\mathbf{x}_1; \mathbf{x}_1') \varrho_1^S(ss'|\mathbf{x}_2; \mathbf{x}_2') + \right.$$
$$+ \quad \left. \varrho_1^S(ss'|\mathbf{x}_1; \mathbf{x}_1') \varrho_1^R(rr'|\mathbf{x}_2; \mathbf{x}_2') \right]$$

where P_{12} interchanges the unprimed variables $\mathbf{x}_1$ and $\mathbf{x}_2$ as the consequence of indistinguishability of electrons and the prime at $\sum'_{R,S}$ denotes that $R = S$ case has been excluded.

The density and transition density functions determine the diagonal and off–diagonal matrix elements of all operators, respectively

$$\langle \Psi_k | \sum_i h(i) | \Psi_k \rangle = \int_{\mathbf{x}_1' = \mathbf{x}_1} h(1) \varrho_1(kk|\mathbf{x}_1; \mathbf{x}_1') d\mathbf{x}_1 \tag{4.10}$$

$$\langle \Psi_k | \sideset{}{'}\sum_{i,j} g(i,j) | \Psi_k \rangle = \int_{\substack{\mathbf{x}_1' = \mathbf{x}_1 \\ \mathbf{x}_2' = \mathbf{x}_2}} g(1,2) \varrho_2(kk|\mathbf{x}_1, \mathbf{x}_2; \mathbf{x}_1', \mathbf{x}_2') d\mathbf{x}_1 d\mathbf{x}_2 \tag{4.11}$$

$$\langle \Psi_k | \sum_i h(i) | \Psi_l \rangle = \int\limits_{\mathbf{x}_1' = \mathbf{x}_1} h(1) \varrho_1(kl | \mathbf{x}_1; \mathbf{x}_1') d\mathbf{x}_1 \qquad (4.12)$$

$$\langle \Psi_k | {\sum_{i,j}}' g(i,j) | \Psi_l \rangle = \int\limits_{\substack{\mathbf{x}_1' = \mathbf{x}_1 \\ \mathbf{x}_2' = \mathbf{x}_2}} g(1,2) \varrho_2(kl | \mathbf{x}_1, \mathbf{x}_2; \mathbf{x}_1', \mathbf{x}_2') d\mathbf{x}_1 d\mathbf{x}_2 \qquad (4.13)$$

where the summations on the left sides run over all electrons under consideration.

Thus, using Eqs 4.9 abc and Eqs 4.10–4.13 the matrix elements of the total Hamiltonian

$$\mathcal{H} = \sum_i h(i) + \frac{1}{2} {\sum_{i,j}}' g(i,j) \qquad (4.14)$$

between the group product functions (Eq.4.1) are readily given as

$$(a) \qquad \mathcal{H}_{kk} = \sum_R \mathcal{H}^R(rr) + \frac{1}{2} {\sum_{R,S}}' \left[J^{RS}(rr, ss) - K^{RS}(rr, ss) \right]$$

$$(b) \qquad \mathcal{H}_{kk'} = \mathcal{H}^R(rr') + \sum_{S(\neq R)} \left[J^{RS}(rr', ss) - K^{RS}(rr', ss) \right] \qquad (4.15)$$

$$(c) \qquad \mathcal{H}_{kk''} = J^{RS}(rr', ss') - K^{RS}(rr', ss')$$

where as above k' and k'' refer to product functions differing from that of k in one group (R) and two groups (R, S), respectively. $J^{RS}(rr', ss')$ and $K^{RS}(rr', ss')$ correspond to generalized two–electron direct and exchange Coulomb integrals

$$J^{RS}(rr', ss') = \int g(1,2) \varrho_1^R(rr' | \mathbf{x}_1; \mathbf{x}_1) \varrho_1^R(ss' | \mathbf{x}_2; \mathbf{x}_2) d\mathbf{x}_1 d\mathbf{x}_2 \qquad (4.16)$$

$$K^{RS}(rr', ss') = \int g(1,2) \varrho_1^R(rr' | \mathbf{x}_2; \mathbf{x}_1) \varrho_1^R(ss' | \mathbf{x}_1; \mathbf{x}_2) d\mathbf{x}_1 d\mathbf{x}_2 \qquad (4.17)$$

and

$$\mathcal{H}^R(rr') = \int\limits_{\mathbf{x}_1' = \mathbf{x}_1} h(1) \varrho_1^R(rr' | \mathbf{x}_1; \mathbf{x}_1') d\mathbf{x}_1 +$$

$$+ \frac{1}{2} \int g(1,2) \varrho_2^R(rr' | \mathbf{x}_1, \mathbf{x}_2; \mathbf{x}_1, \mathbf{x}_2) d\mathbf{x}_1 d\mathbf{x}_2 \qquad (4.18)$$

Since $g(1,2) = 1/r_{12}$ is a factor in the integrand, the primes are dropped.

Under above conditions the expectation value of $\mathcal{H}$ for a group product function (Eq.4.1) may be written in separable form

$$E_k = \langle \Phi_k | \mathcal{H} | \Phi_k \rangle = E_{Aa} + E_{Bb} + \ldots + E_{Rr} + E_{Ss} + \ldots \qquad (4.19)$$

and this is the main facility offered by this approach. It is also evident that $\mathcal{H}^R(rr')$ has the usual form of a matrix element expression of a Hamiltonian for N_R electrons of group R alone taken between states r and r'.

The optimized component group functions can be found just as spin-orbitals in the Hartree–Fock method by starting with a single generalized product and seeking a minimum value of the corresponding variational energy expression (Eq.4.15a), i.e.

$$E_k = \sum_R \mathcal{H}^R(rr) + \sum_{R<S} \left[J^{RS}(rr, ss) - K^{RS}(rr, ss) \right]$$

We require that E_k be stationary for first–order variation. Since each group variation makes its own first–order change in E_k it is enough to consider only one, e.g. $\phi_{Aa} \rightarrow \phi_{Aa} + \delta\phi_{Aa}$ where $\delta\phi_{Aa}$ is a variation constructed from A group orbitals alone so that ϕ_{Aa} will remain automatically strong orthogonal to the other group functions. It leads to the situation in which the stationary condition for E_k reduces to the stationary conditions for each group function separately but in the presence of all other groups. Thus, for the R group functions we have to minimize the expression

$$E_{eff}^R = \mathcal{H}^R(rr) + \sum_{S(\neq R)} \left[J^{RS}(rr, ss) - K^{RS}(rr, ss) \right] \qquad (4.20)$$

and then subsequently for all groups.

In this casting E_{eff}^R is the energy of the R group electrons in the effective field of the nuclei and all electrons of the remaining groups.

The interaction terms in Eq.4.20 can be expressed as matrix elements of actual one–electron operators describing the potential due to the electrons outside group R. For any group S we have

$$J^S(1)\Psi(\mathbf{x}_1) = \left[\int g(1, 2)\varrho_1^S(ss|\mathbf{x}_2; \mathbf{x}_2)d\mathbf{x}_2 \right] \Psi(\mathbf{x}_1) \qquad (4.21)$$

$$K^S(1)\Psi(\mathbf{x}_1) = \int g(1, 2)\varrho_1^S(ss|\mathbf{x}_1; \mathbf{x}_2)\Psi(\mathbf{x}_2)d\mathbf{x}_2 \qquad (4.22)$$

These two operators are the direct and exchange operators for an electron in the effective field due to the electrons of group S. So we have

$$J^{R,S}(rr, ss) = \langle \phi_{Rr} | \sum_{i=1}^{N_R} J^S(i) | \phi_{Rr} \rangle \qquad (4.23)$$

$$K^{R,S}(rr, ss) = \langle \phi_{Rr} | \sum_{i=1}^{N_R} K^S(i) | \phi_{Rr} \rangle \qquad (4.24)$$

The R–group electrons thus behave as if they were quite alone but each in the field described by the one–electron Hamiltonian

$$h_{eff}^R = h + \sum_{S \neq R} (J^S - K^S) \tag{4.25}$$

instead of h. The complete R-group energy expression (Eq.4.20) may be written as

$$E_{eff}^R = \langle \phi_{Rr} | \mathcal{H}_{eff}^R | \phi_{Rr} \rangle \tag{4.26}$$

where

$$\mathcal{H}_{eff}^R = \sum_{i=1}^{N_R} h_{eff}^R(i) + \frac{1}{2} \sum_{i,j=1}^{N_R} {}'g(i,j) \tag{4.27}$$

The stationary value condition for E_{eff}^R subject to ϕ_{Rr} variation is

$$\frac{\delta \langle \phi_{Rr} | \mathcal{H}_{eff}^R | \phi_{Rr} \rangle}{\delta \phi_{Rr}} = 0 \tag{4.28}$$

with

$$\langle \phi_{Rr} | \phi_{Rr} \rangle = 1$$

All groups other than R have been formally eliminated, their presence being absorbed into the effective one–electron Hamiltonian (Eq.4.25) and the optimization of the R–group wave function is essentially an N_R–electron problem.

The variational determination of optimum group functions is a well defined problem, which may be solved iteratively as the SCF problem. From plausible initial choice of the group wave functions ϕ_{Aa}, ϕ_{Bb}, ... the group density matrices are constructed, the matrix elements of the effective one–electron Hamiltonian are found and each group function is then optimized. After improving all functions the cycle is repeated until the computed density matrices agree with those used in the preceding cycle.

The above considerations have concerned construction and optimization of a single group product function. A more general case is wave function of the form

$$\Psi = \sum_k c_k \Phi_k \tag{4.29}$$

where Φ_k is a generalized product function corresponding to a particular assignment of states to the various electron groups. It is apparent that this expansion cannot be complete for every group function is expanded in terms of its own spin–orbitals with fixed partitioning of the electrons into groups. This is an "one–configurational" approach disregarding the possibility of electron transfer between groups. However, on account of degeneracy of any open–shell group functions the expansion (Eq.4.29) establishes the simplified crystal field theory.

The mixing different group product functions resembles the configuration interaction. The energy corrections due to admixture of "excited" configurations may be found perturbationally. There are two kinds of excitations: $k \rightarrow k'$ (single) and $k \rightarrow k''$ (double). It turns out that if the R-group functions have been determined variationally within the subspace of functions belonging to group R it follows that the single–excitation contribution vanishes [79]. Since the admixture modifies the charge density it cannot be further improved to this order of perturbation theory. In other words, polarization of each group by the presence of the others has been fully accounted for by optimizing the one–configuration approximation.

Incompleteness of the expansion (Eq.4.29) is the crucial point of the method. In the situation if validity of the crystal field theory depends critically on the correct choice of orthogonal functions [76] it seems to be a barrier not to be overcome. Several trials of extending the method will be presented in the next chapters.

4.4 Crystal field effect in the product function model

Let us now consider in this formalism a metal ion in crystal lattice from the crystal–field effect viewpoint. The initial trial function may be taken in the form of antisymmetrized and normalized product of chosen groups of spin–orbitals of all free ions forming the cluster (the central ion $+ t$ ligands) in their basic configurations. For the metal ion it is advisable to distinguish its open–shell states and additionally its external closed shell (s, p) states. This group product function has the form

$$\Phi_k = M_k \, \hat{A} \, [\phi_{Ak} \cdot \phi_\Xi \cdot \Pi_t \, \phi_{Xt}] \tag{4.30}$$

where ϕ_{Ak} is a wave function of electrons of the open–shell A, ϕ_Ξ and ϕ_{Xt} wave functions describe the singlet, one–determinant electron states of outer closer shells of the central ion and ligand t, respectively. N_A open–shell electrons in $2(2l_A+1)$ available states give $\begin{pmatrix} 2(2l_A + 1) \\ N_A \end{pmatrix}$ one–determinant wave functions $|a_{m_1} a_{m_2} \ldots a_{m_{N_A}}|$ of N_A order in the basis of orthonormal spin–orbitals a_m. At zero approximation level they are degenerate. Only a perturbation, e.g. the crystal field, splits them according to the secular equation solution.

In practice, the open–shell wave function as a product of a vector coupling the angular and spin momenta of its electrons is usually described with the large quantum numbers. Obviously, this form is reducible to the one–electron spin–orbital basis and this is just in the spirit of the crystal field theory in its both tensor (Wybourne) and operator equivalent (Stevens) approaches.

The Hamiltonian of all N electrons of the considered cluster has the form (in atomic units)

$$
\begin{aligned}
\mathcal{H} &= \sum_{i=1}^{N} h(\mathbf{r}_i) + \sum_{i>j} \frac{1}{r_{ij}} = \\
&= \sum_{i=1}^{N} \left[-\frac{\nabla_i^2}{2} + V^{\mathrm{Mc}}(r_i) + \xi \, \mathbf{l}_i \cdot \mathbf{s}_i + \sum_{t} V_t^{\mathrm{Lc}}(r_{ti}) + V^{\mathrm{ext}}(\mathbf{r}_i) \right] + \\
&\quad + \sum_{i>j} \frac{1}{r_{ij}}
\end{aligned}
\tag{4.31}
$$

where in the square bracket there is its one–electron part including apart from the kinetic energy of the electrons and the spin–orbit coupling, the potentials of the ionic cores: $V^{\mathrm{Mc}}(r_i)$ – that of the central ion and $V_t^{\mathrm{Lc}}(r_{ti})$ – that of ligand t as well as $V^{\mathrm{ext}}(\mathbf{r}_i)$ – the potential of the external surroundings of the cluster, $r_{ti} = |\mathbf{r}_i - \mathbf{R}_t|$ where $\mathbf{R}_t$ is the radius vector of ligand t in the central coordinate system and $\sum_{i>j} \frac{1}{r_{ij}}$ is the two–electron part of the Hamiltonian, i.e. interelectronic direct and exchange Coulomb interaction.

The expectation value of H (Eq.4.31) in the product state (Eq.4.30) is separable relative to the particular groups

$$
\langle \Phi_k | \mathcal{H} | \Phi_k \rangle = \langle \phi_{Ak} | \mathcal{H}_A | \phi_{Ak} \rangle + \langle \phi_\Xi | \mathcal{H}' | \phi_\Xi \rangle + \sum_{t} \langle \phi_{X_t} | \mathcal{H}'' | \phi_{X_t} \rangle
\tag{4.32}
$$

where

$$
\begin{aligned}
\mathcal{H}_A &= \sum_{i=1}^{N_A} \left[h(\mathbf{r}_i) + J^\Xi(\mathbf{r}_i) - K^\Xi(\mathbf{r}_i) + \sum_{t} \left(J^{X_t}(\mathbf{r}_i) - K^{X_t}(\mathbf{r}_i) \right) \right] + \\
&\quad + \sum_{i>j} \frac{1}{r_{ij}}
\end{aligned}
\tag{4.33}
$$

and the summation over i and j refers now to the electrons of the open–shell only, and J^Ξ, K^Ξ, J^{X_t}, K^{X_t} stand for direct and exchange Coulomb operators for Ξ and X_t groups, respectively, $\mathcal{H}'$ and $\mathcal{H}''$ have the same form as $\mathcal{H}$ (Eq.4.31) but the summation in $\mathcal{H}'$ runs over all electrons of the Ξ group and in $\mathcal{H}''$ over all electrons of the X_t group only.

The first component on the right side of Eq.4.32 is equal to energy of ϕ_{Ak} state of the open–shell electrons with taking into account the closer and further surroundings including the effective field produced by groups of outer electrons of the cluster's ions. The stationary value condition for $\langle \phi_{Ak} | \mathcal{H}_A | \phi_{Ak} \rangle$ supplemented with normalizing condition leads to the following expressions for both the optimized one–electron wave functions of the open–shell electrons and the coefficient c_i determining the eigenvectors of $\mathcal{H}_A$

$$\left(-\frac{\nabla^2}{2} + V^M(r) + \Delta V^M(r)\right) a_m = E_a a_m \qquad (4.34)$$

$$\sum_i \sum_j (\langle \phi_{Ai} | \mathcal{H}_A | \phi_{Aj} \rangle - E_A \delta_{ij}) c_i = 0 \qquad (4.35)$$

where E_a and E_A are the eigenenergies of the individual spin–orbitals a and the N_A–electron wave functions, respectively. The one–electron operator $V^M(r)$ defined as

$$V^M(r) = V^{Mc}(r) + \left[J^\Xi(r) - K^\Xi(r)\right] + \left[J^A(r) - K^A(r)\right] \qquad (4.36)$$

is the sum of metal ion core potential, the potential generated by outer central ion electrons (group Ξ) and the potential generated by N_A electrons of the open–shell A averaged over all a_m states. There is no need to remove the one–electron from the A group (the chosen one) since J-K for it cancels.

$\Delta V^M(r)$ is the spherically averaged potential of the central ion surroundings

$$\Delta V^M(r) = \left\{ \sum_t V_t^{Lc}(|\mathbf{r} - \mathbf{R}_t|) + \sum_t \left[J^{X_t}(\mathbf{r}) - K^{X_t}(\mathbf{r})\right] + \right.$$
$$\left. + V^{ext}(\mathbf{r}) \right\}_{\text{sph. av. for M}} \qquad (4.37)$$

In the case of the remaining group functions the procedure is similar. However, since this time we deal with one–determinant wave functions for closed shells we have to solve only the counterparts of Eq.4.34. These are, respectively

$$\left(-\frac{\nabla^2}{2} + V^M(r) + \Delta V^M(r)\right) \xi_m = E_\xi \xi_m \qquad (4.38)$$

$$\left(-\frac{\nabla^2}{2} + V_t^{Lc}(r_t) + J^{X_t}(r_t) - K^{X_t}(r_t) + \Delta V_t^L(r_t)\right) \chi_{tm} = E_{t\chi} \chi_{tm} \quad (4.39)$$

where

$$\Delta V_t^L(r_t) = \left\{ V^M(\mathbf{r}_t + \mathbf{R}_t) + \sum_{t' \neq t} \left[V_{t'}^{Lc}(\mathbf{r}_t + \mathbf{R}_t - \mathbf{R}_{t'}) + J^{X_{t'}}(\mathbf{r}_t + \mathbf{R}_t)\right.\right.$$
$$\left.\left. - K^{X_{t'}}(\mathbf{r}_t + \mathbf{R}_t)\right] + V^{ext}(\mathbf{r}_t + \mathbf{R}_t) \right\}_{\text{sph. av. for L}_t} \qquad (4.40)$$

The averaged self–consistent exchange potentials (K^Ξ, K^A, K^{X_t}) are assumed to be proportional to $\varrho^{1/3}$ [83, 90] and the charge density ϱ is found from the wave functions known.

In this way the initial N–electron problem has been reduced to the system of Eqs 4.34, 4.38 and 4.39 which may be solved exactly with iterative procedure till their self–consistency.

The solutions of one–electron equations (Eqs 4.34, 4.38 4.39) can be obtained by means of standard methods developed for free ions by Slater [84, 85, 86] with the Hartree–Fock method in non–relativistic approximation [48, 87] or its Dirac–Fock [88] and Dirac–Slater [89] versions for heavier ions. In the case of a virtual bound state, e.g. $5d$ one its wave function is available with the interpolation procedure by Herman and Skillman [85]. These one–electron functions are usually described in a convenient for further calculations form of the Slater type functions or their combinations (STO — Slater type orbital). Numerical technics for the functions in tabulated form are also in use.

In the specified Hamiltonian (Eq.4.33) yielding the eigenstates of the open–shell electrons (Eq.4.35) the one–electron terms of non–spherical symmetry contribute to the crystal field potential being sought for. In this approach the potential consists of contributions from the closest neighbors the potentials of their ionic cores treated as point charges and the direct and exchange Coulomb potentials of their outer closed shells, and from further neighbors (from outside the cluster) — as potentials of point charges. It should be underlined that the above result has been gained with the one–configuration approximation and for the strong orthogonality assumption. However, in fact the group functions built up of wave functions being solutions of Eqs 4.34, 4.38 and 4.39 do not obey the strong orthogonality requirement with a sufficient accuracy. Fortunately, the problem can be overcome in the frame of the same formalism. It will be presented thoroughly in chapter 7.

Going beyond the one–configuration approximation is also realizable. You only need to complement the initial trial function with admixtures of suitable functions. For the excited group product functions the analogous optimizing process can be carried out. We will meet the problem again in chapters 8, 9 and 10.

CHAPTER 5

Point Charge Model (PCM)

Within the framework of the point charge model the crystal field is identified with the electrostatic field generated by point charges attributed to the lattice sites. This is the simplest and historically first model. The concept of the crystal field and the PCM was introduced to the solid-state physics by Bethe in 1929 [68] who pointed out at the correspondence between the energy levels of an unfilled electron shell of a metal ion in crystal and the irreducible representations of its point symmetry group. Based on this simplified electrostatic model the interpretation of the magnetic properties of transition metal ions in their salts have been given by Kramers [69], Van Vleck [91, 92], Penney and Schlapp [93, 94], Abragam and Pryce [95]. This model turned out to be also very effective in explanation of optical [1, 96] and EPR [97, 13] spectra.

5.1 PCM potential and its parameters

The first of two main immanent features of the PCM is the additivity of the crystal field potential in the model with respect to its sources which means that the contribution of each lattice site (or ligand) to the global potential can be calculated separately. On the other hand, all unpaired electrons of the central ion are assumed to be equivalent and their potential energies in the crystal field independent, i.e. we deal with an one–electron potential.

Let us denote the central ion i-th electron radius vector and coordinates by $r_i(x_i, y_i, z_i)$ and those of j-th lattice site with $Z_j e$ charge by $R_j(X_j, Y_j, Z_j)$, where j runs over all lattice sites (ligands).

Thus, we have

$$V(x_i, y_i, z_i) = \sum_j \frac{Z_j e}{|R_j - r_i|} =$$

$$= \sum_j \frac{Z_j e}{[(X_j - x_i)^2 + (Y_j - y_i)^2 + (Z_j - z_i)^2]^{1/2}} \quad (5.1)$$

The potential such of a charge lattice can be, trough the expansion of the right side of Eq.5.1 into the Maclaurine series about the central ion position and suitable grouping of the expansion terms, presented in the form of the multipole series which exact to constant coefficients coincides with the expansion with respect to the spherical harmonic functions [39, 98, 35]. The form of the series is imposed by point symmetry of the central ion.

Due to the additivity of the potential it is enough to consider without loss of generality only a binary system – the central ion–ligand one. Such a system is characterized in the PCM by the axial symmetry $C_{\infty v}$ about the line joining the central ion and the ligand. This is the second immanent feature of the PCM which allows a convenient local coordinate frame with the z-axis along the $C_{\infty v}$ axis to be chosen. Then, the orientation of the x and y axes is inessential and optional. In this system the central atom nucleus there is at $(0, 0, 0)$ and the ligand at $(0, 0, R)$ point, respectively. Expansion (Eq.5.1) takes then the form

$$
\begin{aligned}
V(x, y, z) = \quad & V(0, 0, 0) - \frac{Ze}{R^2} z + \frac{1}{2} \frac{Ze}{R^3} \left(2z^2 - x^2 - y^2\right) - \\
& - \frac{1}{2} \frac{Ze}{R^4} \left(2z^3 - 3x^2 z - 3y^2 z\right) \\
& + \frac{1}{8} \frac{Ze}{R^5} \left(8z^4 + 3x^4 + 3y^4 + 6x^2 y^2 - 24x^2 z^2 - 24y^2 z^2\right) \\
& - \frac{1}{8} \frac{Ze}{R^6} \left(8z^5 + 15x^4 z + 15y^4 z - 40z^3 x^2 - 40z^3 y^2 + 30x^2 y^2 z^2\right) \\
& + \frac{1}{16} \frac{Ze}{R^7} \left(16z^6 - 5x^6 - 5y^6 - 135z^4 x^2 - 135z^2 y^2 \right. \\
& \left. \quad + 90z^2 x^4 + 90z^2 y^4 + 180x^2 y^2 z^2 - 15x^4 y^2 - 15y^4 x^2\right) \\
& + \dots
\end{aligned}
\tag{5.2}
$$

in which the uniform polynomials within the parentheses are the Cartesian counterparts of the spherical harmonics and as well the Stevens polynomials [35].

However, it is far more convenient to use spherical coordinates in the system which for the i-th electron and the chosen ligand are $\mathbf{r}_i(r_i, \upsilon_i, \varphi_i)$ and $\mathbf{R}(R, 0, 0)$, respectively. Thus, we have [99]

$$
V(\mathbf{r}_i) = \frac{Ze}{|\mathbf{R} - \mathbf{r}_i|} = \sum_k \frac{Ze \, r_i^k}{R^{k+1}} P_k \left(\cos \left(\frac{\mathbf{R} \cdot \mathbf{r}_i}{R \cdot r_i} \right) \right)
\tag{5.3}
$$

where $P_k \left(\cos \left(\frac{\mathbf{R} \cdot \mathbf{r}_i}{R \cdot r_i} \right) \right)$ is the Legendre polynomial of the k-th degree. The above expansion (Eq.5.3) is valid for $r_i < R$, i.e. if there is no overlapping the electron density with the charge Ze and the potential acting on the i-th electron obeys the Laplace equation $\Delta V(\mathbf{r}_i) = 0$.

Making use of the spherical harmonic addition theorem [100, 101] which now has the form

$$P_k(\cos v_i) = \sum_q (-1)^q C_q^{(k)}(v_i, \varphi_i) C_{-q}^{(k)}(0,0) \tag{5.4}$$

we get

$$V(\mathbf{r}_i) = \sum_k \sum_q \frac{Ze\, r_i^k}{R^{k+1}} (-1)^q C_q^{(k)}(v_i, \varphi_i) C_{-q}^{(k)}(0,0) \tag{5.5}$$

Since $C_{-q}^{(k)}(0,0)$ is different from zero only for $q = 0$ [102] (see also Eq.5.14) and

$$C_0^{(k)}(v, \varphi) = P_k(\cos v) \tag{5.6}$$

Hence

$$C_0^{(k)}(0,0) = P_k(1) = 1 \tag{5.7}$$

and finally

$$V(\mathbf{r}_i) = \sum_k \frac{Ze\, r_i^k}{R^{k+1}} C_0^{(k)}(v_i, \varphi_i) \tag{5.8}$$

Integrating $V(\mathbf{r}_i)$, (Eq.5.8), over the radial distribution of the i-th electron yields the following expression for the potential energy of the electron in the electrostatic field produced by individual lattice site

$$\overline{\mathcal{H}}(i) = \sum_k \overline{B}_{k0} C_0^{(k)}(v_i, \varphi_i) \tag{5.9}$$

where

$$\overline{B}_{k0} = \frac{Ze^2 \langle r_i^k \rangle}{R^{k+1}} \tag{5.10}$$

is the PCM parameter of the crystal field originating from a single lattice site distanced by R, and

$$\langle r_i^k \rangle = \langle \varphi_i | r_i^k | \varphi_i \rangle = \int_0^\infty dr_i P_{nl}^2(r_i) r_i^k \tag{5.11}$$

where φ_i is the i-th electron wave–function and $P_{nl}(r_i)$ its radial distribution.

It is easy to see that only the parameters with $q = 0$ are effective in the expansion (Eq.5.8) in this simple axial reference frame. They are identical with the intrinsic parameters $\overline{B}_{k0}$ in the superposition model [27].

The global crystal field potential can be obtained by means of a proper rotation of each local system performing separately which transforms it to the main system. The global crystal field parameters are equal to

$$B_{kq} = \sum_j C_q^{(k)*}(\theta_i, \phi_i) \overline{B}_k(R_j) \tag{5.12}$$

where (R_j, θ_j, ϕ_j) are the spherical coordinates of the j-th ligand in the main reference system (see chapter 2) where

$$C_q^{(k)^*}(\theta_i, \phi_i) = \left[\frac{(k-q)!}{(k+q)!}\right]^{1/2} P_k^q(\cos\theta_j)\exp(-iq\phi_j) \qquad (5.13)$$

and

$$P_k^q(\cos\theta_j) = \frac{(-1)^k}{2^k k!}(\sin\theta_j)^q \left[\frac{\partial}{\partial(\cos\theta_j)}\right]^{k+q} (\sin\theta_j)^{2k} \qquad (5.14)$$

is the associated Legendre function [102].

The needed values of $C_q^{(k)^*}(\theta, \phi)$ and $P_k^q(\cos\theta)$ functions may be calculated directly from Eq.5.13 and Eq.5.14 or taken from their tables.

The PCM unambiguously defines the potential of the surroundings. This is the pure electrostatic potential. But as for the parameters of the model (Eq.5.10) they additionally contain the $\langle r^k \rangle$ values which are attributes of the state on which the potential acts. These values can be theoretically calculated either in nonrelativistic [103, 115] or in relativistic [104, 114, 115] approaches. However, this is altogether off–PCM problem and one should not maintain that the PCM itself gives an exact recipe (within the model) for the crystal field parameters.

After such aforementioned rotations (see chapter 2) of the local coordinate systems some off–axial components with $q \neq 0$ can occur in the expansion of the global potential.

5.2   Simple partial PCM potentials.

Considering the additivity of the potential the global potential can be presented as a sum of contributions coming from optionally separated parts of the surroundings (lattice sites or ligands).

As an instance the potentials of three basic structural configurations of ligands of symmetry C_{4v}, C_{3v} and $C_{\infty}v$, respectively, are given below. Most of higher symmetry crystal field potentials may easily be obtained by combining the three module potentials.

(i) The PCM potential of four ligands with C_{4v} point symmetry (Fig.5.1).

$$\begin{aligned}
\mathcal{H}_{\mathrm{CF}}(C_{4v}, \mathrm{c.n} = 4) &= \sum_i \left[B_{20}\hat{C}_0^{(2)}\left(\frac{\mathbf{r}_i}{r_i}\right) + B_{40}\hat{C}_0^{(4)}\left(\frac{\mathbf{r}_i}{r_i}\right) + \right. \\
&+ \left. B_{44}\hat{C}_4^{(4)}\left(\frac{\mathbf{r}_i}{r_i}\right) + B_{60}\hat{C}_0^{(6)}\left(\frac{\mathbf{r}_i}{r_i}\right) + \right. \\
&+ \left. B_{64}\hat{C}_4^{(6)}\left(\frac{\mathbf{r}_i}{r_i}\right) \right]
\end{aligned} \qquad (5.15)$$

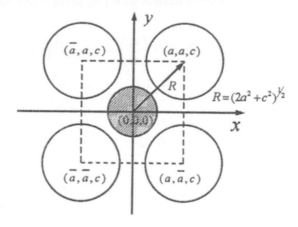

Fig. 5.1 Coordination of C_{4v} symmetry, central ion at $(0,0,0)$, $c.n. = 4$, $c/a = p$

Table 5.1 C_{4v} (c.n.$= 4$) PCM crystal field parameters

Parameter	Roots of $B_{kq}(p) = 0$	Roots of $\frac{dB_{kq}(p)}{dp} = 0$
$B_{20} = 4\left(\frac{Ze^2\langle r^2\rangle}{R^3}\right)\frac{p^2-1}{p^2+2}$	1	0 (min)
$B_{40} = 4\left(\frac{Ze^2\langle r^4\rangle}{R^5}\right)\frac{p^4-6p^2+3/2}{(p^2+2)^2}$	0.51	0 (max)
	2.40	1.22 (min)
$B_{60} = 4\left(\frac{Ze^2\langle r^6\rangle}{R^7}\right)\frac{2p^6-30p^4+456p^2-5}{2(p^2+2)^3}$	0.35	0 (min)
	1.24	0.75 (max)
	3.65	2.11 (min)
$B_{44} = -4\left(\frac{Ze^2\langle r^4\rangle}{R^5}\right)\frac{\sqrt{70}}{4}\frac{1}{(p^2+2)^2}$	—	0 (min)
$B_{64} = -4\left(\frac{Ze^2\langle r^6\rangle}{R^7}\right)\frac{3\sqrt{14}}{4}\frac{5p^2-1}{2(p^2+2)^3}$	0.45	0 (max)
		1.14 (min)

The negative signs of the tetragonal parameters stand for the coordinate system as shown in Fig.5.1. After rotation of the system about the z-axis by $\pi/4$ they are altered. Therefore, these signs depending on the choice of reference system have no physical meaning, but they have to be assigned consequently. Approximate values of zero places of the parameters and positions of their extrema are also given in Table 5.1.

58

(ii) The PCM potential of three ligands with C_{3v} point symmetry (Fig.5.2).

$$\mathcal{H}_{\mathrm{CF}}(C_{3v}, \text{c.n} = 3) = \sum_i \left[B_{20}\hat{C}_0^{(2)}\left(\frac{\mathbf{r}_i}{r_i}\right) + B_{40}\hat{C}_0^{(4)}\left(\frac{\mathbf{r}_i}{r_i}\right) + \right.$$
$$+ B_{43}\hat{C}_3^{(4)}\left(\frac{\mathbf{r}_i}{r_i}\right) + B_{60}\hat{C}_0^{(6)}\left(\frac{\mathbf{r}_i}{r_i}\right) + B_{63}\hat{C}_3^{(6)}\left(\frac{\mathbf{r}_i}{r_i}\right) +$$
$$+ \left. B_{66}\hat{C}_6^{(6)}\left(\frac{\mathbf{r}_i}{r_i}\right) \right] \tag{5.16}$$

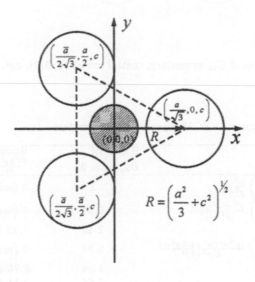

Fig. 5.2 Coordination of C_{3v} symmetry, central ion at $(0,0,0)$, c.n. $= 3$, $c/a = p$

This time the signs of the trigonal terms are unambiguous. Reflection in the horizontal plane alters these signs and as a consequence of this the potential of the regular triangular prism symmetric with respect to the horizontal plane includes no trigonal terms. The same concerns the potential of a regular triangle lying in the horizontal plane.

Table 5.2 C_{3v} (c.n.$= 3$) PCM crystal field parameters

Parameter	Roots of $B_{kq}(p) = 0$	Roots of $\frac{dB_{kq}(p)}{dp} = 0$
$B_{20} = 3\left(\frac{Ze^2\langle r^2\rangle}{R^3}\right)\frac{3p^2-1/2}{3p^2+1}$	0.41	0 (min)
$B_{40} = 3\left(\frac{Ze^2\langle r^4\rangle}{R^5}\right)3\frac{3p^4-3p^2+1/8}{(3p^2+1)^2}$	0.21	0 (max)
	0.98	0.50 (min)
$B_{60} = 3\left(\frac{Ze^2\langle r^6\rangle}{R^7}\right)\frac{1}{16}\frac{432p^6-1080p^4+270p^2-5}{(3p^2+1)^2}$	0.14	0 (min)
	0.51	0.31 (max)
	1.49	0.86 (min)
$B_{43} = -3\left(\frac{Ze^2\langle r^4\rangle}{R^5}\right)\frac{\sqrt{105}}{4}\frac{p}{(3p^2+1)^2}$	0	0.33 (min)
$B_{63} = -3\left(\frac{Ze^2\langle r^6\rangle}{R^7}\right)\frac{3\sqrt{35}}{14}\frac{21p^3-2p}{(3p^2+1)^3}$	0	0.15 (max)
	0.31	0.69 (min)
$B_{66} = 3\left(\frac{Ze^2\langle r^6\rangle}{R^7}\right)\frac{\sqrt{231}}{32}\frac{1}{(3p^2+1)^2}$	—	0 (max)

(iii) The PCM potential of one ligand system with $C_{\infty}v$ point symmetry (Fig.5.3)

$$\mathcal{H}_{CF}(C_{\infty v}, c.n = 1) = \sum_i \left[B_{20}\hat{C}_0^{(2)}\left(\frac{\mathbf{r}_i}{r_i}\right) + B_{40}\hat{C}_0^{(4)}\left(\frac{\mathbf{r}_i}{r_i}\right) + \right.$$
$$\left. + B_{60}\hat{C}_0^{(6)}\left(\frac{\mathbf{r}_i}{r_i}\right) \right] \tag{5.17}$$

$$B_{20} = \frac{Ze^2\langle r^2\rangle}{R^3}, \qquad B_{40} = \frac{Ze^2\langle r^4\rangle}{R^5}, \qquad B_{60} = \frac{Ze^2\langle r^6\rangle}{R^7}$$

The terms of odd orders ineffective in most cases are omitted in the above three potentials (Eq.5.15–5.17) and the signs of the parameters stand for $Ze < 0$ (negative ligands), i.e. $Ze^2 > 0$.

As can be easily checked up the potential of octahedron is a sum of the potential of square (Eq.5.15) for $p = 0$ and double axial potential of single ligand on the z-axis (Eq.5.17) for the same distance R whereas the potential of cube is equal to double potential of square with $p = 1$, etc.

The PCM renders qualitatively all important features of the system which result from its symmetry but it does not work, except some particular cases, as a method of calculating the crystal field parameters and crystal field splittings. It results from an obvious inadequacy of the assumption on the point character of ligands and ignoring all its consequences.

As follows from calculations performed in more general models [27, 105, 80] the contribution originating from point charges to the total crystal field

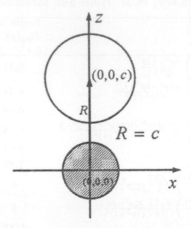

Fig. 5.3 Coordination of $C_{\infty v}$ symmetry, central ion at $(0,0,0)$, $c.n. = 1$

potential is only one of several contributions and not at all that of dominating magnitude.

It happens sometimes that a mutual cancellation of the remaining contributions or their weakness causes the point charge contribution itself to the parameters to be close to their resultant experimental values as it takes place in PrN [106].

In general, PCM values of the second order parameters are overestimated. This means that in spite of contribution coming from further neighbors which is the largest for this long–distance component (R^{-3}), they are considerably reduced by a strong screening (see chapter 9). The model parameters of the fourth order are usually of the same order as those obtained from experiments, and in turn those of sixth order are much more lowered (as regards their absolute values).

Mutual interrelations between phenomenological and model parameters are well illustrated by the two following examples. For Pr^{3+} ion in $PrCl_3$ of C_{3h} point symmetry the parameters calculated in PCM [107] and their experimental values [108] given in parentheses amount to in cm^{-1}, respectively

$$B_{20} = 316 \ (94) \qquad B_{60} = -43 \ (-634)$$
$$B_{40} = -136 \ (-325) \qquad B_{66} = 35 \ (426)$$

In the case of U^{4+} ion in Cs_2UCl_6 [109] for much more stronger crystal field of approximate O_h symmetry the parameters (for z-axis along the four–fold axis) are equal to (in cm^{-1}), respectively

$$B_{40} = 1016 \ (7211) \quad B_{60} = 39 \ (1367)$$
$$B_{44} = 607 \ (4309) \quad B_{64} = -73 \ (-2556)$$

As seen from the comparison the fourth order PCM terms are 7 times lowered whereas those of sixth order as many as 35 times.

Including the trigonal distortion of the U4+ ion surroundings ($O_h \rightarrow D_{3d}$) in Cs_2UCl_6 yields the following PCM crystal field parameters (in cm^{-1}) [80]

$$B_{20} = -1590 \ (-) \quad B_{43} = 960 \ (5750)$$
$$B_{40} = -1000 \ (-4810) \quad B_{63} = 10 \ (1470)$$
$$B_{60} = 230 \ (2430) \quad B_{66} = 140 \ (1540)$$

The experimental values (in parentheses) correspond to the idealized octaedral coordination [109]. In this case the values are given in the reference frame with the z-axis along the three–fold axis.

The above comparisons show that the signs of both the model and experimental parameters are, as a rule, consistent (not always!) but their absolute values are widely divergent.

Phenomenological attempts to reconcile the PCM have been directed either to replacing the point charge Ze by an effective charge $Z_{eff}e$ (even different Z_{eff} for different k) or by replacing R by an effective R_{eff}, so as to fit the calculated and experimentally observed parameters.

So-called modified point charge model (MPCM) proposed by Żołnierek [111, 112] is one of more interesting trials of this type. In this model the areas of overlapping the metal and ligands electron densities play the role of effective ligands.

The potential produced by further surroundings of the central atom, V^{ext}, is in general successfully calculated within the PCM [113]. Including the polarizabilities of the atoms (see chapter 10) leads to slight corrections only.

5.3 Extension of PCM – higher point multipole contribution

Let us consider last of all a generalization of the model, viz. including the contribution of higher point multipoles (beyond the monopole) ascribed to lattice sites to the total crystal field potential.

As a matter of fact, the whole problem of parameterization of crystal field potential resolves itself, de facto, into the expansion of the potential generated at a certain point of space B about a shifted center A, see

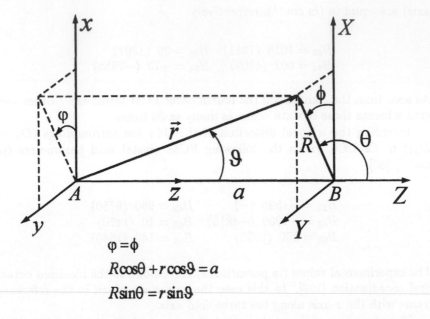

$$\varphi = \phi$$
$$R\cos\theta + r\cos\vartheta = a$$
$$R\sin\theta = r\sin\vartheta$$

Fig. 5.4 Two systems of coordinate axes used for expansion of a function centered at point B in terms of functions centered at point A

Fig.5.4. We can use a relatively simple and close formula for the transformation of the multipole moment potentials of characteristic radial dependence $R^{-(L+1)}Y_L^M(\theta, \phi)$ for a parallel shift of the reference system along z-axis by a [105]

$$\frac{C_M^{(L)}(\theta, \phi)}{R^{L+1}} = \sum_l (-1)^{l+m} \left[\frac{(2l+2L+1)!}{(2l)!(2L)!} \right]^{1/2} \times$$

$$\times \begin{pmatrix} l & L & l+L \\ -m & M & m-M \end{pmatrix} \frac{r^l}{a^{l+L+1}} C_m^{(l)}(v, \varphi) \quad (5.18)$$

where R, θ, ϕ coordinates and L, M indices refer to the initial coordinate system centered at the ligand (point B in Fig.5.4) and r, v, φ and l, m to the system shifted along $Z = z$ axis by a (point A in Fig.5.4), hence $\theta = v$ and $M = m$; the factor within the round parentheses is a 3-j symbol.

The contribution of three basic moments ($L = 0, 1, 2$), i.e. monopole, dipole and quadrupole, respectively, are presented below. The pure multipole potentials become transformed as follows:

$$\frac{Ze}{R} = \frac{Zer^2}{a^3} C_0^{(2)}(v, \varphi) + \frac{Zer^4}{a^5} C_0^{(4)}(v, \varphi) + \frac{Zer^6}{a^7} C_0^{(6)}(v, \varphi)$$

$$\frac{M_0^1 C_0^{(1)}(\theta, \phi)}{R^2} = -\frac{3M_0^{(1)}r^2}{a^4}C_0^{(2)}(v, \varphi) - \frac{9M_0^{(1)}r^4}{a^6}C_0^{(4)}(v, \varphi) -$$
$$- \frac{7M_0^{(1)}r^6}{a^8}C_0^{(6)}(v, \varphi)$$

$$\frac{M_{\pm 1}^1 C_{\pm 1}^{(1)}(\theta, \phi)}{R^2} = \frac{\sqrt{3}M_{\pm 1}^{(1)}r^2}{a^4}C_{\pm 1}^{(2)}(v, \varphi) + \frac{\sqrt{10}M_{\pm 1}^{(1)}r^4}{a^6}C_{\pm 1}^{(4)}(v, \varphi) +$$
$$+ \frac{13\sqrt{21}M_{\pm 1}^{(1)}r^6}{a^8}C_{\pm 1}^{(6)}(v, \varphi)$$

$$\frac{M_0^2 C_0^{(2)}(\theta, \phi)}{R^3} = \frac{6M_0^{(2)}r^2}{a^5}C_0^{(2)}(v, \varphi) + \frac{15M_0^{(2)}r^4}{a^7}C_0^{(4)}(v, \varphi) +$$
$$+ \frac{28M_0^{(2)}r^6}{a^9}C_0^{(6)}(v, \varphi) \tag{5.19}$$

$$\frac{M_{\pm 1}^2 C_{\pm 1}^{(2)}(\theta, \phi)}{R^3} = -\frac{4M_{\pm 1}^{(2)}r^2}{a^5}C_{\pm 1}^{(2)}(v, \varphi) - \frac{2\sqrt{30}M_{\pm 1}^{(2)}r^4}{a^7}C_{\pm 1}^{(4)}(v, \varphi) -$$
$$- \frac{8\sqrt{7}M_{\pm 1}^{(2)}r^6}{a^9}C_{\pm 1}^{(6)}(v, \varphi)$$

$$\frac{M_{\pm 2}^2 C_{\pm 2}^{(2)}(\theta, \phi)}{R^3} = \frac{M_{\pm 2}^{(2)}r^2}{a^5}C_{\pm 2}^{(2)}(v, \varphi) + \frac{\sqrt{15}M_{\pm 2}^{(2)}r^4}{a^7}C_{\pm 2}^{(4)}(v, \varphi) +$$
$$+ \frac{\sqrt{70}M_{\pm 2}^{(2)}r^6}{a^9}C_{\pm 2}^{(6)}(v, \varphi)$$

A formal extension for the higher multipoles is obvious. In the expansion given above only the components which are effective from the crystal field parameterization point of view, i.e. those of $l = 2, 4$ and 6 are shown. Ze denotes the point charge of the ligand and M_0^1, $M_{\pm 1}^1$, M_0^2, $M_{\pm 1}^2$, $M_{\pm 2}^2$ the components of its dipole and quadrupole moments, respectively.

The first row in Eq.(5.19) constitutes the essence of the PCM. The contributions of the higher multipole moments decay with higher powers of a, e.g. by 1 for dipole, by 2 for quadrupole etc. For $L + M$ odd the expansion coefficients are negative, for even they are positive. For fixed L their absolute magnitudes are largest for the axial components and decrease regularly when $|M|$ rises. The radial distribution (within the shifted system) of a component of the angular part $C_m^{(l)}(v, \varphi)$ is simply r^l/a^{l+L+1}.

CHAPTER 6

One-configurational model with neglecting the non-orthogonality. The charge penetration and exchange effects

In the ionic model the ligands electrons are assigned a passive role. The overlap and mixing of the ligand electron wave functions with those of the metal ion are neglected. The effect of the distribution of the electronic charges at the ligand sites leads to the following consequences:

- more general expressions for the intrinsic crystal field potential and its parameters including non–axial ones (according to the local symmetry) as opposed to those of the PCM (see chapter 2),

- possibility of mutual penetration of the metal and ligand ions electronic charge distributions,

- occurring the exchange interaction between the electrons of the metal ion and the ligands.

6.1 Classical electrostatic potential produced by the ligand charge distribution

An element of charge $\varrho(\mathbf{R})d\mathbf{R} = \chi^*(\mathbf{R})\chi(\mathbf{R})d\mathbf{R}$ (in atomic units) enclosed in a volume element $d\mathbf{R}$ at the point P' of the extended ligand charge distribution centered at the point O' produces an electric potential dV at the point P near the central ion located at the point O (Fig.6.1), where dV is given by

$$dV(\mathbf{R}_P) = \frac{\varrho(\mathbf{R})d\mathbf{R}}{|\mathbf{R}_P - \mathbf{R}|} \qquad (6.1)$$

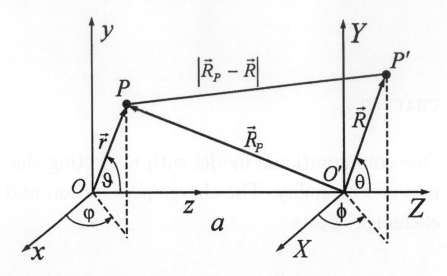

Fig. 6.1 Vector relations in calculating the potential at point P due to the ligand electron at point P' in the linear local coordinate system, the central atom at O, the ligand at O'

$\chi(\mathbf{R})$ represents the wave function of the ligand electrons for which the effect of its extension is being considered. Expanding $\frac{1}{|\mathbf{R}_P - \mathbf{R}|}$, one may write

$$dV(\mathbf{R}_P) = \varrho(\mathbf{R}) \, d\mathbf{R} \sum_{k=0}^{\infty} \sum_{q=-k}^{k} \frac{R_<^k}{R_>^{k+1}} C_q^{(k)}(\theta_{P'}, \phi_{P'}) \, C_q^{(k)^*}(\theta_P, \phi_P) \quad (6.2)$$

where θ_P, ϕ_P and $\theta_{P'}, \phi_{P'}$ are the polar and azimuthal angles of the points P and P' with respect to the coordinate frame XYZ attached to the ligand center O', and $R_<$, $R_>$ are the smaller and larger value from R and R_P, respectively.

The total potential V at P is obtained by integrating dV over the extended charge distribution represented by the wave function $\chi(\mathbf{R})$. Thus one may write

$$\begin{aligned}
V(\mathbf{R}_P) &= \sum_{k=0}^{\infty} \sum_{q=-k}^{k} \left\{ \int_0^{R_P} R^2 dR \int_0^{\pi} \sin\theta d\theta \int_0^{2\pi} d\phi \frac{\varrho(\mathbf{R}) R^k}{R_P^{k+1}} C_q^{(k)}(\theta, \phi) \right\} \times \\
&\quad \times \ C_q^{(k)^*}(\theta_P, \phi_P) + \sum_{k=0}^{\infty} \sum_{q=-k}^{k} \left\{ \int_{R_P}^{\infty} R^2 dR \int_0^{\pi} \sin\theta d\theta \times \right. \\
&\quad \times \left. \int_0^{2\pi} d\phi \frac{\varrho(\mathbf{R}) R_P^k}{R^{k+1}} C_q^{(k)}(\theta, \phi) \right\} C_q^{(k)^*}(\theta_P, \phi_P) \quad (6.3)
\end{aligned}$$

Let us now presume the spherical distribution of the ligand charge density $\varrho(\mathbf{R})$. This is a quite realistic assumption which is valid for undistorted closed $p-$ and $s-$ shells. Then, the only non–vanishing contribution to the angular integral in Eq.6.3 comes from $k = 0$ $q = 0$ term and the angular integrations involving $C_q^{(k)}(\theta, \phi)$ become trivial. Thus, the total potential at the point P due to the extended ligand charge distribution is

$$V(R_P) = \frac{1}{R_P} \int\limits_0^{R_P} 4\pi R^2 \varrho(R) dR + \int\limits_{R_P}^\infty \frac{4\pi R^2 \varrho(R) dR}{R} \qquad (6.4)$$

The potential $V(R_P)$ may be rewritten in the form

$$V(R_P) = \frac{Q}{R_P} + \int\limits_{R_P}^\infty 4\pi R^2 \left[\frac{\varrho(R)}{R} - \varrho(R) \right] dR = \frac{Q}{R_P} + F(R_P) \qquad (6.5)$$

where $Q = \int\limits_0^\infty 4\pi R^2 \varrho(R) dR$ is the total charge connected with the state $\chi(\mathbf{R})$ and $F(R_P)$ is the value of the integral at the sphere of R_P radius. The potential $V(R_P)$ in Eq.6.5 and earlier is expressed with respect to the coordinate system XYZ having origin at the ligand centre. The first term on the right–hand side of Eq.6.5 represents the potential at P if the whole electronic charge of the ligand is located as point charge at the ligand centre. The second term represents just the correction to the point–charge potential due to the extension of the ligand charge distribution. Since the crystal–field potential is expressed with respect to the metal centre O, the potential $V(R_P)$, Eq.6.5, has to be transformed from the ligand centre to the metal centre.

Using the Sharma expansion method of a radial function about a displaced center at the assumption that a is larger than r [116] we get for the $F(R_P)$ function

$$F(R_P) = \sum_k \alpha_k(a, r) C_0^{(k)}(v, \varphi) \qquad (6.6)$$

where $a_k(a, r)$ is the Sharma expansion coefficient, and simple expansion of $1/R$ (Eq.5.19), we have

$$V(r) = \sum_k \left[\frac{Qr^k}{a^{k+1}} + \alpha_k(a, r) \right] C_0^{(k)}(v, \varphi) \qquad (6.7)$$

where a is the separation between the metal centre O and ligand centre O'. Next, averaging $V(r)$ over the radial part of i-th magnetic electron wave function which is assumed to be the same for all the open–shell electrons we get for its intrinsic Hamiltonian

$$h(i) = \sum_k \left[\frac{Q\langle r^k \rangle}{a^{k+1}} + \langle \alpha_k(a,r) \rangle \right] C_0^{(k)}(v_i, \varphi_i) \tag{6.8}$$

or

$$h(i) = \sum_k B_{k0}^{(\mathrm{PCM})}(1 + \eta_k) C_0^{(k)}(v_i, \varphi_i) \tag{6.9}$$

where $B_{k0}^{(\mathrm{PCM})}$ is the intrinsic PCM contribution to the crystal field parameter coming from the specified spherical ligand electron wave function $\chi(\mathbf{R})$ only, and $\eta_k = \langle \alpha_k(a,r) \rangle / B_{k0}^{(\mathrm{PCM})}$ is the penetration correction factor which modifies the PCM contribution to the parameter according to the relation

$$B_{k0}^{(\varrho)} = B_{k0}^{(\mathrm{PCM})}(1 + \eta_k) \tag{6.10}$$

where $B_{k0}^{(\varrho)}$ is the crystal field parameter (intrinsic in this formulation) in the model including the effect of the distribution of the electronic charges at the ligand site.

It is not out of place to remind that the case of individual ligand with a spherical distribution of its electron charge cloud in the simplest local coordinate system has been considered. This is according to the spirit of the superposition model and the parameters obtained are intrinsic in their nature. In the case of a general charge distribution, i.e. that of a non–spherical symmetry, some non–zero components of the potential (Eq.6.3) with k and q being different from zero can occur, and consequently the crystal field potential (Eq.6.7) may loose its simple axial form.

6.2 The charge penetration effect and the exchange interaction in the generalized product function model

The generalized product function model (see chapter 4) allowing some chosen groups of electrons to be treated explicitly may be successfully adopted to descript the charge penetration and exchange effects arising from the spatial distribution of the ligand electronic charges. The one–electron wave functions (spin–orbitals), found by means of the self–consistent iterative procedure, forming the particular group functions enable us to express the interelectronic interaction operators and in consequence the Hamiltonian of the distinguished group of electrons, i.e. the central ion a-open shell electrons.

Explicitly, we have

$$
\mathcal{H}_a = \sum_{i=1}^{Na} \left\{ \frac{\nabla_i^2}{2} + V^{\mathrm{Mc}}(r_i) + \xi\, \mathbf{l}_i \cdot \mathbf{s}_i + \overbrace{\sum_t V_t^{\mathrm{Lc}}(|\mathbf{r}_i - \mathbf{R}_t|) + V^{\mathrm{ext}}(\mathbf{r}_i)}
\right.
$$

$$
\left. + \quad J^{\Xi}(\mathbf{r}_i) - K^{\Xi}(\mathbf{r}_i) + \underbrace{\sum_t \left[J^{X_t}(\mathbf{r}_i) - K^{X_t}(\mathbf{r}_i) \right]} \right\} +
$$

$$
+ \quad \sum_{i>j} \frac{1}{r_{ij}} \tag{6.11}
$$

where the denotations have been introduced in chapter 4, the summations over i and j are confined to the open–shell electrons only, and that over t concerns the ligands. On the other hand, the so–called effective Hamiltonian acting within the same basis of states has the form

$$
\mathcal{H}_{\mathrm{eff}} = \sum_{i=1}^{Na} \left[\frac{\nabla_i^2}{2} + V(r_i) + \xi\, \mathbf{l}_i \cdot \mathbf{s}_i + \overbrace{V_{\mathrm{CF}}(\mathbf{r}_i)} \right] + \sum_{i>j} \frac{1}{r_{ij}} \tag{6.12}
$$

Comparing the two equivalent Hamiltonians, Eq.6.11 and Eq.6.12, and equating their parts of non–spherical symmetry (marked with the frames) we get

$$
V_{\mathrm{CF}}(\mathbf{r}_i) = \sum_t \left[V_t^{\mathrm{Lc}}(\mathbf{r}_i - \mathbf{R}_t) + J^{X_t}(\mathbf{r}_i) - K^{X_t}(\mathbf{r}_i) \right] + V^{\mathrm{ext}}(\mathbf{r}_i) \tag{6.13}
$$

Conventionally, the potential $V_{\mathrm{CF}}(\mathbf{r}_i)$, Eq.6.13, is partitioned into the following terms

$$
V_{\mathrm{CF}}(\mathbf{r}) = V^{\mathrm{PC}}(\mathbf{r}) + V^{\mathrm{Kl}}(\mathbf{r}) + V^{\mathrm{exch}}(\mathbf{r}) + V^{\mathrm{ext}}(\mathbf{r}) \tag{6.14}
$$

in which the two terms $\sum_t V_t^{\mathrm{Lc}}(\mathbf{r}_i - \mathbf{R}_t)$ and $\sum_t J^{X_t}(\mathbf{r}_i)$ in Eq.6.13 are replaced by the two new ones:

- the potential produced by the ligands in the point charge (Z_t) approximation

$$
V^{\mathrm{PC}}(\mathbf{r}) = -\sum_t \frac{Z_t}{|\mathbf{r} - \mathbf{R}_t|} \tag{6.15}
$$

where $\mathbf{R}_t$ is the radius vector of the ligand t in the central coordinate system, and

- the Kleiner correction [73] defined by

$$
V^{\mathrm{Kl}}(\mathbf{r}) = \sum_t \left[V_t^{\mathrm{Lc}}(|\mathbf{r} - \mathbf{R}_t|) - \left(-\frac{Z_t + N_{X_t}}{\mathbf{r} - \mathbf{R}_t|} \right) \right.
$$

$$
\left. + \quad J^{X_t}(\mathbf{r}) - \frac{N_{X_t}}{|\mathbf{r} - \mathbf{R}_t|} \right] \tag{6.16}
$$

where N_{X_t} denotes the number of outer electrons of the ligand t treated explicitly.

It is now clear that the Kleiner correction (Eq.6.14) directly corresponds to the second term in Eq.6.7 and, as it has been stated earlier, this so–called charge penetration effect is a consequence of the spatial distribution of the electronic charges at the ligands.

Since the potentials of the ligand cores, $\sum_t V_t^{\mathrm{Lc}}(\mathbf{r}_i - \mathbf{R}_t)$, may be treated with a satisfactory approximation as those produced by the corresponding point charges the first two components on the right side of Eq.6.16 are practically compensated and the Kleiner correction is

$$V^{\mathrm{Kl}}(\mathbf{r}) = \sum_t \left[J^{X_t}(\mathbf{r}) - \frac{N_{X_t}}{|\mathbf{r} - \mathbf{R}_t|} \right] \qquad (6.17)$$

from which its physical meaning appears immediately.

The second immanent consequence of the charge distribution is the interelectronic exchange interaction, potential of which has the form

$$V^{\mathrm{exch}}(\mathbf{r}) = -\sum_t K^{X_t}(\mathbf{r}) \qquad (6.18)$$

In spite of some early and even later estimations depreciating the meaning of the exchange potential [117, 120] its contribution to the crystal field potential is comparable with that by the classical penetration potential [80]. Calculation of the exchange matrix elements is a tedious numerical problem mainly due to the necessity of integrating slow convergent infinite series. This problem has been managed to be overcome with a suitable grouping the terms in the series [118].

A rough estimation of the exchange matrix elements is available by using two alternative approximation methods. In the first of these a modified form of the Mulliken approximation [47] is used, i.e.

$$\chi a = \frac{1}{2} f \langle \chi | a \rangle [\chi\chi + aa] \qquad (6.19)$$

where the adjustable numerical factor, more appropriate for Coulomb integrals, is obtained by substituting values for the matrix elements in the following relation

$$\langle \chi | \frac{1}{r} | a \rangle = \frac{1}{2} f \langle \chi | a \rangle \left[\langle \chi | \frac{1}{r} | \chi \rangle + \langle a | \frac{1}{r} | a \rangle \right] \qquad (6.20)$$

Consequently,

$$
\begin{aligned}
\langle a\chi | g | \chi a \rangle &= \frac{1}{2} f \langle \chi | a \rangle [\langle a\chi | g | aa \rangle + \langle \chi\chi | g | \chi a \rangle] \\
&= \frac{1}{4} f^2 (\langle \chi | a \rangle)^2 [\langle aa | g | aa \rangle + \langle \chi\chi | g | \chi\chi \rangle + \\
&\quad + 2\langle a\chi | g | a\chi \rangle]
\end{aligned}
\qquad (6.21)
$$

In the second method the exchange integral is evaluated in the approximation that the open–shell wave functions are spherically symmetric $(a \rightarrow a_s)$ and the scaling relation is used

$$\langle a\chi|g|\chi a\rangle = \left(\frac{\langle\chi|a\rangle}{\langle\chi|a_s\rangle}\right)^2 \langle a_s\chi|g|\chi a_s\rangle \qquad (6.22)$$

6.3   The weight of the penetration and exchange effects in the crystal field potential

Both the classical charge penetration and the quantum exchange effects depend on the spatial distribution of interacting electronic densities and in consequence critically depend on the distance between their centers, i.e. the nuclei. Therefore the exact wave functions of the considered electrons particularly in their peripheral areas are requested in calculations of these effects. In the case of heavier ions due to the relativistic contraction of their core states an expansion of their outer states is expected.

The dependence both of these effects on the internuclear distance is especially sensitive in the range of close distances. In order to illustrate the dependence a practical coefficient β_k defined as the ratio of the k-th order crystal field parameters obtained in the penetration model, i.e. for $V^{PC} + V^{Kl}$ but without the exchange term, and in the PCM has been introduced [119]

$$\beta_k = \frac{B_{k0}^{(\varrho)}}{B_{k0}^{(PCM)}} = 1 + \eta_k \qquad (6.23)$$

The above relation is valid for the intrinsic parameters (Eq.6.10). Two representative plots $\beta_k(a)$ are presented in Fig.6.2 and 6.3.

In the ionic compounds in which the ligands are anions the opposed to the PCM values corrections to the parameters are expected and in fact the higher is the order k of the crystal field component the more effective reduction of the corresponding parameters in comparison to the values calculated in the PCM is observed. For large interionic distances $\beta_k \rightarrow 1$ for all k.

As an instance, in Nd_2O_3, where a is equal to about 2.3 Å the calculations yield $\beta_2 = 0.93$, $\beta_4 = 0.59$ and $\beta_6 = -0.10$ [119]. As is seen the correction is always opposite to the PCM contribution, i.e. η_k is negative (Eq.6.23) and may drop below -1 when the parameter sign changes. This is expected indeed, since in the PCM the nucleus of the ligand is perfectly shielded from the open–shell electrons of the central ion by the electrons of the ligands. However, when the magnetic electrons penetrate into the ligand charge cloud, it experiences more strongly the field of the nucleus, which is now somewhat exposed to the magnetic electron. The imperfect screening of the nucleus tends to lower the point charge contribution to the crystal field parameters. For the small distances it may lead to negative values of the

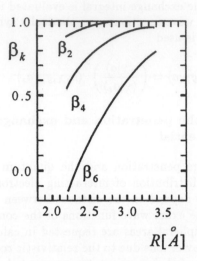

Fig. 6.2 β_k coefficient as a function of internuclear distance a between Nd^{3+} and O^{2-} ions in Nd_2O_3 (acc. to Garcia and Faucher [119])

β_k parameters (see the β_6 curve in Fig.6.2). For $PrCl_3$ $\beta_2 = 0.75$, $\beta_4 = 0$, $\beta_6 = -1.52$ [120, 121]. Note the mutual compensation of the point charge and classical penetration contributions for the fourth–order component of the crystal field potential.

In the actinide compounds the penetration effect is distinctly stronger, e.g. in Cs_2UCl_6 $\beta_2 = 0.75$, $\beta_4 = -0.64$ and $\beta_6 = -2.38$ [80].

In metallic materials where the positively charged atomic cores play a role of ligands a rapid increase of the β_k coefficients with the diminishing of the internuclear distance a is observed (Fig.6.3).

The calculations performed for the Er^{+3} ion in the Au matrix, Er^{+3}:Au [70], in which $a = 5.45$ a.u. give $\beta_4 = 1.21$ and $\beta_6 = 1.76$, whereas for the distance $a = 3.0$ $a.u.$, otherwise unrealistic one, they would reach so large values as 8.0 and 15.0, respectively.

The penetration effect is sometimes presented in the so–called pseudo–point charge model. The pseudo–point charges are simply the effective charges including the penetration effect and they are, in general, different for various k components of the potential [119]. Such a purely formal approach facilities taking into account the dynamical screening mechanism when the conduction electrons are present (see chapter 13).

The exchange potential varies with the interionic distance similarly as the penetration effect. The exchange interaction reinforces distinctly the penetration effect. The generalized $\tilde{\beta}_k$ coefficients including apart from

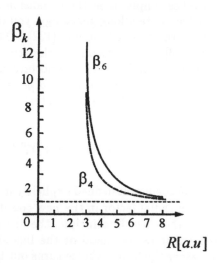

Fig. 6.3 β_k coefficient as a function of internuclear distance a between Er^{3+} ion and Au atomic core in Er^{3+}: Au matrix. Only direct Coulomb contribution included. (acc. to Christodoulos [70])

the point charge and classical penetration effects also the exchange contribution take for Cs_2UCl_6 [80] the following values: $\tilde{\beta}_k = 0.32$ (0.54), $\tilde{\beta}_k = -2.02$ (−0.64) and $\tilde{\beta}_k = -5.62$ (−2.38) where for comparison the previous values of β_k are given in the brackets.

6.4 Calculation of the two–centre integrals

Estimating the effect of distribution of the electronic charges at the ligand sites and the consequent mutual penetration of the metal and the ligand electronic charge densities on the crystal–field potential of the metal ion reduces itself to calculation of appropriate two–centre integrals. Actually, we deal with the two clouds of interacting (directly and exchangely) charge distributions of different origins. In order to use standard two–centre integration procedures it is necessary to align the coordinate systems of the ions so that they have a common z axis and parallel x and y axes.

There are two methods of treating this problem, either with transformation of all functions in the two–electron Coulomb interaction matrix element, i.e. in the integrand, to a common coordinate system or alternatively with a two centre expansions of $\frac{1}{r_{12}}$ operator acting on the functions remaining this time intact in their own local systems. This method may be employed in the direct interaction calculation only.

The first method relies simply upon the expansion of a function about a displaced center. It has quite a long genealogy, you only need to remind yourself the Coulson–Barnett zeta function [122]. The alpha–expansion technique of Sharma [116] is the crowning achievement of the evolution process. In this method, in application to spinorbitals, a wave function of a ligand in its local coordinate system $\chi_{nlm}(R, \theta, \phi)$, where n, l, m are the appropriate quantum numbers, takes in the central ion system (r, v, φ) the following form

$$\chi_{nlm}(R, \theta, \phi) = \sum_k \alpha_k(nml|a, r) Y_k^m(v, \varphi) \tag{6.24}$$

where $\alpha_k(nml|a, r)$ is the coefficient defined explicitly in the Sharma method which plays a role of the radial factor in the new coordinate system, and a is the distance between the sites. In principle, the distinguished coordinate system may be fixed either at the metal or the ligand nucleus. However, from the numerical reasons the former choice turns out to be more profitable [118].

The two–electron operator $\frac{1}{r_{12}}$ keeps then its initial form in the central ion system, i.e.

$$\frac{1}{r_{12}} = \sum_{k=0}^{\infty} \sum_{q=-k}^{k} \frac{r_<^k}{r_>^{k+1}} C_q^{(k)^*}(v_1, \varphi_1) C_q^{(k)}(v_2, \varphi_2) \tag{6.25}$$

where the subscript 1 and 2 numbers the electrons.

The direct Coulomb contribution may also be calculated with the second method employing the expansion of $\frac{1}{r_{12}}$ given by Buehler and Hirschfelder [123, 124] and used by Garcia and Faucher [119]. The two local coordinate systems are now preserved but the operator is expanded to the form

$$\frac{1}{r_{12}} = \sum_{k_1} \sum_{k_2} \sum_q b_{k_1,k_2}^{|q|}(r_1, R_2, a) C_q^{(k_1)^*}(v_1, \varphi_1) C_q^{(k)}(\theta_2, \phi_2) \tag{6.26}$$

where the radius vector $\mathbf{r}_1(r_1, v_1, \varphi_1)$ of electron (1) is relative to one site and the position vector $\mathbf{R}_2(R_2, \theta_2, \phi_2)$ of electron (2) is relative to the other site, and a, as above, is the distance between the sites. The expressions for $b_{k_1,k_2}^{|q|}$ as functions of r_1, R_2 and a can be found in [70, 123, 124].

6.5 Final remarks

The mutual penetration of the charge density clouds should not be identify with the overlap of the corresponding wave functions in the strict sence of the notion, i.e. with their non–orthogonality. There are a lot of convincing experimental data evidencing a mutual compensation of the penetration and exchange effects by the overlap effect, and more strictly by so–called contact covalency effect (see chapter 7). This compensation effect was probably the reason of some earlier striking successes of the PCM [125, 126]. An analogous effect makes the pseudopotential in metals so weak that the almost free–electron model works so well [127]. The effective repulsion due to the Pauli exclusion counteracts the attractive potential acting on the metal ion electrons due to their penetration into the outer ligand shells.

However, in general the crystal field parameters, especially those of higher orders, corrected for these effects are still far from their experimental values. It thus appears that important contributions to these parameters must also arise from mechanisms which perturb the free–ion states.

exchange effects by the overlap effect, and more strictly by so-called contact covalence effect (see chapter ?). This compensation effect was probably the reason of some earlier striking successes of the PCM [125, 126]. An analogous effect makes the pseudopotential in metals so weak that the almost free-electron model works so well [127]. The effective repulsion due to the Pauli exclusion calculated, the attractive potential acting on the metal ion electrons due to their penetration into the outer ligand shells.

However, in general the crystal field parameters, especially those of higher orders, corrected for these effects are still far from their experimental values. It thus appears that important contributions to these parameters must also arise from mechanisms which perturb the free-ion states.

CHAPTER 7

The exclusion model. One–configurational approach with regard to non–orthogonality of the wave functions

The model presented in this chapter differs from the purely electrostatic models considered so far only in that overlap, i.e. the non–orthogonality of the one–electron functions of the complex has been included. Since the orthogonalization effects have their roots in the Pauli exclusion between the free–ion wave–functions this extended approach is called the exclusion model. In the frame of the model all three types of non–orthogonality possible for generalized product functions of the complex and their consequences are treated independently.

7.1 Three types of the non–orthogonality

The one–electron functions entering into the composition of a generalized product function (chapter 4) are not, in fact, the orthogonal ones. The three possible different types of non–zero overlap integrals are

$$S_{t\tau m} = \langle a_m | \chi_{t\tau} \rangle, \qquad S_{t\tau \nu} = \langle \xi_\nu | \chi_{t\tau} \rangle \quad \text{and} \quad S_{t\tau t'\tau'} = \langle \chi_{t\tau} | \chi_{t'\tau'} \rangle \quad (7.1)$$

where a_m, ξ_ν and $\chi_{t\tau}$ are the one–electron component spinorbitals from the group functions ϕ_A, ϕ_Ξ and ϕ_{X_t}, respectively, are followed by the three types of mutually orthogonal linear combinations of the atomic spinorbitals. Their renormalization leads to three independent contributions which in sum compose the so–called orthogonalizing potential — an important part of the effective crystal field potential.

To simplify the matters, separate, intergroup orthogonalization procedures are in use. They lead to the functions which are not exactly orthogonal to one another within a group, nor are they exactly orthogonal to the functions from the other groups. However, this non–orthogonality only occurs to second and higher orders of the overlap integrals and may be ignored in a first approximation. Under assumption that the overlap integrals as for their absolute values are small quantities of order of ε, and since we want to know energy of the states (chapter 4) exact to the terms of ε^2 order, it is enough to assure the orthogonality with accuracy to ε and the new functions (primed) can be described in the form:

$$
\begin{aligned}
a'_m &= \left[1 - \sum_{t,\tau}(S_{t\tau m})^2\right]^{-1/2} \left(a_m - \sum_{t,\tau}S_{t\tau m}\chi_{t\tau}\right) \\
&\approx \left[1 + \frac{1}{2}\sum_{t,\tau}(S_{t\tau m})^2\right] \left(a_m - \sum_{t,\tau}S_{t\tau m}\chi_{t\tau}\right) \\
&\approx a_m - \sum_{t,\tau}S_{t\tau m}\chi_{t\tau} + \frac{1}{2}\sum_{t,\tau}(S_{t\tau m})^2 a_m
\end{aligned} \tag{7.2}
$$

$$
\begin{aligned}
\xi'_\nu &= \left[1 - \sum_{t,\tau}(S_{t\tau\nu})^2\right]^{-1/2} \left(\xi_\nu - \sum_{t,\tau}S_{t\tau\nu}\chi_{t\tau}\right) \\
&\approx \xi_\nu - \sum_{t,\tau}S_{t\tau\nu}\chi_{t\tau} + \frac{1}{2}\sum_{t,\tau}(S_{t\tau\nu})^2\xi_\nu
\end{aligned} \tag{7.3}
$$

$$
\begin{aligned}
\chi'_{t\tau} &= \left[1 - \frac{3}{4}\sum_{\substack{t',\tau' \\ t'\neq t}}(S_{t'\tau't\tau})^2\right]^{-1/2} \left(\chi_{t\tau} - \frac{1}{2}\sum_{\substack{t',\tau' \\ t'\neq t}}S_{t'\tau't\tau}\chi_{t'\tau'}\right) \\
&\approx \chi_{t\tau} - \frac{1}{2}\sum_{\substack{t',\tau' \\ t'\neq t}}S_{t'\tau't\tau}\chi_{t'\tau'} + \frac{3}{8}\sum_{\substack{t',\tau' \\ t'\neq t}}(S_{t'\tau't\tau})^2\chi_{t\tau}
\end{aligned} \tag{7.4}
$$

The ligand wave functions, $\chi_{t\tau}$, have been orthogonalized symmetrically in order to avoid a differentiation among them and hence the coefficient $-\frac{1}{2}$ and in consequence those of $-\frac{3}{4}$ and $\frac{3}{8}$ in Eq.7.4 [113]. The normalization in Eqs 7.2–7.4 is characteristic for overlapping functions. The second order (S^2) renormalization contributions to the primed functions can not be ignored because their contribution to the energy is of the some order.

7.2 The renormalization of the open-shell Hamiltonian $\mathcal{H}_a$ owing to the non-orthogonality of the one-electron functions

To each zero–approximation generalized product function obtained variationally the orthogonality of component one–electron functions is always a priori assumed. Thus, these can not be the free–ions wave functions and rather their orthogonalized, at least exact to ε^2, linear combinations such as these given above, Eqs 7.2 – 7.4.

Return to the convenient basis of the open–shell (A) states leads through an appropriate renormalization of the Hamiltonian of the complex to an additional, corresponding to the accuracy level assumed, contribution into the effective crystal field potential.

The zero–approximation group function of an open–shell electrons is, in general, an one–determinantal function of the form

$$|a_{m_1} a_{m_2} \ldots a_{m_{N_a}}|$$

Let us now consider a typical matrix element of the group Hamiltonian $\mathcal{H}_a$ within the basis, i.e.

$$\langle a_{m_1} a_{m_2} \ldots a_{m_{N_a}} |\mathcal{H}_a| a_{m'_1} a_{m'_2} \ldots a_{m'_{N_a}} \rangle \tag{7.5}$$

This element is different from zero only if the bra and ket functions differ between themselves at most in two spinorbitals a_m. At first, let us analyze the outcomes of the renormalization of the one–electron operator part of $\mathcal{H}_a$, i.e. h_a. It yields non–zero matrix elements only if the functions in the integrand differ at most in one spinorbital. Taking into account the non–orthogonalities corresponds to substitution the a'_m (primed) function (Eq.7.2) in place of a_m, and ξ'_ν (Eq.7.3), $\chi'_{t\tau}$ (Eq.7.4) functions in place of ξ_ν and $\chi_{t\tau}$, respectively. The two latter functions enter into the composition of h_a (Eq.4.33). Consequently, we get

$$\langle a_m|h_a|a_{m'}\rangle \quad \longrightarrow \quad \langle a'_m|h'_a|a'_{m'}\rangle \approx$$

$$\approx \quad \left\langle a_m - \sum_{t,\tau} S_{t\tau m}\chi_{t\tau} + \frac{1}{2}\sum_{t,\tau}(S_{t\tau m})^2 a_m |h'_a|a_{m'} - \right.$$

$$+ \quad \left. \sum_{t,\tau} S_{t\tau m'}\chi_{t\tau} + \frac{1}{2}\sum_{t,\tau}(S_{t\tau m'})^2 a_{m'} \right\rangle \tag{7.6}$$

The corrected Hamiltonian h'_a can be written as a sum of $h_a + \Delta h_a$, where Δh_a is the correction resulting from the non–orthogonalities of $\langle\xi|\chi\rangle$ and $\langle\chi_t|\chi_{t'}\rangle$ types. The Δh_a turns out to be of order of ε^2 and its contribution will be taking into account after considering the main consequences of the $\langle a|\chi\rangle$ non–orthogonality.

Expanding Eq.7.6, exact to terms of $\varepsilon^2 \sim S^2$, we have

$$
\begin{aligned}
\langle a'_m|h'_a|a'_{m'}\rangle &= \langle a_m|h'_a|a_{m'}\rangle + \frac{1}{2}\sum_{t,\tau}\left[(S_{t\tau m})^2 + \right.\\
&+ \left.(S_{t\tau m'})^2\right]\langle a_m|h_a|a_{m'}\rangle - \\
&- \sum_{t,\tau}S_{t\tau m}\langle\chi_{t\tau}|h_a|a_{m'}\rangle - \\
&- \sum_{t,\tau}S_{t\tau m'}\langle\chi_{t\tau}|h_a|a_m\rangle + \\
&+ \sum_{t,\tau}\sum_{t',\tau'}S_{t\tau m}S_{t'\tau' m'}\langle\chi_{t\tau}|h_a|\chi_{t'\tau'}\rangle \qquad (7.7)
\end{aligned}
$$

For the present let us leave the first term on the right side of Eq.7.7 aside and deal with the four remaining terms. Fortunately, introducing a reasonable approximation we may easily make the correction to be independent on the open–shell spinorbitals being occupied. First, the matrix element $\langle a_m|h_a|a_{m'}\rangle$ in the second term of Eq.7.7 is still multiplied by a small factor S^2. Secondly, since the diagonal matrix elements of h_a within the open–shell states are considerably greater than the off–diagonal ones and relatively weakly differentiated as for their energy, we may assume that

$$
\langle a_m|h_a|a_{m'}\rangle = \overline{\langle a|h_a|a\rangle}\,\delta_{mm'} \qquad (7.8)
$$

Where $\overline{\langle a|h_a|a\rangle}$ is the average value of all the diagonal elements $\langle a_m|h_a|a_m\rangle$. Thus, the second term in Eq.7.7 may be approximated by

$$
\sum_{t,\tau}(S_{t\tau m})^2\overline{\langle a|h_a|a\rangle}\,\delta_{mm'} \qquad (7.9)
$$

By way of digression, the ratio of the off–diagonal to diagonal matrix elements corresponds roughly to the ratio of the crystal field splitting to the energy of individual spinorbitals, which is rather small, indeed.

The last four terms of Eq.7.7 including the above approximation may be treated as a matrix element within the unprimed function basis of a certain new one–electron Hamiltonian containing the following renormalization correction

$$
\sum_{t,\tau}|\chi_{t\tau}\rangle\langle\chi_{t\tau}|\overline{\langle a|h_a|a\rangle}\,\delta_{mm'} - 2\sum_{t,\tau}|\chi_{t,\tau}\rangle\langle\chi_{t\tau}|h_a +
$$
$$
+ \sum_{t,\tau}\sum_{t',\tau'}|\chi_{t\tau}\rangle\langle\chi_{t\tau}|h_a|\chi_{t'\tau'}\rangle\langle\chi_{t'\tau'}| \qquad (7.10)
$$

The first term in Eq.7.10 yields only the diagonal contributions.

An analogous approach may be used in the case of matrix elements of $\frac{1}{r_{ij}}$ operator – the two electron part of the open–shell Hamiltonian $\mathcal{H}_a$ (Eq.4.33). Let us write down a general matrix element of this operator in the form

$$\langle a_{m_i} a_{m_j} || a_{m'_i} a_{m'_j} \rangle \tag{7.11}$$

where to simplify the notation a convention is here introduced (only here) that for equal spin functions of both interacting electrons the negative exchange element is automatically included. Number of the non–zero matrix elements, Eq.7.11, is limited by both the spin restriction (there is no change of electron spin for the Coulomb interaction) and the angular momentum conservation requirement

$$m_i + m_j = m'_i + m'_j \tag{7.12}$$

Similarly as above (Eq.7.7), the same substitution (Eq.7.6) allows the non–orthogonality to be included. Thus, substitute the primed a'_m in place of the unprimed a_m ones

$$\langle a_{m_i} a_{m_j} || a_{m'_i} a_{m'_j} \rangle \rightarrow \langle a'_{m_i} a'_{m_j} || a'_{m'_i} a'_{m'_j} \rangle \tag{7.13}$$

Expanding, in turn, the $\langle a'_{m_i} a'_{m_j} || a'_{m'_i} a'_{m'_j} \rangle$ matrix element exact to the terms of order of S^2 yields

$$
\begin{aligned}
\langle a'_{m_i} a'_{m_j} || a'_{m'_i} a'_{m'_j} \rangle &= \langle a_{m_i} a_{m_j} || a_{m'_i} a_{m'_j} \rangle + \\
&+ \frac{1}{2} \sum_{t,\tau} \left[(S_{t\tau m_i})^2 + (S_{t\tau m'_i})^2 + (S_{t\tau m_j})^2 + (S_{t\tau m'_j})^2 \right] \times \\
&\times \langle a_{m_i} a_{m_j} || a_{m'_i} a_{m'_j} \rangle - \\
&- \sum_{t,\tau} \Big[S_{t\tau m_i} \langle \chi_{t\tau} a_{m_j} || a_{m'_i} a_{m'_j} \rangle + S_{t\tau m_j} \langle a_{m_i} \chi_{t\tau} || a_{m'_i} a_{m'_j} \rangle + \\
&+ S_{t\tau m'_i} \langle a_{m_i} a_{m_j} || \chi_{t\tau} a_{m'_j} \rangle + S_{t\tau m'_j} \langle a_{m_i} a_{m_j} || a_{m'_i} \chi_{t\tau} \rangle \Big] + \\
&+ \sum_{t,\tau} \sum_{t',\tau'} \Big[S_{t\tau m_i} S_{t'\tau' m'_i} \langle \chi_{t\tau} a_{m_j} || \chi_{t'\tau'} a_{m'_j} \rangle + \\
&+ S_{t\tau m_j} S_{t'\tau' m'_j} \langle a_{m_i} \chi_{t\tau} || a_{m'_i} \chi_{t'\tau'} \rangle \Big] .
\end{aligned}
\tag{7.14}
$$

In the frame of the same accuracy as for the one–electron part of the Hamiltonian one may, on the analogy of the approximation, Eq.7.8, take

$$(S_{t\tau m_i})^2 \langle a_{m_i} a_{m_j} || a_{m'_i} a_{m'_j} \rangle \approx (S_{t\tau m_i})^2 \langle \overline{a | J(a,a) | a} \rangle \delta_{m_i m'_i} \delta_{m_j m'_j} =$$
$$= (S_{t\tau m_i})^2 \, F_0 \, \delta_{m_i m'_i} \delta_{m_j m'_j} \tag{7.15}$$

where

$$J(a,a) = \frac{1}{2l_a+1} \sum_m J(a_m, a_m) \tag{7.16}$$

is the arithmetic average of all $J(a_m, a_m)$, l_a is the angular momentum quantum number of the open–shell, and $J(a_m, a_m)$ is the direct Coulomb operator (minus the exchange one, if any) related to the charge density $a_m^* a_m$, $F_0 = \langle a|J(a,a)|a\rangle$ is the zero–order Slater integral.

Similarly, we may assume that

$$S_{t\tau m_i}\langle \chi_{t\tau} a_{m_j}||a_{m_i'} a_{m_j'}\rangle \approx S_{t\tau m_i}\langle \chi_{t\tau}|J(a,a)|a_{m_i'}\rangle \delta_{m_j m_j'} \tag{7.17}$$

and

$$S_{t\tau m_i} S_{t'\tau' m_i'}\langle \chi_{t\tau} a_{m_j}||\chi_{t'\tau'} a_{m_j'}\rangle \approx$$
$$\approx S_{t\tau m_i} S_{t'\tau' m_i'}\langle \chi_{t\tau}|J(a,a)|\chi_{t'\tau'}\rangle \delta_{m_j m_j'} \tag{7.18}$$

Introducing the $J(a,a)$ operator reduces the two–electron problem to the one–electron operator matrix elements and the two–electron matrix elements with $m_j =\neq m_j'$ are automatically ignored due to the Kronecker delta $\delta_{m_j m_j'}$.

Thus, the contribution of the two–electron operator $\frac{1}{r_{ij}}$ to the renormalization correction has the approximate form

$$\sum_{t,\tau}|\chi_{t\tau}\rangle\langle\chi_{t\tau}|(N_a - 1)F_0\delta_{mm'} - 2\sum_{t,\tau}|\chi_{t\tau}\rangle\langle\chi_{t\tau}|(N_a - 1)J(a,a) +$$
$$+\sum_{t,\tau}\sum_{t',\tau'}|\chi_{t\tau}\rangle\langle\chi_{t\tau}|(N_a - 1)J(a,a)|\chi_{t'\tau'}\rangle\langle\chi_{t'\tau'}| \tag{7.19}$$

The approximations introduced (Eqs 7.8, 7.15, 7.17, 7.18) have allowed us to eliminate a dependence of the corrections (Eqs 7.10, 7.19) on the open–shell spinorbitals being occupied. It has been worth of trouble. Nevertheless, one should realize that this simple approximate form of the correction is obtained at the price of introducing some errors. Now, combining into a whole Eq.7.10 together with Eq.7.19 the total renormalization correction to the Hamiltonian $\mathcal{H}_a$ resulting from the $\langle a|\chi\rangle$ non-orthogonality is obtained. This is equal to

$$\sum_{t,\tau}|\chi_{t\tau}\rangle\langle\chi_{t\tau}|[\overline{\langle a|h_a|a\rangle} + (N_a - 1)F_0]\delta_{mm'} -$$
$$-2\sum_{t,\tau}|\chi_{t\tau}\rangle\langle\chi_{t\tau}|[h_a + (N_a - 1)J(a,a)] +$$
$$+\sum_{t,\tau}\sum_{t',\tau'}|\chi_{t\tau}\rangle\langle\chi_{t\tau}|h_a + (N_a - 1)J(a,a)|\chi_{t'\tau'}\rangle\langle\chi_{t'\tau'}| \tag{7.20}$$

At last, let us examine the input to the orthogonalizing potential coming from the non–orthogonalities of $\langle \xi | \chi \rangle$ and $\langle \chi_t | \chi_{t'} \rangle$ types. The wave functions ξ and χ take their part in the h_a through the Coulomb interelectronic J and K operators. With the accuracy to S^2 they are

$$J(\xi'_\nu, \xi'_\nu) = J \left[\left(\xi_\nu - \sum_{t,\tau} S_{t\tau\nu} \chi_{t\tau} + \frac{1}{2} \sum_{t,\tau} S^2_{t\tau\nu} \xi_\nu \right), \right.$$

$$\left. \left(\xi_\nu - \sum_{t,\tau} S_{t\tau\nu} \chi_{t\tau} + \frac{1}{2} \sum_{t,\tau} S^2_{t\tau\nu} \xi_\nu \right) \right]$$

$$\approx J(\xi_\nu, \xi_\nu) + \sum_{t,\tau} S^2_{t\tau\nu} J(\xi_\nu, \xi_\nu) + \sum_{t,\tau} -2 \sum_{t,\tau} S_{t\tau\nu} J(\chi_{t\tau}, \xi_\nu) +$$

$$+ \sum_{t,\tau} \sum_{t',\tau'} S_{t\tau\nu} S_{t'\tau'\nu} J(\chi_{t\tau}, \chi_{t'\tau'}) \tag{7.21}$$

and

$$\sum_{t,\tau} J(\chi'_{t\tau}, \chi'_{t\tau}) =$$

$$= \sum_{t,\tau} J \left[\left(\chi_{t\tau} - \frac{1}{2} \sum_{\substack{t',\tau' \\ t' \neq t}} S_{t\tau t'\tau'} \chi_{t'\tau'} + \frac{3}{8} \sum_{\substack{t',\tau' \\ t' \neq t}} (S_{t\tau t'\tau'})^2 \chi_{t\tau} \right), \right.$$

$$\left. \left(\chi_{t\tau} - \frac{1}{2} \sum_{\substack{t',\tau' \\ t' \neq t}} S_{t\tau t'\tau'} \chi_{t'\tau'} + \frac{3}{8} \sum_{\substack{t',\tau' \\ t' \neq t}} (S_{t\tau t'\tau'})^2 \chi_{t\tau} \right) \right]$$

$$\approx \sum_{t,\tau} \left[J(\chi_{t\tau}, \chi_{t\tau}) - \sum_{\substack{t',\tau' \\ t' \neq t}} S_{t\tau t'\tau'} J(\chi_{t'\tau'}, \chi_{t\tau}) \times \right.$$

$$\left. \times \sum_{\substack{t',\tau' \\ t' \neq t}} (S_{t\tau t'\tau'})^2 J(\chi_{t\tau}, \chi_{t\tau}) \right] \tag{7.22}$$

An analogous procedure can be employed in the case of the exchange operator K. Calculating scrupulously their matrix elements of various kinds, e.g. $\langle a'_m | J(\xi'_\nu, \xi'_\nu) | a'_{m'} \rangle$, we are faced with expressions such as

$$S_{t\tau\nu} S_{t'\tau'm} \langle \chi_{t'\tau'} | J(\xi_\nu, \chi_{t\tau}) | a_{m'} \rangle,$$

which seemingly is quadratic in ε but in fact it is a quantity of ε^4 order for the Coulomb interaction of the cross charges $\chi^* a$ and $\xi^* \chi$ is itself of ε^2 order.

The last three terms in Eq.7.21 and the two in Eq.7.22 are the part of the renormalization correction Δh_a due to the $\langle \xi | \chi \rangle$ and $\langle \chi_t | \chi_{t'} \rangle$ non–orthogonalities. This simple part of the correction has been derived from the first term in Eq.7.7.

Summarizing, the procedure of returning to the initial (unprimed) functions leads to an additional orthogonalizing potential V^{ort} which enters into the effective open–shell Hamiltonian $\tilde{\mathcal{H}}_a$

$$\tilde{\mathcal{H}}_a = \mathcal{H}_a + V^{\text{ort}} \tag{7.23}$$

The effective Hamiltonian can be determined from the identity

$$\langle \phi_A' | \mathcal{H}_a' | \phi_A' \rangle \equiv \langle \phi_A | \tilde{\mathcal{H}}_a | \phi_A \rangle \tag{7.24}$$

Each of the three types of the considered non–orthogonality contributes to V^{ort} separately

$$V^{\text{ort}} = V^{\text{c.cov}} + V^{\text{c.sh}} + V^{\text{c.pol}} \tag{7.25}$$

These are the so–called contact or exclusion contributions to the orthogonalizing potential — the contact–covalency $V^{\text{c.cov}}$, the contact–shielding $V^{\text{c.sh}}$ and the contact–polarization $V^{\text{c.pol}}$ according to the $\langle a | \chi \rangle$, $\langle \xi | \chi \rangle$ and $\langle \chi_t | \chi_{t'} \rangle$ non–orthogonality, respectively.

7.3 The contact–covalency — the main component of the crystal field potential

The contact–covalency potential $V^{\text{c.cov}}$ is introduced by the non–orthogonality of the central ion open–shell electron wave functions a_m and the outer closed shell electron wave functions of the ligands $\chi_{t\tau}$. According to Eq.7.20 we have

$$
\begin{aligned}
V^{\text{c.cov}} &= \sum_{t,\tau} |\chi_{t\tau}\rangle \langle \chi_{t\tau} | \langle a | h_a^{\text{ion}} - J(a,a) | a \rangle \delta_{mm'} - \\
&\quad - 2 \sum_{t,\tau} |\chi_{t\tau}\rangle \langle \chi_{t\tau} | [h_a^{\text{ion}} - J(a,a)] + \\
&\quad + \sum_{t,\tau} \sum_{t',\tau'} |\chi_{t\tau}\rangle \langle \chi_{t\tau} | h_a^{\text{ion}} - J(a,a) | \chi_{t'\tau'}\rangle \langle \chi_{t'\tau'}|
\end{aligned} \tag{7.26}
$$

where

$$h_a^{\text{ion}} = h_a + N_a \, J(a,a) \tag{7.27}$$

The one–electron Hamiltonian h_a^{ion} introduced above contains the average potential of all N_a electrons of the ion and hence its denotation h_a^{ion}.

As a matter of facts, h_a^{ion} is the Hartree–Fock Hamiltonian of the open–shell group of electrons obtained in the procedure of minimizing the energy of the generalized product function of the complex (see chapter 4). In practical calculations of $V^{c.cov}$ the free–ion values of both the central ion and ligand energies as well as the Madelung potential energies of the ions in the considered crystal lattice can be exploited [121]. Fortunately, it turns out that this approximation is not a source of serious errors. A quantitative analysis of the contributions to the matrix elements of $V^{c.cov}$ shows that the mutual penetration of the exchange–charge cloud $a^*\chi$ (see further) and the ligand charge distribution plays a dominant role.

Comparison of many experimental values of crystal field parameters with their model counterparts reveals an interesting and somewhat intriguing fact. It turns out namely, that the contact–covalency contribution supplemented with the true covalency contribution in its conventional understanding as a charge transfer (see chapter 8) give in sum the crystal field parameters which are close to those observed experimentally. It means that the remaining contributions, which can not be ignored, are subject to a mutual compensation. Some exemplary results of the calculations for typical representatives of various groups of metals ($3d$, $4f$, $5f$) are compiled in Table 7.1.

This characteristic regularity possesses, on the one hand, a deeper physico-chemical interpretation, and on the other some important practical consequences. It corresponds particularly well with the known simplified approach proposed by Jørgensen, Pappalardo and Schmidtke [78]. The authors assuming that the covalency component dominates among the others in the crystal field potential of the transition metal ions have formulated a simple one–parameter ligand field model, i.e. the so–called overlap model (of σ type), AOM [131]. As a matter of fact, the covalency contribution itself in its conventional meaning, to which the authors refer, does not play the expected dominating role, as seen in Table 7.1. Nevertheless, from the phenomenological point of view both the contact–covalency and the co-valency contributions are indistinguishable. These two contributions are roughly proportional to the square of interionic overlap integrals and the standing out consistence between their sum and the total values of the crystal field parameters is an independent and convincing confirmation of the AOM adequacy, yielding simultaneously a clear physical interpretation of the parameterization. We will return to this problem in chapter 17 which will be devoted to phenomenological aspects of the crystal field model.

Table 7.1. Comparison of the contact–covalency (c.cov.) and the covalency (cov.) contributions to the crystal field parameters with their experimental (exp) values for typical representatives of various groups of metals (all values in cm^{-1}).

Compound	Parameter	c.cov.	cov.	c.cov.+ cov.	exp
$KNiF_3^{(a)}$	B_{40}	$12040^{(b)}$	$6720^{(c)}$	18760	$15225^{(d)}$
(O_h)	B_{44}	7195	4015	11210	9100
$PrCl_3$	B_{20}	$66^{(e)}$	$14^{(f)}$	80	94
(C_{3h})	B_{40}	-256	-112	-368	-328
	B_{60}	-480	-208	-688	-640
	B_{66}	383	168	557	426
$CsUF_6$	B_{20}	$720^{(h)}$	$280^{(h)}$	1000	$880^{(i)}$
$(D_{3d} \approx O_h)$	B_{40}	-12380	-3990	-16370	-15870
	B_{43}	13890	4490	18390	17880
	B_{60}	4140	140	4280	4370
	B_{63}	4100	140	4240	4290
	B_{66}	2900	100	3000	3060

(a) the parameters calculated from $\Delta = E(t_{2g}) - E(e_g) = \frac{5}{3}\bar{B}_4$, where $\bar{B}_4$ is the intrinsic parameter, (b) Watson and Freeman [128], (c) Hubbard et al [81], (d) Knox et al [129], (e) Ellis and Newman [47], (f) Newman [27], (g) Margolis [108], (h) Gajek et al [80], (i) Soulié [130].

The particular importance of the contact–covalency mechanism forces us to a more rigorous estimating the adequacy of the approximations made during derivation of the renormalized potential corresponding to it (Eqs 7.8, 7.15, 7.17, 7.18). By means of relatively simple numerical tests it is not difficult to settle that replacing the self–consistent solutions of the appropriate equations (Eqs 4.34, 4.38, 4.39) with the free–ion data of the electronic structures introduces only inconsiderable quantitative modifications. On the other hand, the averaging certain expressions over all open–shell states a_m such as that used in Eqs 7.8, 7.15, 7.17, 7.18, although formally consistent with the accuracy level assumed, i.e. to ε^2, causes rather serious changes including also qualitative ones.

It is impossible to reduce exactly the expressions occurring on the left side of Eqs 7.8, 7.15, 7.17, 7.18 to the form of matrix elements of some one–electron operators. It would be possible, at most, to say of a renormalization potential being dependent on which state of the open–shell is just occupied or, in other words, of two–electron correlation crystal field. This problem will be taken up in chapter 19.

7.4   The contact–shielding

The contact–shielding potential, $V^{\text{c.sh}}$, resulting from the non–orthogonality of outer closed–shell wave functions ξ_ν of the central ion and those of outer closed–shells of the ligands, $\chi_{t\tau}$, where t runs over the ligands, has, according to Eq.7.21, the form

$$
\begin{aligned}
V^{\text{c.sh}} = & \sum_\nu \sum_{t,\tau} \{ S^2_{t\tau\nu}[2J(\xi_\nu,\xi_\nu) - K(\xi_\nu,\xi_\nu)] - \\
& - 2S_{t\tau\nu}[2J(\chi_{t\tau},\xi_\nu) - K(\chi_{t\tau},\xi_\nu)] + \\
& + \sum_{\substack{t',\tau' \\ t' \neq t}} S_{t\tau\nu}S_{t'\tau'\nu}[2J(\chi_{t\tau},\chi_{t'\tau'}) - K(\chi_{t\tau},\chi_{t'\tau'})]\}
\end{aligned}
\qquad (7.28)
$$

The first term in Eq.7.28 corresponds to a correction of intraionic Coulomb interactions between the electrons of the closed central ion shells ξ_ν which is induced by the above mentioned non–orthogonality. This is an interesting term, indeed, because it concerns, in principle, the second degree terms of the effective crystal field potential only. It appears immediately from the angular distributions of the interacting electronic densities $\xi^*\xi$ and a^*a. To be accurate, in the expressions for the crystal field parameters a 3–j factor occurs in this case with the three numbers in the upper row being equal to l_ξ, k, l_ξ, respectively. As a consequence of this and according to the triangle rule, i.e. $k \leq 2l_\xi$, the only admissible value of k is 2, and additionally the whole contribution comes from the p–electrons ($l_\xi = 1$), whereas the s–electrons are ineffective. If the d–electrons deeper hidden in the central ion electronic cloud were included to its outer electrons and if their wave function overlap with ligand spinorbitals were large enough then also the fourth degree crystal field parameters would undergo a modification. This restriction on k results from the fact that the interacting electrons belong to the same nucleus. It does not refer to the remaining terms in Eq.7.28. The terms in the second and third rows of Eq.7.28 contribute to all B_{kq} parameters which are effective for the central ion symmetry under consideration. The second row describes the interaction of electron density from the $\langle \xi | \chi \rangle$ overlap region. The third, in turn, the renormalization correction to the Coulomb potential of the ligand electrons.

Calculations carried out for the uranium compounds [80] have evidenced that although each of the terms in Eq.7.28 produces itself a fairly large input the total effect is rather weakened due to their mutual compensation. As the usual screening (see chapter 9) so the contact–shielding exerts the strongest influence on the second degree crystal field parameters. Under some special conditions the correction can reach even 40% of the total value of B_{20} parameter [132]. The impact of the contact–shielding on the higher degree parameters is considerably weaker and does not exceed few percents.

7.5   The contact–polarization

The non–orthogonality of the spinorbitals belonging to various ligands leads to the contact – polarization potential, $V^{c.pol}$, which according to Eq.7.22 is

$$V^{c.pol} = \sum_{t,\tau} \{ \sum_{\substack{t',\tau' \\ t' \neq t}} (S_{t\tau t'\tau'})^2 [2J(\chi_{t\tau}, \chi_{t\tau}) - K(\chi_{t\tau}, \chi_{t\tau})]$$

$$- \sum_{\substack{t',\tau' \\ t' \neq t}} S_{t\tau t'\tau'} [2J(\chi_{t'\tau'}, \chi_{t\tau}) - K(\chi_{t'\tau'}, \chi_{t\tau})] \} \qquad (7.29)$$

In spite of rather large values of the overlap integrals between the ligands the weight of the contact–polarization effect is limited because of mutual compensation of the individual term contributions in Eq.7.29, as it has been shown with the model calculations [132]. However, this mechanism plays a specific role as one of those, apart from the usual polarization (see chapter 10), which do not come under the superposition principle. It is easy to notice that the second term in Eq.7.29 is responsible for the departure from the rule.

7.6   Mechanisms of the contact–shielding and contact–polarization in terms of the exchange charge notion

The effects of the contact–shielding and contact–polarization have been taken into account in the crystal field theory by Newman and his co–workers [27, 133, 134, 135]. These effects have been labeled by them as the exchange–charge contributions. This is connected with their interesting physical interpretation. So, both the effects may be explained as result of a change in angular dependence of the electrostatic field which arises from the removal of a charge distribution, the so–called exchange–charge defined as

$$\varrho^{c.sh} = -4 \sum_{t,\tau} \sum_{\nu} S_{t\tau\nu} \chi_{t\tau} \xi_\nu \qquad (7.30)$$

$$\varrho^{c.pol} = -2 \sum_{t,\tau} \sum_{\substack{t',\tau' \\ t' \neq t}} S_{t\tau t'\tau'} \chi_{t\tau} \chi_{t'\tau'} \qquad (7.31)$$

for the screening and polarization, respectively, from each of the regions of overlap between the central ion and ligand ions or between each pair of ligands. In the contact–shielding case $\varrho^{c.sh}$ is being shared equally between the orbitals on the central ion and ligands centers. The former part contributes to $V^{c.sh}$ according to the first row in Eq.7.28 whereas the latter shifted towards the ligands according to the third row in Eq.7.28. In the case of the contact–polarization the charge $\varrho^{c.pol}$ is shifted equally towards both the ligands.

It is worth adding that the formulations given by Newman et al [121, 47, 133, 134, 135] differ from those presented here in a couple of points. First, in the previous approach the exchange contributions occurring in Eq.7.28, 7.29 have been neglected whereas on the ground of our calculations [118] they considerably reduce the direct contributions. Secondly, disregarding the factor $\frac{1}{2}$ at the function $\chi_{t'\tau'}$ in Eq.7.4 causes a change of the normalized primed functions and as a result of this a change in the effective potential. The factor $\frac{1}{2}$ is essential if the functions $\chi_{t\tau}$ are to be orthogonal exact to ε. Thirdly, the contributions corresponding to the first row in Eq.7.28, originally neglected, can appreciably modify the second degree parameters. Therefore, no wonder the results calculated in the original works [133] [135] are overestimated.

In evaluating the exchange–charge terms it is necessary to remember that the field produced by the exchange–charge distribution is to some extent shielding by the central ion outer shell electron excitations. Considering the mutual situation of the exchange–charge and the outer shell charge the screening problem is rather complicated. However, we may hope that the proposed approach provides an upper bound to the effects of shielding.

CHAPTER 8

Covalency contribution, i.e. the charge transfer effect

So far, the contributions into the effective crystal field potential arising directly from interactions of undisturbed electronic densities and from the non–orthogonality of one–electron functions localized at different ions of the complex have been considered. That is tantamount to confining the model to the one–configuration approximation with the basis of functions limited to those of ground configurations of the component ions. In other words, we have dealt with the group product functions from a certain subspace of the state space defined on the ground of available physical premises. As is well known, a more adequate approximation of ground state of the model system can be obtained by taking into account admixtures of states individual electrons of which are excited.

Considering the excitation energy magnitudes one can confine itself, in first approximation, to one–electron excitations only. It should be emphasized that these are formal excitations, i.e. virtual ones allowing the basis of functions needed for description of the states of the complex to be extended. These admixtures bring the initial (one–configurational) states into closer agreement with the actual low–lying states of the complex. The contribution defined as the covalency one is connected with inter–ionic excitations of the charge–transfer type. Intra–ionic excitations determining the screening and polarization effects will be considered later (chapters 9 and 10).

8.1 The one-electron excitations. Group product function for the excited state

The charge transfer from ligand to metal ion, i.e. the transfer of an electron from the outer ligand shell (X) into the open–shell (A) of the central ion, $\chi \to a$, is the most important excitation from the crystal field theory point of view. Notably, this is a consequence of a rather typical donor role assigned to ligands and acceptor character of metal ions. Other possible excitations are: $\xi \to a, \xi \to \zeta, \chi \to \zeta, a \to \zeta$, where apart from the symbols introduced previously (chapter 4), ζ has been used to denote an empty–shell orbital on either centre. These excitations can only produce secondary contributions to the crystal field splitting. Those of the type $\xi \to a$ and $a \to \zeta$ give expressions of exactly the same form as those for $\chi \to a$ charge–transfer and, in turn, the electron transfer from ligand to an empty outer orbital of the central ion, $\chi \to \zeta$, is, in general, still less effective for due to the spherical symmetry of the empty shells (mainly s and p) it may yield, in principle, only an uniform shift of all energy levels. Admittedly, there is a possibility that the empty shell suffers from a splitting in the crystal field, e.g. the p-shell in the field of symmetry lower than the cubic one, or the $5d$ shell for the lanthanides in whichever field, which might lead to a renormalization correction. Obviously, the weight of the particular excitations is conditioned by their energies. Thus, let us focus our attention on the leading covalency mechanism. The electron transfer $\chi \to a$ brings about a modification of partition of the electrons into the groups which is followed by the corresponding changes in their electron configurations.

For each of the new configurations (the group functions) the optimizing procedure can be performed analogously to that described for ground configurations (chapter 4). The generalized product functions for the excited states (after an electron transfer) are set up according to the scheme

$$\Psi_A^0 = \mathcal{A}\left[\phi_A^0 \phi_\Xi^0 \prod_t \phi_{X_t}^0\right] \quad \begin{matrix} \phi_A^0 \xrightarrow{+e^-} \phi_{A'} \\ \phi_{X_t}^0 \xrightarrow{-e^-} \phi_{X_t'} \end{matrix} \quad \Psi_{A'X_t'} = \mathcal{A}\left[\phi_{A'} \phi_\Xi^0 \phi_{X_t'} \prod_{t' \neq t} \phi_{X_{t'}}^0\right] \tag{8.1}$$

where Ψ_A^0 is the initial group product function defined earlier in chapter 4 and $\Psi_{A'X_t'}$ – the excited function composed of $\phi_{A'}$ – the optimized group function of $N_A + 1$ electrons of the open–shell A of the central ion and $\phi_{X_t'}$ – the optimized group function of $N_{X_t} - 1$ outer electrons of ligand t, as well as, the same as in Ψ_A^0, ϕ_Ξ^0 and $\phi_{X_{t'}}^0$ functions for outer closed–shell of the central ion and for ligands t' being different from t, respectively. The subscripts at the group functions denote both their electron configurations as well as all quantum numbers necessary for their unambiguous description. Since $\phi_{X_t}^0$ are the closed–shell one–determinantal functions the parent spin–orbital of ligand t of the transferred electron may be pointed out specifically.

On the other hand, one can not be explicit about the open–shell spin-orbital which accepts the electron, except the reference state model [27]. It results just from multi–determinantal structure of the open–shell group functions ϕ_A^0 and $\phi_{A'}$. Now, it is a due time to remind the three different approaches employed in the process of arriving at the group product functions of the excited states. They are arranged in order of reducing their restrictions.

— In the simplest model based on the reference state [27] only one open-shell electron is treated explicitly, i.e. we have a trivial one–electron group function of the shell. Both correlation effects inside the open shell and polarization ones arising from the electron transfer and a change of ionization pattern are in this approximation intentionally ignored. This model leads to results being related to those of simplified version of the molecular orbital or Hartree–Fock approach.

— In the second approach a many–electron group function of the open shell (according to its population) set up with the optimizing procedure is applied. However, the optimization of the excited states is carried out for the initial ionization pattern. This approach, after some assumptions reducing the problem to the one–electron description, is convergent with the MO method with characteristic for all one–electron approximations neglecting the correlation effects. Strong assumptions of the model making the covalency contribution to be independent of occupation of the open–shell [121] reduces, in practice, this approach to the first one.

— In the third approach the optimizing the excited state is carried out for the new ionization pattern, i.e. for $M^{(n-1)+} - L^{(m-1)-}$ pair of ions instead of the initial one $M^{n+} - L^{m-}$. This is so–called configuration interaction scheme [128, 81] including important correlation and polarization effects omitted previously. These effects play an important part in determining the energy of the excited levels. The assumption of an instantaneous polarization leads to the adoption of a Heitler–London model like that employed by Hubbard et al [81] in their $KNiF_3$ calculation.

If we allow the remaining electrons to relax in response to the electron transfer the energy denominator (see further) which represents the difference in energy of the initial and final configurations is reduced compared with that for the MO method and therefore the covalency contribution increases in this approach. A useful feature of the configuration interaction approach is that the MO approach can be obtained as a special case of it by making suitable approximations. A further simplification of the calculation is obtained by collecting together those terms which correspond to the energies of the electrons on the free metal and ligand ions and substituting for them the known free–ion Hartree–Fock ground state energies as well as the electron Madelung potentials. These substitutions also improve the calculations by including the effects of the other electrons although they have been omitted from our model.

The cross–matrix element of the Hamiltonian between the initial Ψ_A^0 and final $\Psi_{A'X_l'}$ states will turn out to be the crucial quantity for the considered covalency mechanism. But since both the functions are superpositions of products of one–determinantal functions the total matrix element is the sum of the partial matrix elements for all pairs of the product functions. After including some approximations, which will be postulated further, the cross–matrix elements will be different from zero only when the final $\Psi_{A'X'_t}$ functions originate directly from the initial Ψ^0_A function, i.e. if the N_A electron parts of both open–shell determinants are identical.

To simplify the procedure applied it can be assumed that the initial and excited functions are mutually orthogonal. This feature may always be formally ensured.

The contribution each of the charge–transfer excitations yielding the covalency effect is determined by the ratio of square of the cross–matrix element between the ground and excited states to the difference of their energies. This ratio should be scrupulously estimated for all virtual electron transitions. The result of a such review can be sometimes unexpected. For example Dexter [136] has suggested that the first excited state in NaCl is $4s$ state fairly well localized on the Cl^- ion. Similarly, one of inter–ionic mechanisms of the so–called zero–splitting of the $^8S_{7/2}$ of the gadolinium ion in lanthanium ethyl sulphate matrix is based on analogous excited state of the oxygen anion [70].

8.2 The renormalization of the open–shell Hamiltonian due to the covalency effect

The configuration interaction leads, from the very nature of things, to admixtures of the excited states from higher configurations into the ground state. In other words, covalency effects may be introduced by making allowance for some admixture of these states into the trial wave functions of the complex. Thus, the extended wave function of the ground state Ψ_A including the essential admixtures has the form

$$\Psi_A = \Psi^0_A + \sum_{A'} \sum_{X'_t} \gamma^A_{A'X'_t} \Psi_{A'X'_t} \qquad (8.2)$$

where $\gamma^A_{A'X'_t}$ are the coefficients (amplitudes) of the excited wave functions admixtured, the summations run over all possible $\phi_{A'}$ and $\phi_{X'_t}$ group functions and t numbers the ligands.

Strictly, taking into account the optimizing procedure applied Ψ^0_A function in Eq.8.2 should be replaced by a superposition of the initial functions $\sum_i k_i \Psi^0_{Ai}$. The corresponding variational calculation [81] leads, in principle, to the results being parallel to those for the individual Ψ^0_{Ai} functions. What we loose is a secular character of the variational equations and more exact dependence on energy of the open–shell states.

Applying the Ritz variational principle with the function Ψ_A (Eq.8.2) as a trial function, remembering of the assumed orthogonality of the initial and excited state wave functions and neglecting all terms of orders higher than γ^2 one gets for the energy of Ψ_A

$$
E_A = \frac{\langle\Psi_A|\mathcal{H}|\Psi_A\rangle}{\langle\Psi_A|\Psi_A\rangle} = (1 - \sum_{A'}\sum_{X_t'}|\gamma_{A'X_t'}^A|^2) \times \Big\{ \langle\Psi_A^0|\mathcal{H}|\Psi_A^0\rangle +
$$

$$
+ 2\sum_{A'}\sum_{X_t'}\gamma_{A'X_t'}^A\langle\Psi_A^0|\mathcal{H}|\Psi_{A'X_t'}\rangle +
$$

$$
+ \sum_{A'}\sum_{X_t'}|\gamma_{A'X_t'}^A|^2\langle\Psi_{A'X_t'}|\mathcal{H}|\Psi_{A'X_t'}\rangle\Big\} \tag{8.3}
$$

From the stationary condition $\partial E_A/\partial\gamma_{A'X_t'}^A = 0$, we have

$$
\gamma_{A'X_t'}^A = -\frac{\langle\Psi_A^0|\mathcal{H}|\Psi_{A'X_t'}\rangle}{\langle\Psi_{A'X_t'}|\mathcal{H}|\Psi_{A'X_t'}\rangle - \langle\Psi_A^0|\mathcal{H}|\Psi_A^0\rangle} \tag{8.4}
$$

and consequently

$$
E_A = \langle\Psi_A^0|\mathcal{H} + \Delta\mathcal{H}_A|\Psi_A^0\rangle \tag{8.5}
$$

where

$$
\Delta\mathcal{H}_A = -\sum_{A'}\sum_{X_t'}\frac{\mathcal{H}|\Psi_{A'X_t'}\rangle\langle\Psi_{A'X_t'}|\mathcal{H}}{\langle\Psi_{A'X_t'}|\mathcal{H}|\Psi_{A'X_t'}\rangle - \langle\Psi_A^0|\mathcal{H}|\Psi_A^0\rangle} \tag{8.6}
$$

At the cost of developing Hamiltonian by the renormalization correction $\Delta\mathcal{H}_A$ reversion to the initial basis of the one–configurational approximation has been attained (Eq.8.5). The whole optimizing procedure can now be repeated with the renormalized Hamiltonian $\mathcal{H} + \Delta\mathcal{H}_A$. Fortunately, the corresponding modifications of the potentials occurring in the Schrödinger equations for the one–electron functions (Eqs 4.34, 4.38, 4.39) as those in the orthogonalizing potential (Eq.7.25) arising from the presence of the $\Delta\mathcal{H}_A$ term turn out to be negligible. However, the secular equation (Eq.4.35) will undergo a radical change. The modified Hamiltonian of the open–shell electrons will contain an additional term V_A^{cov}, so that

$$
\widetilde{\widetilde{\mathcal{H}}}_A = \mathcal{H}_A + V^{\text{ort}} + V_A^{\text{cov}} \tag{8.7}
$$

V_A^{cov} can be determined from the identity

$$
\langle\Psi_A^0|\Delta\mathcal{H}_A|\Psi_A^0\rangle \equiv \langle\phi_A^0|V_A^{\text{cov}}|\phi_A^0\rangle \tag{8.8}
$$

Unfortunately, there is one inconvenient feature of this approach. The renormalization term $\Delta\mathcal{H}_A$ and thereby V_A^{cov}, according to Eqs 8.6 and 8.8, depend on the electron eigenstate ϕ_A energy of which is calculated. It means that each electron term and moreover each crystal field level possesses its own set of crystal field parameters. This effect is known in literature as the term dependent crystal field (TDCF) [137]. Using the parameters averaged over all open–states the effect can be eliminated. Fortunately, an almost universal adequacy of phenomenological one–electron parameterization of the crystal field potential supports the approach.

Thus, the fundamental requirement towards the model constructed to be compatible with the classical phenomenological approach needs making the Hamiltonian $\widetilde{\widetilde{\mathcal{H}}}_A$ in Eq.8.7 independent on individual states of the open–shell, i.e. on ϕ_A functions explicitly. This problem arose earlier during derivation of the orthogonalizing potential V^{ort} in the effective $\tilde{\mathcal{H}}$ Hamiltonian (chapter 7). However, in the case of $\Delta\mathcal{H}_A$ or V_A^{cov} it is more complex problem. As seen from Eq.8.6, $\Delta\mathcal{H}_A$ depends on ϕ_A both openly through the energy denominator, and indirectly through limiting the summation over the excited states only to those available via an one–electron transfer from the initial state. The averaging the energy denominator over all ϕ_A states excludes both the dependencies. It is therefore since the cross–matrix elements $\langle \Psi_A^0 | \mathcal{H} | \Psi_{A'X'_t} \rangle$ vanish automatically unless the ϕ_A^0 state is the parent state for the $\phi_{A'}$ one in the sense given earlier, and therefore the summation in Eq.8.6 can be formally extended to all $\phi_{A'}$ states. The averaging the energy denominator is equivalent to the assumption that the energy difference between the excited and initial states depends only on the $\chi_{t\tau}$ spinorbital of the ϕ_{X_t} state, i.e. on the initial state of the ligand electron which is transferred, e.g. n_s, np_σ, np_π etc. The vanishing of the cross-matrix elements of the Hamiltonian if the N_A–electron part of the $\phi_{A'}$ wavefunction differs from the ϕ_A^0 function is, in turn, tantamount to their independence on the occupation of the open–shell and suggests that the proportion of the crystal field due to covalency will not vary significantly along e.g. the lanthanide series provided of course the crystallographic symmetry is settled. This shows that the approximations eliminating the dependence of the crystal field potential on the particular eigenstates of the open–shell correspond to the approximation made in representing the charge–transfer effect in terms of a one–electron crystal field [121]. The remaining two– and many–electron parts form the correlation crystal field which we have already met during formulating the contact–covalency contribution (chapter 7). The developed form of V^{cov}, useful in actual calculations, will be presented after introducing certain further approximations ample with consequences of both fundamental and numerical nature.

8.3   Basic approximations

The considerations we have led hitherto, at least those which are confined to the one–electron part of the covalency effect have had an exact character. Starting from the initial many–electron problem an effective Hamiltonian for the specified group of electrons has been defined and expressed in the form a priori postulated (Eq.3.4). This effective Hamiltonian is a projection of the total Hamiltonian into the open–shell basis of functions.

Despite of unquestionable advantages arising from employing the concept of generalized product functions, the model obtained is still too complex and cumbersome to play a role of a practical model on a wide scale. It requires, among others, solution of several sets of self–consistent equations and orthogonalization of wave functions originating from various configurations. Both these drawbacks of the approach are eliminated by making allowances for the two following approximations:

(i) all the one–electron wave functions occurring in the secular equation (Eq.4.35), with Hamiltonian $\tilde{\tilde{\mathcal{H}}}_A$ instead of $\mathcal{H}_A$, are the free–ion functions,

(ii) the radial functions of excited configuration states are the same as those of the ground configuration.

The first assumption is equivalent with setting back the iterative procedures for the one–electron functions of the ground state (Eqs 4.34, 4.38, 4.39) and their counterparts in excited states at the zero step. This is nothing more but taking the surroundings potential in the crystal ion area as a constant one. In reality it is a slow–changing function of the space coordinates.

The second approximation means that after the electron transfer from ligand to central ion the radial function of the remaining electrons is not changed. The extra a electron entering the a^{N_a+1} configuration is described in this approximation by a function of the identical radial distribution as that in the configuration a^{N_a}. Undoubtedly, this is rather a coarse approximation but referring only to covalency contribution which, as it has been already stated, is not at all a dominating one. It is unquestionable that after the electron transfer from ligand to metal ion the orbitals of the metal open–shell expand from those for M^{n+} to those for $M^{(n-1)+}$ ion (the ligand orbitals contract) and the process gives a relaxation energy.

The problem of taking into account the covalency effects in the crystal field potential has always evoked a lot of controversies. Various approaches to this problem are known in literature, e.g. that by Watson and Freeman [128], Hubbard et al [81], Moskovitz et al [138], Newman [27]. Each of them is a compromise between requirement of an appointed accuracy from the one hand and simplicity of the model from the other. The model presented here, having its original in the Newman formulation, is closer to the latter concept.

8.4 The one–electron covalency potential V^{cov}

In order to obtain the explicit form of the covalency potential V^{cov} the diagonal and cross–matrix elements $\langle \Psi_A^0 | \mathcal{H} | \Psi_A^0 \rangle$, $\langle \Psi_{A'X_t'} | \mathcal{H} | \Psi_{A'X_t'} \rangle$ and

$\langle \Psi_A^0 | \mathcal{H} | \Psi_{A'X_t'} \rangle$ are needed according to Eqs 8.6 and 8.8.

To this end simple rules which are consequences of the orthogonality and antisymmetry of the generalized product functions are used [139]. Let us investigate the calculation method taking by way of example a cross–matrix element $\langle \Psi_A^0 | \mathcal{H} | \Psi_{A'X_t'} \rangle$. First of all, let us notice that each group product function of the complex (antisymmetrized ex definitione, Eq.4.30) is a superposition of products of determinantal functions of particular groups of electrons. Thus, it is enough to examine, without loss of generality, a matrix element of the type

$$\left\langle a_{m_1} a_{m_2} \ldots a_{m_{N_a}} \prod_\nu \xi_\nu \prod_{t\tau} \chi_{t\tau} \middle| \mathcal{H} \middle| a_{m_1'} a_{m_2'} \ldots a_{m_{N_a}'} \prod_\nu \xi_\nu \prod_{\tau'\neq\tau} \chi_{t\tau'} \prod_{\substack{t'\neq t \\ \tau'}} \chi_{t'\tau'} \right\rangle$$

$$(8.9)$$

where all the denotations introduced previously remain in force. Let us notice that $\prod_{\tau'\neq\tau} \chi_{t\tau'}$, i.e. $\phi_{X_t'}$ (see Eq.8.1) defines indirectly but unambiguously the state of the electron of ligand t transferred to the metal ion.

Expanding the matrix element (Eq.8.9) into the matrix elements of the one– and two– electron operators in the Hamiltonian

$$\mathcal{H} = \sum_i h(\mathbf{r}_i) + \sum_{\substack{i,j \\ i>j}} \frac{1}{r_{ij}} \qquad (8.10)$$

and using approximation (ii) from the former section we get

$$\langle \chi_{t\tau} | h | a_{m_{N_a+1}'} \rangle \prod_{i=1}^{N_a} \delta_{m_i m_i'} +$$

$$+ \sum_i \langle \chi_{t\tau} a_{m_i} || a_{m_{N_a+1}'} a_{m_i'} \rangle \prod_{\substack{j=1 \\ j\neq i}}^{N_a} \delta_{m_j m_j'}$$

$$+ \sum_\nu \langle \chi_{t\tau} \xi_\nu || a_{m_{N_a+1}'} \xi_\nu \rangle \prod_{i=1}^{N_a} \delta_{m_i m_i'}$$

$$+ \sum_{\substack{t'\tau' \\ \neq t\tau}} \langle \chi_{t\tau} \chi_{t'\tau'} || a_{m_{N_a+1}'} \chi_{t'\tau'} \rangle \prod_{i=1}^{N_a} \delta_{m_i m_i'} \qquad (8.11)$$

where $\langle \cdot || \cdot \rangle$ denotes shortly a matrix element of the two electron operator $1/r_{ij}$.

For clarity the exchange terms in Eq.8.11 have been omitted. Nevertheless, if the two interacting electrons possess identical spin functions the exchange terms are formally effective although frequently they can be negligible.

The second term in Eq.8.11 after averaging over all occupied spin-orbitals of the open shell A takes the form

$$\langle \chi_{t\tau} \bar{a} || a_{m'_{N_a+1}} \bar{a} \rangle N_a \prod_{i=1}^{N_a} \delta_{m_i m'_i} \tag{8.12}$$

where $\bar{a}$ is the radial part of a_m functions. Owing to this approximation the covalency contribution becomes independent on occupation of the open–shell. Substituting the approximation (Eq.8.12) into the expansion (Eq.8.11) and adding the non–revealed exchange terms enables us to describe Eq.8.11 in the compact form

$$\langle \chi_{t\tau} | h_{t\tau} | a_{m'_{N_a+1}} \rangle \prod_{i=1}^{N_a} \delta_{m_i m'_i} \tag{8.13}$$

where $h_{t\tau}$ is the effective one–electron operator which, using the ionic Hamiltonian notion, h^{ion}, introduced in the previous chapter, can be presented as

$$h_{t\tau} = h^{\text{ion}} - J(\chi_{t\tau}, \chi_{t\tau}) \tag{8.14}$$

where the much smaller term $K(\chi_{t\tau}, \chi_{t\tau})$ is omitted this time indeed.

The above derivation has been founded on assumption of orthogonality of the component one–electron functions of the complex. Passing on, in compliance with approximation (i), to the free ion functions, i.e. replacing a_m by a'_m functions

$$a'_m = \left[1 - \sum_{t',\tau'} |\langle \chi_{t'\tau'} | a_m \rangle|^2 \right]^{-1/2} \left(a_m - \sum_{t',\tau'} \langle \chi_{t'\tau'} | a_m \rangle \chi_{t'\tau'} \right) \tag{8.15}$$

we get instead of Eq.8.13 the modified expression

$$\langle \chi_{t\tau} | \tilde{h}_{t\tau} | a_{m'_{N_a+1}} \rangle \prod_{i=1}^{N_a} \delta_{m_i m'_i} \tag{8.16}$$

in which the new effective one–electron Hamiltonian $\tilde{h}_{t\tau}$ has the form

$$\tilde{h}_{t\tau} = h_{t\tau} - \sum_{t'\tau'} \sum_{t''\tau''} |\chi_{t'\tau'}\rangle \langle \chi_{t'\tau'} | h_{t\tau} | \chi_{t''\tau''} \rangle \langle \chi_{t''}| \tag{8.17}$$

Expanding analogously the diagonal matrix elements $\langle \Psi_A^0 | H | \Psi_A^0 \rangle$ and $\langle \Psi_{A'} x'_i | H | \Psi_{A'} x'_i \rangle$ and using Eqs 8.6 and 8.8 the full explicit form of V^{cov} is obtained

$$V^{\text{cov}} = \sum_{t,\tau} \frac{\tilde{h}_{t\tau} |\chi_{t\tau}\rangle \langle \chi_{t\tau} | \tilde{h}_{t\tau}}{\Delta_{t\tau}} \tag{8.18}$$

where

$$\Delta_{t\tau} = \langle \bar{a} | h^{\text{ion}} | \bar{a} \rangle - \langle \bar{a} | J(\chi_{t\tau}, \chi_{t\tau}) | \bar{a} \rangle - \langle \chi_{t\tau} | h^{\text{ion}} | \chi_{t\tau} \rangle \qquad (8.19)$$

The last equation gives the energy difference between the initial and excited states after the electron transfer $\chi \to a$ for the assumed accuracy (to the zero order of S). The derivation of Eqs 8.17 and 8.19 and finally Eq.8.18 corresponds to expanding the cross–matrix elements of the group product functions into the one–electron function matrix elements exact to the terms of order of S whereas the diagonal ones exact to zero–order terms with respect to the S where S is the interionic overlap integral. The terms independent on particular open–shell states, i.e. those producing only a uniform shift of all energy levels are omitted. In the calculations some further facilities can be introduced by collecting together the terms which correspond to the electron energy in the free ligand and metal ions and replacing them by the known values of the Hartree–Fock and Madelung energies as in the contact–covalency problem. So, we may use the following substitutions [121, 47]

$$\langle a_m | h | \chi_{t\tau} \rangle = (\epsilon_A + U^+) \langle a_m | \chi_{t\tau} \rangle - \sum_{t'} \langle a_m | V_{t'}^{\text{Lc}} | \chi_{t\tau} \rangle \qquad (8.20)$$

and

$$\langle \chi_{t\tau} | h | \chi_{t\tau} \rangle + \sum_{\tau' \neq \tau} \langle \chi_{t\tau} \chi_{t\tau'} || \chi_{t\tau} \chi_{t\tau'} \rangle =$$

$$= \epsilon_X - U^- - \frac{1}{r_0} - \sum_{t' \neq t} \langle \chi_{t\tau} | V_{t'}^{\text{Lc}} | \chi_{t\tau} \rangle \qquad (8.21)$$

where ϵ_A, ϵ_X are, respectively, the free-ion Hartree-Fock energies of A–open–shell electrons and the ligand X–shell electrons, U^+, U^- denote the electron Madelung potential energies at the central ion and ligand sites in the crystal lattice, $V_{t'}^{\text{Lc}}$ is the core potential of ligand t' and r_0 is the metal-ligand distance (in atomic units). The summation index τ' in Eq.8.21 runs over all spin–orbitals of the X_t shell excluding the $\chi_{t\tau}$ spin–orbital empty after the electron transfer. Effective exchange terms are in this sum included automatically.

Detailed analysis of the covalency mechanism [121] shows that the dominant covalency contribution comes from the mutual penetration of the exchange–charge cloud $a\chi$ and the ligand charge distribution, similarly as in the exclusion model. These substitutions, Eqs 8.20 and 8.21, together with the previous approximations reduces, in practice, the presented approach to that by Ellis and Newman [121, 47] based on the reference function concept (see chapter 4).

8.5 The one–electron covalency potential V^{cov} in the molecular– orbital formalism

As it has been already mentioned, on account of the approximations introduced, similar results can be obtained by means of the molecular orbital method (MO). In this method a certain one–electron effective Hamiltonian h^{MO} is postulated. Its function basis is limited to linear combinations of atomic orbitals (LCAO) set up in accordance with symmetry requirements of the system [141]. In the case of axial symmetry molecular orbitals assume exceptionally simple form

$$a'_m = N_m(a_m - \lambda_{m\tau}\chi_\tau)$$
$$\chi'_\tau = N_\tau(\chi_\tau + \gamma_{m\tau}a_m) \tag{8.22}$$

where N_m and N_τ are normalizing factors. The orthogonality of orbitals a'_m and χ'_t (Eq.8.22) is ensured exact to $\varepsilon \approx S_{m\tau} \approx \lambda_{m\lambda} \approx \gamma_{m\gamma}$ if

$$\lambda_{m\tau} = \gamma_{m\tau} + S_{m\tau} \tag{8.23}$$

where $\gamma_{m\tau}$ can be recognized as the covalency coefficient defining the admixture of a_m function in the ligand's orbital χ_τ, and $S_{m\tau}$ is the overlap integral. If from the symmetry selection rules results that $S_{m\tau} = \langle a_m|\chi_\tau\rangle \equiv 0$ then inevitably $\lambda_{m\tau} = \gamma_{m\tau} = 0$, otherwise from the stationary condition for $\langle a'_m|h^{\text{MO}}|a'_m\rangle$ one gets

$$\lambda_{m\tau} = -\frac{\langle a_m|h^{\text{MO}}|\chi_\tau\rangle - S_{m\tau}\langle a_m|h^{\text{MO}}|a_m\rangle}{\langle a_m|h^{\text{MO}}|a_m\rangle - \langle\chi_\tau|h^{\text{MO}}|\chi_\tau\rangle} \tag{8.24}$$

or correspondingly

$$\gamma_{m\tau} = -\frac{\langle a_m|h^{\text{MO}}|\chi_\tau\rangle - S_{m\tau}\langle\chi_\tau|h^{\text{MO}}|\chi_\tau\rangle}{\langle a_m|h^{\text{MO}}|a_m\rangle - \langle\chi_\tau|h^{\text{MO}}|\chi_\tau\rangle} \tag{8.25}$$

and the covalency correction to the energy of a_m spin–orbital (see Eq.8.3) amounts to

$$\Delta E_m^{\text{cov}} = \gamma_{m\tau}^2\left(\langle a_m|h^{\text{MO}}|a_m\rangle - \langle\chi_\tau|h^{\text{MO}}|\chi_\tau\rangle\right) \tag{8.26}$$

Equally, it is easy to get, within the frame of the same approach, the renormalization correction arising from the non–orthogonality $\langle a_m|\chi_\tau\rangle$. It has the form

$$\Delta E_m^{\text{c.cov}} = S_{m\tau}^2\langle a_m|h^{\text{MO}}|a_m\rangle - 2S_{m\tau}\langle a_m|h^{\text{MO}}|\chi_\tau\rangle + S_{m\tau}^2\langle\chi_\tau|h^{\text{MO}}|\chi_\tau\rangle \tag{8.27}$$

Considering the conditions for non–zero overlap integrals in the axial (local) coordinate system it is directly seen that if the ligand active electrons are of type s and p, i.e. for $\tau = s, p_0, p_{\pm 1}$, the covalency contributions both the contact and ordinary ones refer to a_0 and $a_{\pm 1}$ states only. Since these are dominating contributions to the crystal field potential this explains also a dominating role of e_σ and e_π parameters in comparison to e_δ ones in the angular overlap model (AOM), (see chapter 17).

The MO theory itself, leaving out of account some general suggestions does not answer to the basic question — what is the best one–electron Hamiltonian in this approach? The variational approximation to this one–electron but multicentre eigenvalue problem leads to a system of integro–differential equations which can not easily be solved except in the one centre problem [140]. However, comparing the MO covalency correction (Eq.8.26) with the result obtained within the more general configuration interaction approach (Eq.8.18) shows that the effective h^{MO} Hamiltonian corresponds roughly to h^{ion}. To reconcile both the quantities one should include in Eq.8.26 the correlation term $\langle \bar{a} | J(\chi_{t\tau}, \chi_{t\tau}) | \bar{a} \rangle$ which is present in denominator of Eq.8.18. Similarly, Eq.8.27 bears a resemblance to the previously found expression for $V^{c.cov}$ (Eq.7.26) but this time h^{MO} should correspond to $h^{ion} - J(\bar{a}, \bar{a})$. The revealed divergences are not accidental ones. They are a manifestation of the so–called electron with itself interaction energy which is characteristic for any averaged treatment of inter–electron interactions by means of one–electron self–consistent Hamiltonian. It is proper to add that neglecting the correlation terms leads to considerable underestimating the covalency contribution [81].

8.6 Remarks on the covalency mechanism

Measurements of the unpaired spin density based on the resonance techniques both the electron paramagnetic and nuclear magnetic resonance supported by those with the neutron radiography have evidenced a general appearing of the covalency effect even in compounds considered as typical ionic ones, e.g. in $KNiF_3$ [81]. Thus, there is a reliable argument to expect an analogous generality and importance of this mechanism for the crystal field potential. Indeed, the covalency contribution, as seen in Table 7.1 from chapter 7 is one of the largest term yielding only to the contact-covalency contribution.

Nevertheless, the pivotal question still remains – in what extent the interionic charge transfer, as a many-electron process in its nature, can be described with an effective one-electron Hamiltonian? Complexity of the phenomenon can be properly appreciated against a background of difficulties in exact quantitative description of the polarization and screening effects which are related to the covalency one as for their micromechanisms but better understood. Fortunately, some corrections can be introduced without extending the formalism. An adequate description of distortion of the electron density distribution in the vicinity of both ions induced by the electron transfer is a critical point of the method [128, 74]. These are mainly the polarization effects which although formally correspond to higher order terms in perturbation expansion produce appreciable changes in energy of electronic levels, e.g. in $KNiF_3$ the polarization correction calculated from the free–ion polarizabilities reaches about 20% of the total covalency effect [81]. In practice, some experimental data like the ionization energy can be exploited. They include quite naturally the charge–redistribution effects associated with the excitations.

Considering the importance of the covalency contribution the so-called covalo-electrostatic crystal field model has been postulated [142] as a simplified but representative model of the total crystal field effect. In this model the crystal field potential is evaluated as the sum of an electrostatic and a covalent contribution only. When the electrostatic part is created by extended charges of the first neighbours (charge penetration effect) the model yields often quite satisfactory results. The weight of covalency contribution is so significant that including the interaction of the second neighbour electrons with the first neighbour (ligand) valence electrons improves the model appreciably [143, 144]. The distant of the second neighbours to the central ion is the crucial parameter of the influence. For example, in rare earths compounds, second neighbours are allowed to interfere in the covalent process if their distance to the central ion is lower than 3.5 $\mathring{A}$. The second neighbour effect is seen in plots of one-electron ionization energy of ligand versus the shortest central ion – ligand distance for various second neighbours. The chemical status (position in the periodic table) of the second neighbours determines their impact and sometimes may lead to a distinctly original behaviour [143, 144].

CHAPTER 9

Schielding and antishielding effect:
contributions from closed electron shells

Interatomic excitations of closed shell electrons of the central atom (paramagnetic ion or atomic core in metals) induced by external crystal field lead
to the screening or antiscreening effect according to whether an attenuation
or amplification of the crystal field potential giving rise to it at the location
of the open–shell electron density takes place [146, 152, 153, 154, 155, 156].
The screening comes of specific distortion of the closed shell orbitals which
is the consequence of admixing some excited states into the spherical initial
states. The distorted closed shells produce an additional potential which
superimposing on the initial one modifies it. In other words, the crystal–
field shielding terms arise from the interaction between the distorted charge
distribution and the open–shell electrons [145, 146, 147]. Susceptibility to
the distortion, i.e. to the polarization of closed atomic shells depends on
their position in the atom and primarily refers to filled outer shells which
are weaker bound by their nucleus. In a considerably less extent the effect
concerns closed inner shells in the atom [146, 147, 148]. Hence a particular
importance of this effect in the case of the lanthanide ($4f$) and actinide
($5f$) ions, in which the $5s^2p^6$ and $6s^2p^6$ closed shells are situated outside
the $4f$ and $5f$ open–shells, respectively. Thus, the shielding (or antishielding) in an atom is directly connected with its polarizability with respect to
all multipole components of an external potential [145, 149, 150, 151]. An
analogous mechanism determines the hyperfine structure of electronic states
but here the role of the external potential source is played by nuclear multipole moments. The so–called quadrupole antishielding γ_∞ factor [145, 151]
is defined as the ratio of the quadrupole moment induced in the electron
charge distribution to the nuclear quadrupole moment which give rise to it.

The screening phenomenon can also be perceived as a certain redistribution of energy within the considered closed shell, namely, as a loss of a part
of the kinetic energy of the electrons in favour of their correlation energy.

In the simplified approach, under the assumption that the crystal field interacts only with open–shell electrons and closed shells are completely rigid, the effect is ignored. The physical nature of the screening effect may be better understood by realizing that it is the atomic in scale counterpart of the dielectric polarization effect in macroscale. There is also an analogy between the diamagnetism induced by magnetic field and the screening effect produced by the electrostatic crystal field.

9.1 Phenomenological quantification of the screening effect

From the retrospective point of view taking into account the screening effect has originated from ascertaining the inadequacy of the point charge model in the crystal field theory. At that stage of development of the theory this mechanism seemed to be responsible for reducing the second order crystal field parameters and possibly for increasing the sixth order ones, i.e. for antiscreening in the latter case (see discussion in section 4). It is worth noting that the supposition has found a confirmation in later theoretical analyses but unfortunately only a qualitative one. Nevertheless the concept of the screening initiated inquiries about the effective crystal field potential. Consequently, in the phenomenological approach, characteristic for this period of evolution of the crystal field theory, the crystal field term in the Hamiltonian of a bound ion, was assumed in the following form

$$\mathcal{H}_{\mathrm{CF}} = \sum_i \sum_k \sum_q B_{kq}(1 - \sigma_k)C_q^{(k)}\left(\frac{\mathbf{r}_i}{r_i}\right) \tag{9.1}$$

where B_{kq} stand for the PCM parameters, and σ_k are the screening factors. These factors were taken to be independent on q and played the role of free (phenomenological) parameters. Their values have been calculated from the experimental data by way of fitting procedures.

As seen from Eq.9.1 the screening of linear character has been postulated, i.e. that it can be presented in the form of product of screening factor σ_k times the appropriate component of the unscreened potential, or otherwise, that the screening is proportional to the crystal field strength. More precise model calculations [147, 148] sketched further have clearly evidenced that the simple formula (Eq.9.1) is, in principle, well justified from the theoretical point of view.

The phenomenological description of screened crystal field potential according to Eq.9.1 has turned out to be very effective in its parameterization and has been commonly employed or rather misemployed. However, the phenomenological screening factors obtained from Eq.9.1 do not correspond to their values found theoretically. Nowadays we know that the striking efficiency of the simple parameterization has its more general theoretical grounds. This parameterization is formally equivalent with that obtained for a relatively exact microscopic model of the effect (section 4). It means that the phenomenological screening factors σ_k can be well fitted to the experimental data but their values obtained in this way have in fact a weak or no connection with the actual screening itself.

9.2 Microscopic model of the screening effect

There exist several methods of solving the screening problem. However, their common feature is that independently of the formalism applied determination of all possible distortions or admixtures of excited states to the states subject to the crystal field action is indispensable and crucial for the effect.

These states have to be solutions of the Schrödinger equation with the Hamiltonian supplemented with the disturbing potential $\mathcal{H}_{\mathrm{CF}}$. According to symmetry of the perturbing potential the admixtured excited states are related to the initial ones with simple selection rules. Since the perturbing potential is usually expanded into the multipole series (Eq.3.1)

$$\mathcal{H}_{\mathrm{CF}}(\mathbf{r}) = \sum_k \sum_q \beta_{kq}(r) C_q^{(k)} \left(\frac{\mathbf{r}}{r}\right) \tag{9.2}$$

the contribution of any individual excitation mode, $\mathcal{H}_{\mathrm{CF}}^{(kq)}$, may be examined separately.

The angular momentum quantum numbers of the initial and disturbed states, l and l', respectively, and the multipolarity index k of the crystal field component have to obey the triangle rule whereas the magnetic quantum numbers m and m' and the index q the relation $m' - m = q$. As it will be proved in the next section, the dominating part of the screening effect, i.e. the so-called linear screening, does not depend on q of the excitation mode and one may restrict oneself only to the simplest case of $q = 0$ or $m' = m$.

From the symmetry requirements (the selection rules) two types of the excitation can occur, the so-named, after Sternheimer, radial excitations, i.e. those preserving the l quantum number and the angular ones connected with a change of l (Table 9.1). Some of potentially important excitations, as for example $s \rightarrow p$ one, do not occur due to the triangle rule.

Permissible radial excitations for the second order crystal field terms are those of $np \rightarrow n'p$, $nd \rightarrow n'd$ and $nf \rightarrow n'f$ types whereas for the fourth order term — $nd \rightarrow n'd$ and $nf \rightarrow n'f$ and for the sixth order $nf \rightarrow n'f$ only.

Obviously, the contribution to the screening coming from excitations to states of larger and larger values of l' become lesser and lesser since they are penalized by the high energies of the excited orbitals.

Table 9.1 Allowed interacting configurations $(l \to l')$ for 2^k-pole components of the crystal field potential [146]

k=2	k=4	k=6
$s \to d$	$s \to g$	$s \to i$
$p \to p$	$p \to f$	$p \to h$
$p \to f$	$p \to h$	$p \to j$
$d \to s$	$d \to d$	$d \to g$
$d \to d$	$d \to g$	$d \to i$
$d \to g$	$d \to i$	$d \to k$
$f \to p$	$f \to p$	$f \to f$
$f \to f$	$f \to f$	$f \to h$
$f \to h$	$f \to h$	$f \to j$
$\cdots$	$f \to j$	$f \to l$
	$\cdots$	$\cdots$

We might suspect that in the case of crystal field potentials including terms of odd k (of no inversion centre) ineffective within a specified shell (for constant l) they could be effective in angular interconfigurational excitations. Undoubtedly, they can produce large cross–densities, e.g. for $s \to p$ excitation but as can be directly shown based on the expansion (Eq.9.15) of the two–electron interaction these densities do not interact with the open-shell electrons neither in the direct nor in the exchange way.

However, finding the complete wave functions of the disturbed states remains the key step of the screening problem since it is just the electron density resulting from overlap of the initial and disturbed wave functions, i.e. the cross density, which is responsible for the screening effect [145, 147].

Determination of the angular distribution of a disturbed state is a simpler task and it resolves itself into calculation of the appropriate Gaunt or Condon–Shortley coefficient (Eq.9.9).

The cross electron density which determines the screening effect interacts with the open–shell electrons in both the direct and exchange ways. The direct interaction, as it will be evidenced, for some weak assumptions turns out to be diagonal with respect to the excitation mode which means that the response of the system is of the same multipolarity as the perturbation. In the case of the exchange interaction the reaction of several different multipolarities can correspond to each 2^k-pole perturbation mode (Eq.9.25).

The next step is determination of the radial distribution of the disturbed functions. This more difficult task can be accomplished, in principle, by means of two methods split to several variants — either through direct numerical integrating the appropriate inhomogeneous Schrödinger equation, i.e. the so–called Sternheimer equation [145, 147], or with variational-perturbational method in a manner similar to the method of configuration interaction (CI) used in approach to the correlation problem in atoms and molecules [146, 149].

In the latter method we search for the disturbed function in the form of superposition of the initial and excited states treated as admixtures. The combining coefficients on the admixtured states are determined as the solution of a secular equation formed by applying the variational principle to the total energy. This method is conceptually simpler than the former one and consistent with the formalism of the generalized product functions. Individual excitation modes can be analyzed separately and in addition, in the case of radial excitations, one can exploit the convenient method based on the so–called unrestricted Hartree–Fock functions (UHF). In this method the electrons in a given nl shell having different m_l values are allowed to have different radial wave functions. However, from the numerical point of view the Sternheimer approach is more convenient and most of the results have been obtained by this method.

In the next section when deriving general expression for the screening factors the dominating contribution of the linear screening, i.e. that its part which can be cast into the form of a shielding factor times the crystal field strength will be demonstrated. The linear shielding leads to the level patterns being homothetic with those for the unscreened potential. Simultaneously, the independence of the screening factors on the q index will be proved. Besides, the effect of population of the open–shell on the screening will be shortly discussed.

9.3 General expressions for the screening factors

In this section we shall follow the perturbational procedure of calculating the screening factors given by Sternheimer, Blume and Peierls [147]. This very instructive method reveals the microscopic mechanism of the screening effect. Assuming absolute independence of individual electrons of the open–shell (A) in relation to the crystal field potential, i.e. its one–electron character the initial wave functions have been taken in the form of the reference functions (chapter 4). These one–determinantal N–electron functions, which are a special type of generalized product functions, consist of an individual open shell orbital a specified by the quantum numbers l_a and m_a (the spin quantum number is omitted here which does not lead to a confusion) and the closed shell orbitals of the central ion, mainly ξ orbitals. The functions are solutions of the central field Schrödinger equation

$$\mathcal{H}_0 \Psi_{m_a}^0 = \left[\sum_{i=1}^{N} \frac{p_i^2}{2m} + \sum_{i=1}^{N} \frac{Z}{r_i} + \sum_{i=1}^{N} V_0(r_i) \right] \Psi_{m_a}^0 = E^0 \Psi_{m_a}^0 \qquad (9.3)$$

where $\mathcal{H}_0$ is the zero–approximation Hartree–Fock Hamiltonian of the N–electron system, and $Z/r_i + V_0(r_i)$ is the average one–electron central field potential in which the electrons are assumed to move. As the perturbation both the crystal field potential (Eq.9.5) as well as the difference between the actual Hamiltonian (with $1/r_{ij}$ two–electron interaction) and its Hartree–Fock approximation $\mathcal{H}_0$ (Eq.9.3) (with the averaged central potential) has been taken. Let us consider, for clarity, the perturbation produced by one of the crystal field components, say the (k, q) component. Then, the perturbation has the form

$$\mathcal{H}' = \mathcal{H}_{\text{CF}}^{(kq)} + \sum_{i>j}^{N} \frac{1}{r_{ij}} - \sum_{i=1}^{N} \frac{Z}{r_i} - \sum_{i=1}^{N} V_0(r_i) \tag{9.4}$$

where Z denotes the nuclear charge of the central ion and all quantities are expressed in the atomic units. Since the third and fourth terms on the right side of Eq.9.4 possess the spherical symmetry only the first and second terms, i.e. $\mathcal{H}_{\text{CF}}^{(kq)} + \sum_{i>j}^{N} \frac{1}{r_{ij}}$, are essential from the splitting point of view in calculating the matrix elements of the perturbation $\mathcal{H}'$. The perturbation calculation has been performed with accuracy to the first order with respect to $\mathcal{H}_{\text{CF}}$.

Besides, let us assume the PCM crystal field potential described with the usual spherical harmonic operators, i.e. in the form

$$\mathcal{H}_{\text{CF}}^{(kq)} = \frac{4\pi}{2k+1} \sum_{t} \sum_{i=1}^{N} Y_k^{q^*}\left(\frac{\mathbf{R}_t}{R_t}\right) \frac{Z_t}{R_t^{k+1}} r_i^k Y_k^q\left(\frac{\mathbf{r}_i}{r_i}\right)$$

$$= \sum_{i=1}^{N} A_k^q r_i^k Y_k^q\left(\frac{\mathbf{r}_i}{r_i}\right) \tag{9.5}$$

where the summation over t refers to the ligands, Z_t stands for t-ligand charge, R_t for t-ligand–central ion distance, r_i is the radius vector from the central ion nucleus, and A_k^q is the geometry factor determined by the distribution of ions in the lattice.

This is not a necessary assumption but it has been introduced here from instructive reasons. The final results obtained under the assumption possesses however a more general meaning. In the first step let us consider the effect of $\mathcal{H}_{\text{CF}}^{kq}$ on the wave function. We have

$$(\mathcal{H}_0 + \mathcal{H}_{\text{CF}}^{(kq)})\Psi_{m_a} = E_{m_a}\Psi_{m_a} \tag{9.6}$$

or, upon substituting

$$\Psi_{m_a} = \Psi_{m_a}^0 + \Psi_{m_a}^1$$
$$E_{m_a} = E^0 + E_{m_a}^1$$

we get

$$(\mathcal{H}_0 - E^0)\Psi_{m_a}^1 = (E_{m_a}^1 - \mathcal{H}_{\text{CF}}^{(kq)})\Psi_{m_a}^0 \tag{9.6'}$$

The first order perturbed wave functions $\Psi_{m_a}^1$ may be written in the form

$$\Psi_{m_a}^1 = \sum_{nlm} \sum_{l'm'} \Psi_{m_a}^0 (nlm \to l'm') \tag{9.7}$$

where $\Psi_{m_a}^0 (nlm \to l'm')$ is a determinant function identical to $\Psi_{m_a}^0$ except that the single–electron wave function $\xi(nlm)$ is replaced by the perturbed function characterized by the $l'm'$ quantum numbers. In summations of Eq.9.7 all the initial states (nlm) and all allowed excited states $(l'm')$ are to be included. The nml quantum numbers refer primarily to the closed outer shells Ξ of the central ion.

For each individual excitation mode (k, q) a chosen one–electron wave function $\xi_0(nlm)$ is replaced in the determinant by the perturbed function

$$\xi_1(nlm \to l'm')^{(kq)} = D \cdot A_k^q \frac{1}{r} u_1(nl \to l')^{(k)} Y_{l'}^{m'} \tag{9.8}$$

where A_k^q has been defined previously, the constant D, i.e. the so–called Gaunt or Condon–Shortley coefficient is given by

$$
\begin{aligned}
D &= \int_0^{\pi} \sin v dv \int_0^{2\pi} d\varphi Y_l^{m^*}(v, \varphi) Y_k^q(v, \varphi) Y_{l'}^{m'}(v, \varphi) = \\
&= (-1)^{m'} \left[\frac{(2l+1)(2l'+1)(2k+1)}{4\pi} \right]^{1/2} \times \\
&\times \begin{pmatrix} l' & k & l \\ 0 & 0 & 0 \end{pmatrix} \begin{pmatrix} l' & k & l \\ -m' & q & m \end{pmatrix}
\end{aligned} \tag{9.9}
$$

and $u_1(nl \to l')^{(k)}$ is the radial distribution of the perturbed function $\xi_1(nlm \to l'm')^{(kq)}$ (Eq.9.8). This function, as results from the first order perturbation theory (Eq.9.6) has to satisfy the Sternheimer equation [145, 147]

$$
\begin{aligned}
\left[-\frac{d^2}{dr^2} + \frac{l'(l'+1)}{r^2} + V_0(r) - E_0 \right] u_1(nl \to l')^{(k)} = \\
= u_0(nl)[r^k - \langle r^k \rangle_{nl} \delta_{ll'}]
\end{aligned} \tag{9.10}
$$

where $u_0(nl)$ stands for the radial distribution of the unperturbed $\xi_0(nlm)$ function.

The centrifugal potential $l'(l'+1)/r^2$ on the left side of Eq.9.10 is worth noting. For large l' and small r it can dominate in the Hamiltonian and then $u_1 \to 0$. This equation is usually solved by direct numerical integration.

The effective potential $V_0(r)$ can be obtained directly from the unperturbed radial wave function $u_0(nl)$ by a procedure described by Sternheimer [157]

$$V_0(r) - E_0 = \frac{1}{u_0}\frac{d^2 u_0(r)}{dr^2} - \frac{l(l+1)}{r^2} \tag{9.11}$$

Eq.9.11 is the fundamental equation of the Sternheimer theory.

Knowing the disturbed radial distribution $u_1^{(k)}$ the matrix elements of the total perturbation $\mathcal{H}'$ can be calculated. With accuracy to the first order in $\mathcal{H}_{CF}^{(kq)}$ we have

$$\langle \Psi_{m_a'} | \mathcal{H}_{CF}^{(kq)} + \sum_{i>j}^{N} \frac{1}{r_{ij}} | \Psi_{m_a} \rangle \approx \langle \Psi_{m_a'}^0 | \mathcal{H}_{CF}^{(kq)} | \Psi_{m_a}^0 \rangle$$

$$+ \quad \langle \Psi_{m_a'}^1 | \sum_{i>j}^{N} \frac{1}{r_{ij}} | \Psi_{m_a}^0 \rangle + \langle \Psi_{m_a'}^0 | \sum_{i>j}^{N} \frac{1}{r_{ij}} | \Psi_{m_a}^1 \rangle \tag{9.12}$$

where $\Psi_{m_a'}^1$ and $\Psi_{m_a}^1$ differ in one spinorbital of the open–shell, i.e. in one m_a quantum number.

After simple modifications we will see that Eq.9.12 may be transformed to the form $\langle a_{m_a'} | \tilde{\mathcal{H}}_{CF}^{(kq)} | a_{m_a} \rangle$ where $\tilde{\mathcal{H}}_{CF}^{(kq)}$ corresponds to the effective screened crystal field potential acting within the open–shell spinorbitals only. The first term on the right side of Eq.9.12 is the well known matrix element of the unscreened crystal field

$$\langle \Psi_{m_a'}^0 | \mathcal{H}_{CF}^{(kq)} | \Psi_{m_a}^0 \rangle = \int a_{m_a'}^*(\mathbf{r}) A_k^q \, r^k \, Y_k^q(\mathbf{r}/r) \, a_{m_a}(\mathbf{r}) \, d\mathbf{r} =$$

$$= A_k^q \langle r^k \rangle (-1)^{m_a'} (2l_a + 1) \left(\frac{2k+1}{4\pi} \right)^{1/2} \times$$

$$\times \begin{pmatrix} l_a & k & l_a \\ 0 & 0 & 0 \end{pmatrix} \begin{pmatrix} l_a & k & l_a \\ -m_a' & q & m_a \end{pmatrix} \tag{9.13}$$

The next two terms describe the screening effect. Since their contributions are identical it is enough to double the results obtained for one of them. Thus, we have

$$\langle \Psi_{m_a'}^1 | \sum_{i>j}^{N} \frac{1}{r_{ij}} | \Psi_{m_a}^0 \rangle =$$

$$= 2 \sum_{nml} \sum_{l'm'} \int\int \xi_1^*(\mathbf{r}_1) a_{m_a'}^*(\mathbf{r}_2) \frac{1}{r_{12}} \xi_0(\mathbf{r}_1) a_{m_a}(\mathbf{r}_2) d\mathbf{r}_1 d\mathbf{r}_2 -$$

$$- \sum_{nml} \sum_{l'm'} \int\int \xi_1^*(\mathbf{r}_1) a_{m_a'}^*(\mathbf{r}_2) \frac{1}{r_{12}} a_{m_a}(\mathbf{r}_1) \xi_0(\mathbf{r}_2) d\mathbf{r}_1 d\mathbf{r}_2 \tag{9.14}$$

The first term in Eq.9.14 corresponds to the direct Coulomb interaction of the cross electron density $\xi_1^* \xi_0$ with the open–shell electron density $a_{m'_a}^* a_{m_a}$ and the second term to the exchange interaction. For clarity, in Eq.9.14 and the further expressions the indices k, q at the perturbed function are omitted.

Using the expansion

$$\frac{1}{r_{12}} = 4\pi \sum_{k'} \sum_{q'} \left(\frac{1}{2k'+1} \right) \frac{r_<^{k'}}{r_>^{k'+1}} Y_{k'}^{q'*} \left(\frac{\mathbf{r}_1}{r_1} \right) Y_{k'}^{q'} \left(\frac{\mathbf{r}_2}{r_2} \right) \qquad (9.15)$$

where $r_<$ and $r_>$ denote the lesser and larger distance from both r_1 and r_2, respectively, we get the following expression for the direct term

$$8\pi \sum_{nlm} \sum_{l'm'} \sum_{k'} \sum_{q'} \frac{1}{2k'+1} D A_k^q \times$$

$$\times \left[\int \int u_1(r_1) w_0^2(r_2) u_0(r_1) \frac{r_<^{k'}}{r_>^{k'+1}} dr_1 dr_2 \right] \times$$

$$\times \left[\int Y_{l'}^{m'*} Y_{k'}^{q'*} Y_l^m d\Omega_1 \right] \cdot \left[\int Y_{l_a}^{m'_a*} Y_{k'}^{q'} Y_{l_a}^{m_a} d\Omega_2 \right] \qquad (9.16)$$

where $w_0(r_2)$ is the radial distribution of the open–shell a function and Ω_i stands for the pair of angles v_i, φ_i. After substituting the value of D, $Eq.9.9$, and evaluating the integrals over spherical harmonics in terms of 3-j symbols, Eq.9.16 becomes

$$8\pi \sum_{nlm} \sum_{l'm'} \sum_{k'} \sum_{q'} (-1)^{m'_a + q'} \times$$

$$\times \frac{(2l+1)(2l'+1)(2l_a+1)(2k+1)^{1/2}}{(4\pi)^{3/2}} A_k^q \times$$

$$\times \left[\int \int u_1(r_1) w_0^2(r_2) u_0(r_1) \frac{r_<^{k'}}{r_>^{k'+1}} dr_1 dr_2 \right] \times$$

$$\times \begin{pmatrix} l' & k' & l \\ 0 & 0 & 0 \end{pmatrix} \begin{pmatrix} l_a & k' & l_a \\ 0 & 0 & 0 \end{pmatrix} \begin{pmatrix} l' & k & l \\ 0 & 0 & 0 \end{pmatrix} \times$$

$$\times \begin{pmatrix} l' & k' & l \\ -m' & -q' & m \end{pmatrix} \begin{pmatrix} l_a & k' & l_a \\ -m'_a & q' & m_a \end{pmatrix} \begin{pmatrix} l' & k & l \\ -m' & q & m \end{pmatrix} \qquad (9.17)$$

Now, we can exploit one of the orthogonality properties of 3-j symbols [158] which is the key relation in the presented derivation

$$\sum_{mm'} \begin{pmatrix} l' & k' & l \\ -m' & -q' & m \end{pmatrix} \begin{pmatrix} l' & k & l \\ -m' & q & m \end{pmatrix} = \frac{\delta_{kk'} \delta_{-qq'}}{2k+1} \qquad (9.18)$$

It means that if there are no limitations in summation over the m and m' indices the response of the perturbed system is diagonal with respect to the perturbation $(k = k')$.

So, transforming consequently the direct term (Eq.9.17) we finally find

$$A_k^q \langle r^k \rangle (-1)^{m_a} (2l_a + 1) \left(\frac{2k+1}{4\pi} \right)^{1/2} \times$$

$$\times \begin{pmatrix} l_a & k & l_a \\ 0 & 0 & 0 \end{pmatrix} \begin{pmatrix} l_a & k & l_a \\ -m_a' & -q & m_a \end{pmatrix} \times$$

$$\times \sum_{nl} \sum_{l'} 2 \begin{pmatrix} l' & k & l \\ 0 & 0 & 0 \end{pmatrix}^2 \frac{(2l+1)(2l'+1)}{(2k+1)} \frac{1}{\langle r^k \rangle} \times$$

$$\times \int \int u_1(r_1) w_0^2(r_2) u_0(r_1) \frac{r_<^k}{r_>^{k+1}} dr_1 dr_2 \qquad (9.19)$$

Since usual spherical harmonic operators (Y_k^q) has been used in the expansion of $\mathcal{H}_{CF}$ a doubt may arise about the meaning of $(-1)^{m_a}$ sign. This is an apparent problem only because for even q, $(-1)^{m_a} = (-1)^{m_a'}$, and for odd q $(-1)^{m_a} = -(-1)^{m_a'}$ but simultaneously the associated matrix element changes its sign due to reality and hermiticity of $\mathcal{H}_{CF}$. Anyway, the screening effect is always proportional to $\langle \Psi_{m_a'}^0 | \mathcal{H}_{CF}^{(kq)} | \Psi_{m_a}^0 \rangle$. Note that the above expression may be factorized in such a way that the first factor (the first row) is the unscreened crystal field term whereas the second corresponds to the screening factor σ_k. In other words, the screening effect is proportional to the matrix element of the unscreened term. Thus, the linearity of the screening has been demonstrated but under the assumption of unconstraint of the summation. This condition is not always fulfilled, as for example in the case of excitations to the open–shell states. The independence of σ_k on q is also proved (Eq.9.19).

The exchange term is evaluated similarly. It amounts to

$$-4\pi \sum_{nlm} \sum_{l'm'} \sum_{k'} \sum_{q'} \frac{1}{2k'+1} DA_k^q \times$$

$$\times \left[\int \int u_1(r_1) w_0(r_1) u_0(r_2) w_0(r_2) \frac{r_<^{k'}}{r_>^{k'+1}} dr_1 dr_2 \right] \times$$

$$\times \left[\int Y_{l'}^{m'*} Y_{k'}^{q'*} Y_{l_a}^{m_a} d\Omega_1 \right] \cdot \left[\int Y_{l_a}^{m_a*} Y_{k'}^{q'} Y_l^m d\Omega_2 \right] \qquad (9.20)$$

Transforming Eq.9.20 in the same way as the direct term (Eq.9.16 → Eq.9.17) we find

$$-4\pi \sum_{nlm} \sum_{l'm'} \sum_{k'} \sum_{q'} (-1)^{m_a' + q'} \times$$

$$\times \quad \frac{(2l+1)(2l'+1)(2l_a+1)(2k+1)^{1/2}}{(4\pi)^{3/2}} A_k^q K_{\mathrm{E}}(nl \to l'; k')^{(k)} \times$$

$$\times \quad \begin{pmatrix} l' & k & l \\ 0 & 0 & 0 \end{pmatrix} \begin{pmatrix} l' & k' & l_a \\ 0 & 0 & 0 \end{pmatrix} \begin{pmatrix} l_a & k' & l \\ 0 & 0 & 0 \end{pmatrix} \times$$

$$\times \quad \begin{pmatrix} l' & k' & l_a \\ -m' & -q' & m_a \end{pmatrix} \begin{pmatrix} l_a & k' & l \\ -m_a & q' & m \end{pmatrix} \begin{pmatrix} l' & k & l \\ -m' & q & m \end{pmatrix} \quad (9.21)$$

where

$$K_{\mathrm{E}}(nl \to l'; k')^{(k)} \equiv \int_0^\infty u_0(r) w_0(r) G(r) dr \quad (9.22)$$

and

$$G(r) \equiv \frac{1}{r^{k'+1}} \int_0^r u_1(r') w_0(r')(r')^{k'} dr' \times$$
$$+ \quad r^{k'} \int_r^\infty u_1(r') w_0(r')(r')^{-k'-1} dr' \quad (9.23)$$

Contrary to the direct contribution the response of the system can be off-diagonal, $k' \neq k$, where k' is the multipolarity of the exchange interaction.

We can perform in Eq.9.21 the summation over m, m' and q using the following mixed formula connecting 3-j and 6-j symbols [159]

$$\begin{pmatrix} j_1 & j_2 & j_3 \\ m_1 & m_2 & m_3 \end{pmatrix} \begin{Bmatrix} j_1 & j_2 & j_3 \\ l_1 & l_2 & l_3 \end{Bmatrix} =$$

$$= \sum_{\text{over all } n} (-1)^{l_1+l_2+l_3+n_1+n_2+n_3} \times$$

$$\times \begin{pmatrix} j_1 & l_2 & l_3 \\ m_1 & n_2 & -n_3 \end{pmatrix} \begin{pmatrix} l_1 & j_2 & l_3 \\ -n_1 & m_2 & n_3 \end{pmatrix} \begin{pmatrix} l_1 & l_2 & j_3 \\ n_1 & -n_2 & m_3 \end{pmatrix} (9.24)$$

where the following assignment takes place: $j_1 = l_a$, $j_2 = k$, $j_3 = l_a$, $l_1 = l'$, $l_2 = k'$ and $l_3 = l$.

Using this relation we find for the exchange term

$$A_k^q \langle r^k \rangle (-1)^{m_a} (2l_a+1) \left(\frac{2k+1}{4\pi} \right)^{1/2} \times$$

$$\times \begin{pmatrix} l_a & k & l_a \\ 0 & 0 & 0 \end{pmatrix} \begin{pmatrix} l_a & k & l_a \\ -m'_a & -q & m_a \end{pmatrix} \times$$

$$\times \sum_{nl} \sum_{l'} \sum_{k'} (-1)^{l+l'+k'+1} \frac{K_{\mathrm{E}}}{\langle r^k \rangle} (2l+1)(2l'+1) \times$$

$$\times \begin{pmatrix} l' & k & l \\ 0 & 0 & 0 \end{pmatrix} \begin{pmatrix} l' & k' & l_a \\ 0 & 0 & 0 \end{pmatrix} \begin{pmatrix} l_a & k' & l \\ 0 & 0 & 0 \end{pmatrix} \times$$

$$\times \begin{Bmatrix} l_a & k & l_a \\ l' & k' & l \end{Bmatrix} \left[\begin{pmatrix} l_a & k & l_a \\ 0 & 0 & 0 \end{pmatrix} \right]^{-1} \quad (9.25)$$

Upon combining these results of evaluating the direct (Eq.9.19) and exchange (Eq.9.25) interactions and using the symmetry properties of the 3-j and 6-j symbols we find that the effect of the perturbation of the wave function by $\mathcal{H}_{\mathrm{CF}}$ results in a screening of the crystal field, and that we may replace $\mathcal{H}_{\mathrm{CF}}$ by $\tilde{\mathcal{H}}_{\mathrm{CF}}$ according to

$$
\mathcal{H}_{\mathrm{CF}}^{(kq)} \to \tilde{\mathcal{H}}_{\mathrm{CF}}^{(kq)} = \sum_i^{N_a} A_k^q r_i^k Y_k^q(\mathbf{r}_i/r_i) \left\{ 1 + 4 \sum_{nl} \sum_{l'} \begin{pmatrix} l' & k & l \\ 0 & 0 & 0 \end{pmatrix}^2 \times \right.
$$

$$
\times\ \frac{(2l+1)(2l'+1)}{2k+1} \frac{1}{\langle r^k \rangle} \int \int u_1(r_1) w_0^2(r_2) u_0(r_1) \frac{r_<^k}{r_>^{k+1}} dr_1 dr_2
$$

$$
-\ 2 \sum_{nl} \sum_{l'} \sum_{k'} (-1)^{l+l'+k'} \frac{K_E}{\langle r^k \rangle} (2l+1)(2l'+1) \begin{pmatrix} l & k & l' \\ 0 & 0 & 0 \end{pmatrix} \times
$$

$$
\times\ \begin{pmatrix} l & k' & l_a \\ 0 & 0 & 0 \end{pmatrix} \begin{pmatrix} l' & k' & l_a \\ 0 & 0 & 0 \end{pmatrix} \times
$$

$$
\times\ \left. \begin{Bmatrix} k & l_a & l_a \\ k' & l & l' \end{Bmatrix} \left[\begin{pmatrix} l_a & k & l_a \\ 0 & 0 & 0 \end{pmatrix} \right]^{-1} \right\}
\tag{9.26}
$$

This time the summation over i is limited to N_a open–shell electrons only. The two terms in the brace after 1 give in sum the screening factor σ_k. As demonstrated, in the first order perturbation approach the screening effect is linear indeed and independent on q which justifies the phenomenological formula (Eq.9.1). An extended approach in relation to that presented above which includes corrections up to the third order in perturbation calculation with setting off the role of electronic correlation effects has been presented by way of example for Pr^{3+} ion by Ahmad [148]. These correlation effects have been found to be important, they reach about 17% of the lowest–order contribution but of opposite sign.

9.4   The screening factors

Speculatively, based on Eq.9.26 we are in a position to foresee the sign of the screening factor σ_k, i.e. to forejudge whether we deal with the screening or antiscreening effect. So, as seen, the σ_k sign depends primarily on the excitation mode which determines signs of the radial integrals in both the direct and exchange contributions as well as signs of the 3-j and 6-j symbols and in consequence the mutual proportion of the two contributions which, in turn, finally determines the resultive sign.

The sign of the direct contribution (Eq.9.19) depends solely on the sign of the radial integral. It is negative as a rule which corresponds to a true screening attenuating the crystal field effect. However, in the case of radial excitations, e.g. $5p \to p$ for the lanthanide ions [145, 147, 148] the corresponding radial integral is positive leading to an antishielding contribution. Nevertheless, the algebraic sum of all direct terms is usually negative.

As seen from Eq.9.25 the question of the sign of the exchange contribution is more complex mainly due to its multipolar nature. Generally, it is opposite to that for the direct term, i.e. positive and compensates partly or even overcome the direct contribution. Consequently, the exchange contributions for the radial excitations are negative and reduce somewhat the direct antishielding. It has been found that the exchange terms play a progressively more important role with increasing k.

Concluding, the typical radial excitations lead to antiscreening. In the case of lanthanides the most important radial excitation is $5p \rightarrow p$, the remaining such as $4p \rightarrow p$ and $4d \rightarrow d$ produce the contributions about ten times smaller [146, 148]. This antiscreening reaches about 15% of the unscreened second order crystal field effect.

Among the angular excitations the most important in lanthanides are $5p \rightarrow f^*$, where the star denotes one of the open–shell $(4f)$ orbitals, and $5s \rightarrow d$ one, and they are in fact responsible for the screening effect. The $5p \rightarrow f^*$ excitation is responsible for about 25% shielding of the $\mathcal{H}_{CF}^{(20)}$ potential for Ce^{3+} ion. There is also good experimental evidence that σ_2 decreases somewhat with increasing the population of the $4f$ shell from Ce^{3+} $(4f^1)$ to Yb^{3+} $(4f^{13})$ [146]. Investigating the procedure of deriving Eq.9.26 the reasons of deviation from the linear relationship (Eq.9.1) can be pointed out. These may be differences in radial distributions of particular orbitals belonging to the same shell (UHF functions) and differences in energy associated with them as well as the necessity of compliance with the exclusion principle when the summation over m' is not fully free (Eq.9.18).

It should be emphasized that the non–linear screening effect for one of the crystal field modes can be separated into linear contributions of other modes. It leads to the potential of deformed modal structure.

The experimental egzamplification of the model is limited to the lanthanide ions for which the screening effect is clear cut and best evidenced. The magnitudes of the total screening factors decrease with rise in the degree of crystal field mode. Their typical values are: σ_2 from 0.6 to 1.0, σ_4 from 0.05 to 0.10 and $\sigma_6 \approx -0.05$ [147, 148, 160]. In the case of σ_2 the direct interaction exceeds several times the exchange interaction. The total values of σ_4 and σ_6 are the result of a considerable amount of cancellation between the direct and exchange terms. For σ_6 the exchange terms actually predominate and give rise to a small net antiscreening. A general decrease of σ_k factors associated with increasing the atomic number of the central ion nucleus is caused by decreasing the polarizability of the stronger bound electronic states. The above model which refers to a system with a simple open–shell electron beyond the closed shells can be simply generalized to any many–electron system by using the tensor method and $3n$-j symbols [44].

CHAPTER 10

Electrostatic crystal field contributions with consistent multipolar effects. Polarization

The problem of including all electrostatic crystal field contributions involves the calculation of the electrostatic field due to point charges and induced moments of any order on neighbouring ions at the particular lattice site under consideration. To this end having a generalized expression for the interaction of two groups of the moments (including point charges) with different origins is necessary.

The crystal lattice may be always treated as a spatially arranged distribution of electron density and nuclear charges. From point of view of electrostatic balance in the crystal it corresponds, in the model description, with co-existing a series of permanent and induced multipole moments on ions, atoms or atomic cores forming the lattice according to the type of the crystal.

Typically, considering the diminishing range of multipole interactions with their degree the multipole expansion of lattice potential can be, with a good approximation, truncated after the first three terms corresponding to point charges, dipole and quadrupole moments, respectively [161, 162, 163, 105]. Nevertheless, it should be noticed that the problem of convergency of the multipole expansions is not always so evident and, in principle, requires every single inspection [105]. The polarization effects are more important for crystals where one or more of the consistent ions occupies a site of low symmetry since they may possess a dipole moment along with higher multipole moments as well as a single point charge.

10.1 Expansion of the electrostatic potential of point charge system into the multipole series

The electrostatic potential produced by a system of point charges q_t being distant by r_t from their centre B at an outer point A (Fig.10.1) can be in

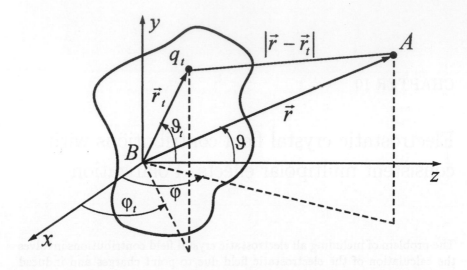

Fig. 10.1 Electrostatic potential at point A generated by a system of point charges centered about point B; vectors and angles in the multipole expansion

the most convenient way presented in the form of its expansion into the spherical function series

$$V(\mathbf{r}) = \sum_t \frac{q_t}{|\mathbf{r} - \mathbf{r}_t|} = \sum_t \frac{q_t}{(r^2 + r_t^2 - 2rr_t \cos \kappa_t)^{1/2}} =$$

$$= \sum_t \sum_{l=0}^{\infty} \frac{q_t r_t^l}{r^{l+1}} P_l(\cos \kappa_t) \qquad (10.1)$$

where $P_l(\cos \kappa_t)$ is the spherical function, i.e. the Legendre polynomial of l degree and κ_t is the angle between the $\mathbf{r}$ and $\mathbf{r}_t$ vectors.

The above expansion form is valid for $r_t < r$. Introducing the spherical angles υ, φ and υ_t, φ_t spanned between the vectors $\mathbf{r}$ and $\mathbf{r}_t$ and the coordinate axes z, x, respectively (Fig.10.1) and using the spherical harmonic addition theorem [164] one gets

$$V(\mathbf{r}) = \frac{4\pi}{2l+1} \sum_t \sum_{l=0}^{\infty} \sum_{m=-l}^{l} \frac{q_t r_t^l}{r^{l+1}} Y_l^{m^*}(\upsilon, \varphi) Y_l^m(\upsilon_t, \varphi_t) \qquad (10.2)$$

Thus, the l-th term of the expansion has the form

$$V^{(l)}(\mathbf{r}) = \left(\frac{4\pi}{2l+1}\right)^{1/2} \frac{1}{r^{l+1}} \sum_{m=-l}^{l} Q_m^{(l)} Y_l^{m^*}(\upsilon, \varphi) \qquad (10.3)$$

and

$$Q_m^{(l)} = \left(\frac{4\pi}{2l+1}\right)^{1/2} \sum_t q_t \, r_t^l \, Y_l^m(v_t, \varphi_t) = \sum_t q_t \, r_t^l \, C_m^{(l)}(v_t, \varphi_t) \quad (10.4)$$

The set of $2l + 1$ quantities $Q_m^{(l)}$ form the 2^l–pole moment of the charge system.

10.2 Extended formula for the crystal field parameters including all multipole moments of the surroundings

The spatial electron charge distribution in crystal lattice, i.e. $\langle \Psi^*(\mathbf{r})\Psi(\mathbf{r})\rangle$ where $\Psi(\mathbf{r})$ is the global electronic eigenfunction, responding to the presence of electric fields and their derivatives, undergoes polarization which leads to appearance of induced multipole moments.

As in the case of the screening effect, the polarization changes in the electronic states result from intraatomic excitations. It means that the corrected electronic states including the polarization effects are composed of appropriate excited states of the ion as the admixtures. Both the screening and polarization effects are represented by means of certain coefficients which are the characteristic quantities for each ion. Previously these were the screening factors. Now these are various multipole polarizabilities. The corresponding induced multipole moments are proportional to the polarizabilities and can be determined from the requirement of the electrostatic equilibrium in the crystal. It turns out that the effective total electrostatic Hamiltonian, $\mathcal{H}_{CF}^E$ acting on electrons of the open–shell A can be described, as the point charge (monopoles) potential, with the standard parametric formula

$$\mathcal{H}_{CF}^E = \sum_i \sum_k \sum_q B_{kq}^E \, C_q^{(k)}(\mathbf{r}_i/r_i) \quad (10.5)$$

with modified crystal field parameters, B_{kq}^E, including, apart from the point charge, also the dipole, quadrupole etc. contributions. The summation over i, as usual for the effective Hamiltonian, runs over a electrons, and the origin of the coordinate system is obviously located at the central ion nucleous.

Unfortunately, in spite of some hopes [162, 163, 105] this complete electrostatic Hamiltonian does not describe the crystal field effect correctly. What is even more surprising, the generalized electrostatic model itself can give even worse results than the simple model. Only the complex attempt taking into account several other mechanisms and in the first place the covalency, overlap and screening effects leads to approximate interpretations of experimental data.

Now, the general expression for the B_{kq}^E parameters modified in consequence of the polarization will be derived. Following the procedure used in chapter 5 we are starting with the generalized potential produced by the charge density $\varrho(\mathbf{r})$ throughout the lattice at $\mathbf{r}_i$ point (Fig.10.2).

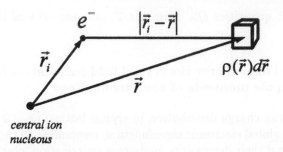

central ion
nucleus

Fig. 10.2 Vector scheme according to Eq.10.6

$$\mathcal{H}_{\text{CF}}^{\text{E}}(\mathbf{r}_i) = -e \int \frac{\varrho(\mathbf{r})d\mathbf{r}}{|\mathbf{r}_i - \mathbf{r}|} \tag{10.6}$$

The right hand side of Eq.10.6 is expanded, as previously, into the series of products of complex conjugate spherical harmonics under the same assumption $r > r_i$. The problem of charge penetration, i.e. the case of $r < r_i$ has been discussed in chapter 6.

$$\mathcal{H}_{\text{CF}}^{\text{E}}(\mathbf{r}_i) = -e \int \sum_k \sum_q \frac{4\pi}{2k+1} \, \varrho(\mathbf{r}) \, \frac{r_i^k}{r^{k+1}} \, Y_k^{q^*}(\mathbf{r}/r) \, Y_k^q(\mathbf{r}_i/r_i) d\mathbf{r} =$$

$$= -e \sum_k \sum_q \left(\frac{4\pi}{2k+1}\right)^{1/2} r_i^k \, C_q^{(k)}(\mathbf{r}_i/r_i) \times$$

$$\times \int \frac{Y_k^{q^*}(\mathbf{r}/r) \, \varrho(\mathbf{r})}{r^{k+1}} d\mathbf{r} \tag{10.7}$$

where

$$C_q^{(k)}(\mathbf{r}_i/r_i) = \left(\frac{4\pi}{2k+1}\right)^{1/2} Y_k^q(\mathbf{r}_i/r_i) \tag{10.8}$$

The electron density is not evenly distributed in the lattice but rather gathered in form of clusters t, R_t being the center of the charge of t–th cluster corresponding to position of the t lattice site. For each such charge region introducing the local radius vector $\mathbf{r}_t$ obeying the relation $\mathbf{r} = \mathbf{R}_t + \mathbf{r}_t$ and the local density $\varrho_t(\mathbf{r}_t) \equiv \varrho(\mathbf{R}_t + \mathbf{r}_t)$, Fig.10.3, allows the direct analytical approach to the be applied. Then, for $r_t < R_t$ the integrand in Eq.10.7 may be expanded into the sum over all the charge regions and within each of the regions into the power series with respect to $\mathbf{r}_t$. Consequently, $\mathcal{H}_{\text{CF}}^{\text{E}}(\mathbf{r}_i)$ takes the form

$$\mathcal{H}_{\text{CF}}^{\text{E}}(\mathbf{r}_i) = -e \sum_k \sum_q \left(\frac{4\pi}{2k+1}\right)^{1/2} r_i^k \, C_q^{(k)}(\mathbf{r}_i/r_i) \times$$

$$\times \sum_t \sum_n \int \frac{1}{n!} \varrho(\mathbf{r}_t)\mathbf{r}_t^{(n)} \cdot \nabla_t^{(n)} \left[\frac{Y_k^{q^\bullet}(\mathbf{R}_t/R_t)}{R_t^{k+1}}\right] d\mathbf{r}_t \quad (10.9)$$

where $\nabla_t^{(n)}$ is the differential tensor operator of rank n acting on the coordinates of the t–ion radius vector $\mathbf{R}_t(R_t, v_t, \varphi_t)$ in the central coordinate system, and $\mathbf{r}_t^{(n)}$ is the n–rank tensor constructed from the components of the local radius vector $\mathbf{r}_t$.

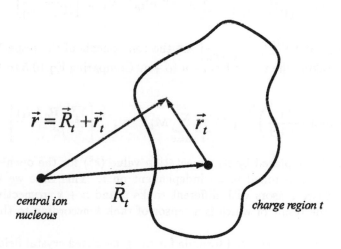

Fig. 10.3 Charge region t and the central ion, the radius vector $\mathbf{r}$ and $\mathbf{r}_t$

Defining the monopole, dipole, quadrupole etc. moments of the t–ion as

$$\mathbf{M}_t^{(0)} = \int \varrho(\mathbf{r}_t)d\mathbf{r}_t$$

$$\mathbf{M}_t^{(1)} = \int \varrho(\mathbf{r}_t)\mathbf{r}_t^{(1)}d\mathbf{r}_t$$

$$\mathbf{M}_t^{(2)} = \int \varrho(\mathbf{r}_t)\mathbf{r}_t^{(2)}d\mathbf{r}_t \quad (10.10)$$

one can rewrite Eq.10.9 in the form

$$\mathcal{H}_{CF}^{E}(\mathbf{r}_i) = -e \sum_k \sum_q \left(\frac{4\pi}{2k+1}\right)^{1/2} r_i^k \, C_q^{(k)}(\mathbf{r}_i/r_i) \times$$

$$\times \sum_t \sum_n \mathbf{M}_t^{(n)} \cdot \nabla_t^{(n)} \left[\frac{Y_k^{q^*}(\mathbf{R}_t/R_t)}{R_t^{k+1}}\right] \qquad (10.11)$$

or in the more compact form

$$\mathcal{H}_{CF}^{E}(\mathbf{r}_i) = -e \sum_t \sum_n \mathbf{M}_t^{(n)} \cdot \nabla_t^{(n)} \left[\frac{1}{|\mathbf{R}_t - \mathbf{r}_i|}\right] \qquad (10.12)$$

The right hand side of Eq.10.12 is the scalar product of two tensor operators of the same rank (n) defined conventionally by [164]

$$\mathbf{M}_t^{(n)} \cdot \nabla_t^{(n)} \left[\frac{1}{|\mathbf{R}_t - \mathbf{r}_i|}\right] = \sum_\mu (-1)^\mu M_{t\mu}^{(n)} \cdot \nabla_{t-\mu}^{(n)} \left[\frac{1}{|\mathbf{R}_t - \mathbf{r}_i|}\right] \qquad (10.13)$$

where $M_{t\mu}^{(n)}$ and $\nabla_{t-\mu}^{(n)} \left[\frac{1}{|\mathbf{R}_t-\mathbf{r}_i|}\right]$ are the components of the respective tensors for μ taking the values from $-n$ to $+n$. Comparing Eq.10.5 to Eq.10.11 one gets

$$B_{kq}^{E} = -e \left(\frac{4\pi}{2k+1}\right)^{1/2} \langle r^k \rangle \sum_t \sum_{n=0}^{\infty} \mathbf{M}_t^{(n)} \cdot \nabla_t^{(n)} \left[\frac{Y_k^{q^*}(\mathbf{R}_t/R_t)}{R_t^{k+1}}\right] \qquad (10.14)$$

in which r_i^k is replaced by its expectation value $\langle r^k \rangle$ for the open–shell orbitals which is assumed to be independent on i. This time we have the product of two tensors with different ranks n and $n + k$, respectively (see Eq.10.16) the result of which is a tensor of rank k according to the expectation.

This is just the extended formula for the generalized crystal field parameters which includes all multipole moments of the surroundings.

The first term in the sum over n (for $n = 0$) is equal to the point charge contribution

$$B_{kq}^{PC} = -e \left(\frac{4\pi}{2k+1}\right)^{1/2} \langle r^k \rangle \sum_t M_t^{(0)} \frac{Y_k^{q^*}(\mathbf{R}_t/R_t)}{R_t^{k+1}} \qquad (10.15)$$

where the factor $\left(\frac{4\pi}{2k+1}\right)^{1/2} Y_k^{q^*}(\mathbf{R}_t/R_t)$ can be recognized as the coordination factor of ligand t (chapter 2). For the axial system, with the z–axis joining the central ion and the ligand, Eq.10.15 reduces to Eq.5.10 in chapter 5. The second term of Eq.10.14 (for $n = 1$) corresponds to the contribution of dipole moments, and the third (for $n = 2$) to that of quadrupole moments etc. It is worth noting that the $\langle r^k \rangle\, C_q^{(k)}(\mathbf{r}_i/r_i)$, Eq.10.11, are the open–shell multipolar (2^k–pole) moments in normal coordinates.

Based on the fundamental relations concerning the gradient notion [164] and using the tensor algebra methods the following expression can be found

$$
\nabla_{t\mu}^{(n)} \left[\frac{Y_k^{q^*}(\mathbf{R}_t/R_t)}{R_t^{k+1}} \right] = (-1)^{k+q+\mu} \left(\frac{2k+1}{4\pi} \right)^{1/2} \left[\frac{(2k+2n+1)!}{2^n (2k)!} \right]^{1/2} \times
$$

$$
\times \begin{pmatrix} k & n & k+n \\ q & \mu & -q-\mu \end{pmatrix} \frac{1}{R^{k+n+1}} \times
$$

$$
\times \; C_{q+\mu}^{(k+n)}(\mathbf{R}_t/R_t) \qquad\qquad (10.16)
$$

where μ runs over all $2l+1$ normal components of the $\nabla_t^{(n)}$ tensor, $-n \leq \mu \leq n$. The expression for the n–rank tensor operator $\nabla_t^{(n)}$, Eq.10.16, has been derived in the recurring from starting from the gradient formula ($\nabla^{(1)}$) [164]. The mathematical induction proof was given by Gajek [113]. The normal components of $\mathbf{M}^{(n)}$ or otherwise the spherical components entering into the composition of the scalar product in Eqs 10.11, 10.12 and 10.14 can be expressed as combinations of their Cartesian components in the way presented in Table 10.1.

Table 10.1 Normal components of $\mathbf{M}^{(0)}$, $\mathbf{M}^{(1)}$ and $\mathbf{M}^{(2)}$ tensors as combinations of the Cartesian components

$M^{(0)}$	$M^{(0)}$
$M_{-1}^{(1)}$	$\frac{1}{\sqrt{2}}(M_x^{(1)} - iM_y^{(1)})$
$M_0^{(1)}$	$M_z^{(1)}$
$M_1^{(1)}$	$-\frac{1}{\sqrt{2}}(M_x^{(1)} + iM_y^{(1)})$
$M_2^{(2)} = M_1^{(1)} M_1^{(1)}$	$\frac{1}{2}(M_{xx}^{(2)} - M_{yy}^{(2)} + 2iM_{xy}^{(2)})$
$M_1^{(2)} = \frac{1}{\sqrt{2}}(M_1^{(1)} M_0^1 + M_0^{(1)} M_1^{(1)})$	$-(M_{xz}^{(2)} + iM_{yz}^{(2)})$
$M_0^{(2)} = \frac{1}{\sqrt{6}}(2M_0^{(1)} M_0^{(1)} + M_1^{(1)} M_{-1}^{(1)} + M_{-1}^{(1)} M_1^{(1)})$	$\frac{1}{\sqrt{6}}(2M_{zz}^{(2)} - M_{xx}^{(2)} - M_{yy}^{(2)})$
$M_{-1}^{(2)} = \frac{1}{\sqrt{2}}(M_{-1}^{(1)} M_0^{(1)} + M_0^{(1)} M_{-1}^{(1)})$	$M_{xz}^{(2)} - iM_{yz}^{(2)}$
$M_{-2}^{(2)} = M_{-1}^{(1)} M_{-1}^{(1)}$	$\frac{1}{2}(M_{xx}^{(2)} - M_{yy}^{(2)} - 2iM_{xy}^{(2)})$

The spherical components of $\mathbf{M}^{(1)}$ possess the well known form and transformational properties of Y_1^1, Y_1^0 and Y_1^{-1}, whereas the components of $\mathbf{M}^{(2)}$ have been constructed by means of coupling the spherical vectors [164] according to

$$M_\mu^{(2)} = \sum_{\mu_1}\sum_{\mu_2} M_{\mu_1}^{(1)}M_{\mu_2}^{(1)}(1\mu_1 1\mu_2|112\mu) =$$

$$= \sum_{\mu_1}\sum_{\mu_2}(-1)^\mu M_{\mu_1}^{(1)}M_{\mu_2}^{(1)}\sqrt{5}\begin{pmatrix} 1 & 1 & 2 \\ \mu_1 & \mu_2 & -\mu \end{pmatrix} \quad (10.17)$$

where $(1\mu_1 1\mu_2|112\mu)$ is the vector coupling coefficient.

10.3 The self–consistent system of permanent and induced multipole moments in crystal lattice

The multipole moments $M_t^{(n)}$ induced on an ion at any lattice site t can be directly calculated from requirement of the electrostatic balance in the system of point charges and permanent and induced multipole moments on all sites of the lattice. In the following course the multipole series will be restricted to dipoles and quadrupoles only, which, in general, is a well justified truncation. From the very definition the induced dipole and quadrupole moments are proportional respectively to the electric field intensity E_t and gradient of the field $\nabla_t^{(1)}E_t$ according to the formula

$$M_t^{(n)} = \alpha_t^{(n)}\nabla_t^{(n-1)}E_t \quad (10.18)$$

where $\alpha_t^{(n)}$ is the tensor of respective polarizability which in the case of isotropic electron density about the t site reduces to the scalar.

The electric field intensity at R_t generated by a 2^p–pole moment at $R_{t'}$ amounts to (see Eq.10.12)

$$E_t = -\nabla_t^{(1)}\left[M_{t'}^{(p)}\cdot\nabla_{t'}^{(p)}\frac{1}{|R_{t'} - R_t|}\right] \quad (10.19)$$

Hence, the 2^p–pole moment at $R_{t'}$ induces at R_t a 2^n–pole moment given by

$$M_t^{(n)} = -\alpha_t^{(n)}\nabla_t^{(n)}\left[M_{t'}^{(p)}\cdot\nabla_{t'}^{(p)}\frac{1}{|R_{t'} - R_t|}\right]$$

$$= (-1)^{n+1}\alpha_t^{(n)}\nabla_{t'}^{(n)}\left[M_{t'}^{(p)}\cdot\nabla_{t'}^{(p)}\frac{1}{R_{t'}}\right] \quad (10.20)$$

where in the last expression the differentiation with respect to coordinates of R_t has been replaced by that with respect to coordinates of $R_{t'}$ causing the corresponding change of the sign. Next, $R_t = 0$ has been substituted, i.e. locating the origin of the coordinate system at the central ion position. The state of electrostatic equilibrium between the $M_t^{(n)}$ moment and the remaining multipole moments of the crystal lattice may be described in the form of equations

$$\mathbf{M}_t^{(n)} = \sum_{t'} \sum_{p=0,1,2} (-1)^{n+1} \, \alpha_t^{(n)} \, \mathbf{I}^{(2n)} \, \nabla_{t'}^{(n)} \left[\mathbf{M}_{t'}^{(p)} \cdot \nabla_{t'}^{(p)} \frac{1}{\mathbf{R}_{t'}} \right] \qquad (10.21)$$

where $\mathbf{I}^{(2n)}$ is the diagonal unit tensor of rank $2n$ introduced to reconcile the transformational properties of both sides of the equation. For the point charge contributions $(n = 0)$ $\alpha^{(0)} = 1$, $\mathbf{I}^{(0)} = 1$.

The system of 12 N linear equations in the unknown moments components (Eq.10.21) for three dipole and nine quadrupole components of the induced moments on N different ions in the unit cell describes the self–consistent state of the electrostatic balance in the whole lattice.

10.4 The off–axial polarization terms in local coordinate systems

Let us return to the expression for B_{kq}^E including the induced multipole moments (Eq.10.14). Substituting Eq.10.16 into Eq.10.14 one gets

$$\begin{aligned} B_{kq}^E &= -e\langle r^k \rangle \sum_t \sum_n \sum_\mu (-1)^{k+q+\mu} \times \\ &\times \left[\frac{(2k + 2n + 1)!}{2^n(2k)!} \right]^{1/2} \left(\begin{array}{ccc} k & n & k+n \\ q & \mu & -q-\mu \end{array} \right) \times \\ &\times \frac{1}{R^{k+n+1}} \, M_{t\mu}^{(n)} \, C_{q+\mu}^{(k+n)}(\mathbf{R}_t/R_t) \end{aligned} \qquad (10.22)$$

where μ runs over the components of the 2^n–pole moments $\mathbf{M}_t^{(n)}$, i.e. in our approximation for $n = 0, 1$ and 2, expressed in the central coordinate system being anchored at the central ion nucleous.

In a local coordinate system associated with the individual lattice site, with the z–axis lying along the central ion – the chosen ligand direction, two kinds of components of the multipole moments of the ion can occur – those lying along the z–axis or the axial components and the off–axial ones. Only the axial components fulfill the requirements of the superposition model, i.e. are characterized by the specified tranformational properties under rotations of the coordinate system. In order to demonstrate this differentiation the moments $\mathbf{M}_t^{(n)}$ ought to be expressed in their local coordinate systems. Between the components of the moment $\mathbf{M}_t^{(n)}$ in the central coordinate system $(\mathbf{M}_{t\mu}^{(n)})$ and its components in the local system $(\tilde{\mathbf{M}}_{t\mu}^{(n)})$ the following relation takes place

$$\mathbf{M}_{t\mu}^{(n)} = \sum_{\nu} D_{\nu\mu}^{(n)}(\varphi_t, \upsilon_t, 0) \, \tilde{\mathbf{M}}_{t\nu}^{(n)} \tag{10.23}$$

where $D_{\nu\mu}^{(n)}(\varphi_t, \upsilon_t, 0)$ is the matrix element of rotation transforming the local system into the central one, and υ_t, φ_t are the spherical coordinates of the point t in the central coordinate system. Substituting Eq.10.23 into Eq.10.22 and using, in succession, the relation [164]

$$C_m^{(l)}(\upsilon, \varphi) = D_{0m}^{(l)}(\varphi, \upsilon, 0) \tag{10.24}$$

and the expression for product of two $D_{m'm}^{(j)}(\alpha, \beta, \gamma)$ matrix elements [165] with employing the orthogonality properties of 3-j symbols [158]

$$\sum_{\mu} \begin{pmatrix} k & n & k+n \\ q & \mu & -q-\mu \end{pmatrix} D_{\nu\mu}^{(n)}(\varphi_t, \upsilon_t, 0) \, D_{0,-\mu-q}^{(k+n)}(\varphi_t, \upsilon_t, 0) =$$

$$= \begin{pmatrix} k & n & k+n \\ -\nu & \nu & 0 \end{pmatrix} D_{-\nu q}^{(k)^*}(\varphi_t, \upsilon_t, 0) \tag{10.25}$$

the new formula for B_{kq}^{E} can be written as

$$B_{kq}^{E} = -e\langle r^k \rangle \sum_{t} \sum_{n} \sum_{\nu} (-1)^k \times$$

$$\times \left[\frac{(2k+2n+1)!}{2^n(2k)!} \right]^{1/2} \begin{pmatrix} k & n & k+n \\ -\nu & \nu & 0 \end{pmatrix} \times$$

$$\times \frac{1}{R_t^{k+n+1}} \, \tilde{M}_{t\nu}^{(n)} \, D_{-\nu q}^{(k)^*}(\varphi_t, \upsilon_t, 0) \tag{10.26}$$

where $\tilde{M}_{t\nu}^{(n)}$ stands for the ν-th component of the 2^n-pole moment of t ion in the local coordinate system.

The off-axial terms are those with index $\nu \neq 0$. Since the polarization contributions rapidly decay with the distance the most important input comes from the nearst neighbours. Fortunately, in most cases the axial contribution predominates and the off–axial terms are of secondary meaning. However, they frequently are of the same order as the total contribution originating from the further neighbours, but often of the reverse sign. The sketched formalism is quite rigorous, the truncation of the multipole series after the quadrupole term is the only approximation introduced. The main reason of observed divergences bewtween the model description and the reality is our inaccurate knowledge on the polarizability coefficients.

10.5 Typical examples of dipole and quadrupole polarization contributions to the crystal field potential

The polarization contributions to the crystal field parameters are particularly large, regarding their absolute values, in the case of second order terms of the potential, when they frequently exceed many times the resultant values of the parameters. For the fourth and sixth order potential terms their magnitudes considerably decrease as it can be seen in Tables 10.2 and 10.3 in which the polarization inputs for representative lanthanide and actinide ions, respectively, are presented.

Table 10.2 Polarization contributions to the crystal field parameters of Pr^{3+} ion in $PrCl_3$ of C_{3h} point symmetry [162] compared with the experimental data [108], in cm^{-1}

	B_{20}	B_{40}	B_{60}	B_{63}
Dipole polarization	938	24	−4.8	5.3
Quadrupole polarization	−1410	−112	16.0	−5.3
Experiment	94	−325	−634	426

Table 10.3 Polarization contributions to the crystal field parameters of U^{4+} in Cs_2UCl_6 of D_{3d} point symmetry [80] compared with the experimental data [109], in cm^{-1}

	$B_{20}^{a)}$	B_{40}	B_{43}	B_{60}	B_{63}	B_{66}
D P of NL	−1330	−1190	1750	400	10	240
DP of FN	390	60	−10	0	0	0
Q P of NL	−1390	−1330	2010	640	−20	370
Q P of FN	−40	−10	−10	0	0	0
OAC	−670	88	−2	42	63	10
Experiment	—	−4810	5750	2430	1470	1540

a) non–typically small parameter due to the almost O_h symmetry
Dipole polarization of nearest ligands = DP of NL
Dipole polarization of further neighbours = DP of FN
Quadrupole polarization of nearest ligands = QP of NL
Quadrupole polarization of further neighbours = QP of FN
Off-axial contributions = OAC

A natural tendency to mutual cancellation of all electrostatic contributions including the screening covers up the individual role each of them emphasizing simultaneously significance of the complex approach.

CHAPTER 11

Crystal field effect in the Stevens perturbation approach

So far, we have discussed models based on effective Hamiltonians where a great deal of attention has been paid to the local symmetry by treating open–shell electrons as localized about a particular site. The remaining electrons are also individualized and distinguished by their localization at other particular sites. Some of them, e.g. those on diamagnetic ions, are also distinguished by being ignored. However , a substantial part of the crystal field potential at a chosen lattice site is due to electrons at other sites and the electrons are in no way distinguished in a general Hamiltonian.

The correct description of crystal field effect in a periodic lattice with taking into account the band states exacts rejecting the conventional one–ion crystal field model which has no regard for translational invariance of electronic states and indistinguishability of electrons. And although the efficiency of the traditional crystal field model (generalized in various ways) is in many cases quite satisfactory (semi–phenomenologically rather) the model is based, in fact, on incoherent and false assumptions and therefore one can hardly expect quantitatively correct results particularly for typical band and metallic systems.

This criticism of the CF theory was raised in the early fifties by Slater in 1953 [166]. The difficulties have been overcome by Stevens who proposed a new method of looking at crystal field which is compatible with the lattice periodicity. This was done in an article entitled *"Exchange interactions in magnetic insulators"* [76]. In this approach the crystal field potential emerges in a natural way from general Hamiltonian defined for the whole system.

This is a perturbation scheme for degenerate systems employing the projection operators and the second quantization method, i.e. the occupation number representation. A novelty of this approach relies on specific construction of the unperturbed Hamiltonian $\mathcal{H}_0$. Inversely to a normal order, i.e. from operator to its eigenstates, $\mathcal{H}_0$ is built up based on its approximate eigenstates anticipated and their projection operators. In general, there is a sufficient number of data allowing the eigenstates, especially those low–lying, to be defined enough precisely without solving any Schrödinger equation. A special attention is paid, this time, to preserve the whole, and not local only, symmetry of the unperturbed Hamiltonian particularly that resulting from indistinguishability of electrons and translational invariance.

To assure hermiticity of $\mathcal{H}_0$ and to use the second quantization method conveniently the orthogonal states are needed. This is the reason the Wannier functions are introduced for description of localized states (or generalized Wannier functions in the case of doped ions) which in turn are orthogonal to the Bloch functions of conduction band electrons.

11.1 The Wannier functions

The Wannier function $W_n(\mathbf{r} - \mathbf{l})$ associated with the band of index n and the lattice site $\mathbf{l}$ [167, 168] may be obtained from a Fourier transform of the Bloch functions $B_{\mathbf{k}n}(\mathbf{r})$ according to the expression

$$W_n(\mathbf{r} - \mathbf{l}) = \frac{1}{\sqrt{N}} \sum_{\mathbf{k}} e^{i\mathbf{k} \cdot \mathbf{l}} B_{\mathbf{k}n}(\mathbf{r}) \tag{11.1}$$

where the summation is taken over the $\mathbf{k}$ vectors (N is number) of the first Brillouin zone.

The Bloch function

$$B_{\mathbf{k}n}(\mathbf{r}) = U_{\mathbf{k}n}(\mathbf{r}) \exp(i\mathbf{k} \cdot \mathbf{r}) \tag{11.2}$$

as a solution of Schrödinger equation for a periodic potential is a plane wave of wave-vector $\mathbf{k}$ modulated with periodic factor $U_{\mathbf{k}n}(\mathbf{r})$. The $W_n(\mathbf{r}-\mathbf{l})$ have the following properties:

 (i) they are exponentially localized about particular sites, i.e. $W_n(\mathbf{r} - \mathbf{l})$ is localized about the site $\mathbf{l}$ but contrary to atomic functions they can have a finite amplitude at distances considerably exceeding the lattice constant,

 (ii) $W_n(\mathbf{r} - \mathbf{l})$ is orthogonal to any $W_{n'}(\mathbf{r} - \mathbf{l})$ for $n \neq n'$,

 (iii) Wannier functions centered at different sites are always orthogonal,

 (iv) they are orthogonal to Bloch functions of different bands,

 (v) the Wannier functions for a given band but with a different cell vector $\mathbf{l}$ are identical in form but relatively displaced by $\mathbf{l}$,

 (vi) as superpositions of Bloch functions of different energy they are not themselves eigenfunctions of Hamiltonian.

When the periodicity of the potential is lost, as for instance by the presence of an admixture or isolated defect, it has been shown [169, 170, 171, 77] that there exists a set of orthonormal Wannier–like functions, called generalized Wannier functions for the perturbed lattice which have properties similar to those for the ideal lattice.

Using in the above manner defined functions of both localized and free electrons the many electron functions may be described in the form of global Slater determinants:

$$|n\rangle = \{W_{n_1}^A, W_{n_2}^A, \ldots, W_{m_1}^B, W_{m_2}^B, \ldots, B_{\lambda_1}, B_{\lambda_2}, \ldots\} \tag{11.3}$$

where the indices A, B show the lattice sites whereas the n, m and λ indices individualize the Wannier and Bloch functions exploited, respectively. Since all the component functions are orthogonal the antisymmetrized many-electron states are also mutually orthogonal, if different. Consequently, for any state $|n\rangle$ its projection operator $P_n = |n\rangle\langle n|$ is defined by

$$P_n|k\rangle = \begin{cases} |n\rangle & \text{if } n = k \\ 0 & \text{otherwise} \end{cases}$$

11.2 The perturbation scheme for degenerate systems employing projection operators

In his construction of the perturbation scheme Stevens starts from the general actual Hamiltonian $\mathcal{H}$ of the whole system, which for a rigid lattice has the known form:

$$\mathcal{H} = \sum_i \left[\frac{p^2}{2m}\right]_i - \sum_{i,M} \frac{Z_M e^2}{r_{Mi}} + \frac{1}{2} \sum_{i \neq j} \frac{e^2}{r_{ij}} + \mathcal{H}' \tag{11.4}$$

The Hamiltonian consists of the sum of the kinetic energies of the electrons, their electrostatic energies in the fields of the nuclei, their mutual Coulomb repulsion energies and a wide variety of other energy terms in the $\mathcal{H}'$. Having at our disposal the well defined one–electron functions and knowing all important interactions in the system we are in a position to construct on the basis of physical knowledge suitable many–electrons states being a good approximation to actual stationary states.

Thus, there is a possibility of approximate reproducing the energy spectrum of a system (at least its low–lying part) without solving any relevant Schrödinger equation, i.e. not easy many–body problem. The unperturbed Hamiltonian $\mathcal{H}_0$ is just defined based on this set properly constructed states as

$$H_0 = \sum_n \lambda_n P_n = \sum_n |n\rangle\langle n|H|n\rangle\langle n| \tag{11.5}$$

where $P_n = |n\rangle\langle n|$ is the projection operator of state $|n\rangle$ from the set of the found states, and λ_n is a scalar, which, from natural reasons, is equaled to $\langle n|H|n\rangle$, i.e. to the expectation value of H taken for the state $|n\rangle$. That is $\lambda_n = \langle n|H|n\rangle$, for if $|n\rangle$ approximates to an eigenstate of H, then $\langle n|H|n\rangle$ should approximate to an eigenvalue. The perturbation will be $H - H_0$ and it yields only off–diagonal elements, whereas H_0 is, ex definitione, purely diagonal operator.

Since the states $|n\rangle$ (Eq.11.3) are all mutually orthogonal, H_0 is a hermitean operator. Additionally, H_0 possesses most of the symmetry properties of H. Its definition (Eq.11.5, Eq.11.3) ensures indistinguishability of electrons and keeps the translational symmetry of H. This is a very favourable characteristic of H_0 and its important feature from the point of view of convergence of perturbation calculation. However, such an unperturbed Hamiltonian does not have the rotational symmetries of H. The difficulty stems from the fact that in defining the Slater determinants direction of quantization for orbit and spin are introduced.

Another problem results from degeneracy or pseudo-degeneracy of the states $|n\rangle$. In fact, we have to manipulate rather with manifolds (families) r of subtly differing states with energies enclosed within certain quasi–bands which can lead to divergences in second and higher orders of perturbation calculation. The reason being that higher order terms involve energy differences like $\langle n|H|n\rangle - \langle m|H|m\rangle$ as denominators.

Suppose now that we constructed by using the above procedure such a set of mutually orthogonal many–electron wavefunctions $|n\rangle$. By way of example, one manifold (the basis one) can now be the states in which, say, each rare earth atom in the system has an occupancy $4f^2$, with the conduction electrons in the lowest conduction band and all core electrons filled. Another could be that in which, say, two sites have $4f^1$, the rest $4f^2$ and the displaced $4f$ electrons are accommodated in the lowest conduction band with core orbitals given an occupancy so that the overall number of electrons is conserved. Obviously, the states in a given family have different energies.

Both above shortcomings can be eliminated by easily made re–defining the unperturbed Hamiltonian, so that the spreading is removed. The $|n\rangle$ states retain the eigenstates of H_0 whereas their eigenvalues are now not $\langle n|H|n\rangle$ but their mean values over the families

$$E_r = \frac{\sum_{n_r} \langle n_r|H|n_r\rangle}{\sum_{n_r} \langle n_r|n_r\rangle} \tag{11.6}$$

Then the modified unperturbed Hamiltonian is given by

$$\overline{H}_0 = \sum_r E_r P_r \tag{11.7}$$

where the projection operator for the r-th family is now defined by

$$P_r = \sum_r |n_r\rangle\langle n_r| \qquad (11.8)$$

After this modification $\overline{H}_0$ possesses the full symmetry of the actual Hamiltonian H including the rotational invariance (any sub–space P_r is rotationally invariant) [76]. All that has happened is that whereas before H_0 was entirely diagonal and $H - H_0$ was entirely off–diagonal, now $\overline{H}_0$ is entirely diagonal but $H - \overline{H}_0$ has both diagonal and off–diagonal elements. The diagonal elements of $H - \overline{H}_0$ are just those needed to bring the diagonal elements of $\overline{H}_0$ up to those of H_0. A standard perturbation method with the projection operator technics given fully by Bates et al [172] leads to the following perturbation expansion of the ground state effective Hamiltonian

$$H \longrightarrow P_0 \left\{ H - \sum_{r \neq 0} \frac{H P_r H}{E_r - E_0} + \ldots \right\} P_0 \qquad (11.9)$$

where P_0 is the projection operator into the ground manifold, r is summed over all excited manifolds. In all orders the effect is always to have to evaluate an operator between two P_0, thus $P_0\{\ldots\}P_0$. In the second order perturbation term in Eq.11.9 $H - H_0$ is replaced by H. This follows because $P_0\overline{H}_0 P_r = 0$ for $r \neq 0$.

The expansion (Eq.11.9) will be efficiently convergent if off–diagonal matrix elements coupling the ground manifold states $|n_0\rangle$ with those of excited manifolds $|n_r\rangle$, $r \neq 0$, are small enough. The first order energy levels for the ground manifold can be obtained by diagonalizing the matrix $P_0 H P_0$ (the first term of Eq.11.9). This corresponds to the one–configuration approximation of crystal field effect since all the ground family states possess definite numbers of electrons in the various shells and conduction bands. The second order corrections are obtained by diagonalizing the second term in Eq.11.9 within the ground manifold

$$\sum_{r \neq 0} \frac{P_0 H P_r H P_0}{E_r - E_0} \qquad (11.10)$$

The higher order contributions are more and more complicated. The magnitude of second order corrections (Eq.11.10) will decide both necessity of their including or possibility of their rejecting. Usually one confines itself to the second order term.

The next step is transforming H in Eq.11.9 into second quantized form using as a basis the complete set of one–electron orthonormal Wannier and Bloch functions. The methods of second quantization automatically take care of the required antisymmetry of the many–electron wavefunctions and are directly related to the Slater determinant representation.

The second quantization creation α^+ and annihilation α operators for electrons obey the following commutation (anticommutation) rules:

$$\alpha^+_{k_1}\alpha^+_{k_2} + \alpha^+_{k_2}\alpha^+_{k_1} = 0$$

$$\alpha_{k_1}\alpha_{k_2} + \alpha_{k_2}\alpha_{k_1} = 0$$

$$\alpha^+_{k_1}\alpha_{k_2} + \alpha_{k_2}\alpha^+_{k_1} = \delta_{k_1 k_2} \tag{11.11}$$

It will be now convenient to denote by a^+, a; b^+, b ... etc. creation and annihilation operators of open shell electrons at site A, B, respectively, f^+, f those of filled orbitals and by e^+, e those for empty (excited) orbitals. The f^+, f and e^+, e operators have not been given site indices so their localization or delocalization is not prejudged.

In general, four types of operators, one–, and two–electron ones, independent or dependent on spin, i.e. h, g, h_s, g_s, respectively, as well as their various combinations in second and higher orders may occur. From the crystal field point of view the operators independent on spin, viz. h (electrostatic energy in the field of nuclei, kinetic energy), and g (direct and exchange Coulomb interactions between electrons) are mainly effective.

In turn, replacing the second quantized forms of the operators by expressions in angular (and spin) momentum variables according to the following relationships [76, 173, 174]:

$$\alpha^+_{m+}\alpha_{m'+} = \frac{1}{2}\sum_{k,i}\Delta_{mm'k}O^{(k)}_{m-m'}(l_i)(1+2s^i_z)$$

$$\alpha^+_{m-}\alpha_{m'-} = \frac{1}{2}\sum_{k,i}\Delta_{mm'k}O^{(k)}_{m-m'}(l_i)(1-2s^i_z)$$

$$\alpha^+_{m+}\alpha_{m'-} = \sum_{k,i}\Delta_{mm'k}O^{(k)}_{m-m'}(l_i)s^i_+$$

$$\alpha^+_{m-}\alpha_{m'+} = \sum_{k,i}\Delta_{mm'k}O^{(k)}_{m-m'}(l_i)s^i_- \tag{11.12}$$

where the summations is over electrons i and the rank k of the Smith and Thornley operators $O^{(k)}_{m-m'}(l_i)$ [33] (see chapter 2), $\Delta_{mm'k}$ are defined in the next section, $+$ and $-$ are identified as $m_s = 1/2$ and $m_s = -1/2$, respectively, the required equivalent form of so–called spin–Hamiltonian is gained [76].

Further, taking use of the Wigner–Eckart theorem the appropriate spin–Hamiltonians for any type of vector couplings in atoms, i.e. for large quantum numbers L or J may be obtained.

11.3 The crystal field effect

Let us now examine more thoroughly how does the crystal field theory look in this formulation? Within the second quantization formalism only these interactions (of any order) the operator form of which after all possible reductions ("contractions" of operators) takes, in general, the form

$$\sum_{\sigma} a^+_{m\sigma} a_{m'\sigma} \tag{11.13}$$

where $a^+_{m\sigma}$ denotes the creation operator of an open shell $(l)m\sigma$ electron and $a_{m'\sigma}$ the annihilation operator of an electron from the same shell but not necessarily from the same orbital, can contribute to the crystal field potential.

By means of $a_{m\sigma}$ symbol the corresponding spinorbital may be denoted too. This is a practical manner and its right interpretation results directly and unambiguously from the context.

Below, an equivalence between the classical crystal field Hamiltonian (in the Wybourne notation [3], see chapter 2)

$$\mathcal{H}_{CF} = \sum_i \sum_{k,q} B_{kq} C^{(k)}_q(i) \tag{11.14}$$

and its second quantization form

$$(\mathcal{H}_{CF})_{s.q} = \sum_{m,m'} \sum_{\sigma} \langle m|\mathcal{H}_{CF}|m'\rangle a^+_{m\sigma} a_{m'\sigma} \tag{11.15}$$

will be demonstrated.

The summation in Eq.11.14 is over N electrons whereas that in Eq.11.15 is over an infinite number of states characterized by m and m'. The antisymmetry requirement (with respect to interchange of electrons) which is a consequence of the exclusion principle has to be in the first case imposed on the mathematics (Slater determinants) whereas in the second case it is automatically ensured by the commutation rules of the creation and annihilation operators [76].

The matrix element of $\mathcal{H}_{CF}$ (Eq.11.14) between one–electron states $|lm\rangle$ and $|lm'\rangle$ is

$$\langle m|\mathcal{H}_{CF}|m'\rangle = \sum_k \sum_q B_{kq} \langle lm|C^{(k)}_q|lm'\rangle =$$

$$= \sum_k \sum_q B_{kq}(-1)^{l-m} \begin{pmatrix} l & k & l \\ -m & q & m' \end{pmatrix} \langle l\|C^{(k)}\|l\rangle =$$

$$= \sum_k \sum_q B_{kq}(-1)^{l-m} \begin{pmatrix} l & k & l \\ -m & q & m' \end{pmatrix} (-1)^l (2l+1) \begin{pmatrix} l & k & l \\ 0 & 0 & 0 \end{pmatrix}$$

Multiply both sides of the last equality by $(-1)^m \begin{pmatrix} l & k & l \\ -m & q & m' \end{pmatrix}$ and sum over all m and m'. Using the orthogonality relation [175]

$$\sum_m \sum_{m'} \begin{pmatrix} l & k & l \\ -m & q & m' \end{pmatrix} \begin{pmatrix} l & k' & l \\ -m & q' & m' \end{pmatrix} = \frac{\delta(k,k')\delta(q,q')}{2k+1}$$

we get

$$\sum_m \sum_{m'} (-1)^m \langle m|\mathcal{H}_{\mathrm{CF}}|m'\rangle \begin{pmatrix} l & k & l \\ -m & q & m' \end{pmatrix} = B_{kq} \frac{2l+1}{2k+1} \begin{pmatrix} l & k & l \\ 0 & 0 & 0 \end{pmatrix}$$

and hence

$$B_{kq} = \frac{2l+1}{2k+1} \left[\begin{pmatrix} l & k & l \\ 0 & 0 & 0 \end{pmatrix} \right]^{-1} \sum_m \sum_{m'} (-1)^m \langle m|\mathcal{H}_{\mathrm{CF}}|m'\rangle \times$$

$$\times \begin{pmatrix} l & k & l \\ -m & q & m' \end{pmatrix} \tag{11.16}$$

This is an important expression relating the crystal field parameters B_{kq} with the matrix elements of $\mathcal{H}_{\mathrm{CF}}$.

In turn, let us transform Eq.11.15. Using the tensor operator properties [174] each pair of $a_{m\sigma}^+ a_{m'\sigma}$ operators being diagonal with respect to spin orientation and summed over σ can be expressed with the Smith and Thornley operators $O_q^{(k)}(\mathbf{l}_i)$ (equivalent to $C_q^{(k)}\left(\frac{\mathbf{r}_i}{r_i}\right)$ ones) (see chapter 2) in the form

$$\sum_\sigma a_{m\sigma}^+ a_{m'\sigma} = \sum_k \Delta_{mm'k} \sum_i O_q^{(k)}(\mathbf{l}_i) =$$

$$= \sum_k \frac{\langle m|O_q^{(k)}|m'\rangle}{N_k^2} \sum_i O_q^{(k)}(\mathbf{l}_i) \tag{11.17}$$

where [176]

$$N_k^2 = \frac{(\langle l||O^{(k)}||l\rangle)^2}{2k+1} \tag{11.18}$$

And since [176]

$$\langle l||O^{(k)}||l\rangle = \frac{1}{2^k} \left[\frac{(2l+k+1)!}{(2l-k)!} \right] \tag{11.19}$$

(note the difference between $\langle l||O^{(k)}||l\rangle$ and $\langle l||C^{(k)}||l\rangle$), N_k^2 for $l = 2$ and 3 takes values presented in Table 11.1 [173]

Table 11.1 N_k^2 coefficients according to Eq.11.18 (and Eq.11.19) for $k = 0, 1, \ldots, 6$, and $l = 2, 3$

k	l=2	l=3
0	5	7
1	10	28
2	31.5	189
3	90	1350
4	157.5	8662.5
5	0	42525
6	0	116943.75

Thus, from Eq.11.17 substituting Eq.11.18 we have

$$
\sum_\sigma a_{m\sigma}^+ a_{m'\sigma} = \sum_k \frac{(2k+1)(-1)^{l-m} \begin{pmatrix} l & k & l \\ -m & q & m' \end{pmatrix}}{(-1)^l (2l+1) \begin{pmatrix} l & k & l \\ 0 & 0 & 0 \end{pmatrix}} \sum_i O_q^{(k)}(\mathbf{l}_i) =
$$

$$
= \sum_k \frac{2k+1}{2l+1} \left[\begin{pmatrix} l & k & l \\ 0 & 0 & 0 \end{pmatrix} \right]^{-1} \times
$$

$$
= (-1)^m \begin{pmatrix} l & k & l \\ -m & q & m' \end{pmatrix} \sum_i O_q^{(k)}(\mathbf{l}_i) \tag{11.20}
$$

Consequently,

$$
(\mathcal{H}_{\mathrm{CF}})_{\mathrm{s.q}} = \sum_{m,m'} \langle m|\mathcal{H}_{\mathrm{CF}}|m'\rangle \sum_\sigma a_{m\sigma}^+ a_{m'\sigma} =
$$

$$
= \sum_k \frac{2k+1}{2l+1} \left[\begin{pmatrix} l & k & l \\ 0 & 0 & 0 \end{pmatrix} \right]^{-1} \times
$$

$$
\times (-1)^m \langle m|\mathcal{H}_{\mathrm{CF}}|m'\rangle \times
$$

$$
\times \begin{pmatrix} l & k & l \\ -m & q & m' \end{pmatrix} \sum_i O_q^{(k)}(\mathbf{l}_i) \tag{11.21}
$$

Since acting $O_q^{(k)}(\mathbf{l}_i)$ on components of i-th electron angular momentum is equivalent to acting $C_q^{(k)} \left(\frac{\mathbf{r}_i}{r_i} \right)$ on space coordinates of the electron, Eq.11.21 after substitution of Eq.11.16 turns out to be equivalent with Eq.11.14. So, both forms (Eq.11.14 and Eq.11.15) are fully equivalent indeed.

It should be once more emphasized that there are two identification attributes of the second quantized form of crystal field Hamiltonian — one-electron character of its operator part with the open shell states involved only and identity of spin indices of both creation and annihilation operators. However, a matrix element or a product of matrix elements (with their energy denominators) accompanying $a^+_{m\sigma} a_{m'\sigma}$ may take various complicated forms depending on the interaction mechanism.

From the construction of H_0 (Eq.11.5) clearly results that it contains most, if not all, of the main energy contributions to $\mathcal{H}$ — the kinetic energy of electrons, the potential energy of each electron in the field of its own nucleus and other nuclei, inter–electronic direct and exchange interactions and spin–orbit coupling.

The crystal field effect is mainly derived from interaction of the central ion unpaired electrons with foreign nuclei and electrons belonging to other ions. However, intra–ionic interactions, although being out of $\mathcal{H}_{CF}$, can generate some "zero"–splitting. The point is, in a solid the Wannier functions describing the localized states $a_{m\sigma}$ are orthogonal but they are not still eigenfunctions of h. No wonder, there is no longer spherical symmetry and in consequence of that, e.g. $\langle a_{m\sigma}|h|a_{m'\sigma}\rangle \neq 0$. This is an effect of the orthogonalization making a variation of the free ion energy spectrum. H_0 is a diagonal operator (Eq.11.5) so the interaction of j-th nucleus at $\mathbf{R}_j$ with i-th electron (of the central ion) at $\mathbf{r}_i$ has the form

$$\left\langle a_{m\sigma} \left| \frac{Z_j e^2}{|\mathbf{R}_j - \mathbf{r}_i|} \right| a_{m\sigma} \right\rangle a^+_{m\sigma} a_{m\sigma} \tag{11.22}$$

whereas the direct and exchange Coulomb interactions of i-th electron occupying $a_{m\sigma}$ state with any foreign j-th electron in one of f, e, b, $ldots$ etc. states gives the contribution, as an instance

$$\text{(D)} \qquad \left\langle a_{m\sigma} f_{m'\sigma'} \left| \frac{e^2}{r_{ij}} \right| a_{m\sigma} f_{m'\sigma'} \right\rangle a^+_{m\sigma} b^+_{m'\sigma'} b_{m'\sigma'} a_{m\sigma}$$

$$\text{(Ex)} \qquad - \left\langle a_{m\sigma} e_{m''\sigma} \left| \frac{e^2}{r_{ij}} \right| a_{m''\sigma} a_{m\sigma} \right\rangle a^+_{m\sigma} e^+_{m''\sigma} e_{m''\sigma} a_{m\sigma} \tag{11.23}$$

Note the negative sign before the exchange term (after rearrangement of the operators) and identity of spins required due to orthogonality of the states. To sum up, H_0 contains a large part of the crystal field effect including its enormous spherical component (the Madelung energy).

At the first order perturbation level apart from the diagonals operators (Eq.11.22, 11.23) the off–diagonals ones of the type $\sum_\sigma a^+_{m\sigma} a_{m'\sigma}$, where $m \neq m'$, are worthy of notice as for the crystal field effect. They represent the same interactions as above in H_0.

Both the diagonal and off–diagonal first order effects are usually considerably smaller than those of the zero approximation diagonals terms. However, they are essential considering the non–spherical part of the crystal field, i.e. the crystal field potential sensu stricto. Diagonalization of the first order interaction matrix corresponds to the conventional crystal field theory.

In the second order excited manifolds $|n_r\rangle$ differing from the ground manifold $|n_0\rangle$ in distribution of electrons over different shells and bands are taken into account. The renormalizing correction (Eq.11.10) will contain several double interactions, $P_0 \, H \, P_r \, H \, P_0$, corresponding to all allowed combinations following from the scheme $\{h + h_s + g + g_s\} : \{h + h_s + g + g_s\}$, e.g. $h : h$ or $g : g$. Only these processes the operator of which can be reduced to the form as in Eq.11.13 are allowed. By way of example, two the most important of them will be considered. $h : h$ process of the type $a^+_{m'\sigma} e_\sigma : e^+_\sigma a_{m\sigma} \rightarrow a^+_{m'\sigma} a_{m\sigma}$, where h contains all the one–electron spin-independent terms and in which an open–shell electron at A site is promoted to an empty orbital e_σ. To return to P_0 via the second h the electron in e must be removed and restored in the open shell at A but not necessarily into the same orbital. $\Delta(e, a_m)$ is the energy required to excite the electron from a to e (it is an energy difference of $\overline{H}_0$). This process contributes as

$$- \frac{\langle a_{m'}|h|e\rangle\langle e|h|a_m\rangle}{\Delta(e, a_m)} a^+_{m'\sigma} a_{m\sigma} \tag{11.24}$$

The terms are identified as free–ion–like if e belongs to A or as crystal–field–like from empty states on other centers (from filled states also in another process $a^+_{m'\sigma} f_\sigma : f^+_\sigma a_{m\sigma}$. There are already terms of similar form coming from the first order theory so the new ones can be regarded as renormalizing the coefficients of the first order ones. Among other operators h contains the electron kinetic energy and this interaction does not make us think of the crystal field effect. So, the above result sounds a note of warning for electron kinetics energy terms (and others) can, in second order, give contributions which are indistinguishable from crystal–field–like terms in first order.

$g : g$ process, where $g = e^2/r_{ij}$, of the type

$$a^+_{m'\sigma} f^+_{\sigma'} e_{1\sigma'} e_{2\sigma} : e^+_{2\sigma} e^+_{1\sigma'} f_{\sigma'} a_{m\sigma} \longrightarrow a^+_{m'\sigma} a_{m\sigma}$$

in which two electrons one from an open shell and the other from a filled shell (of other ion) are excited to empty orbitals e_1 and e_2 and in the second step they are removed with simultaneous restoring them in the same f orbital and in the open shell at A. Its contribution is

$$- \frac{\langle a_{m'\sigma} f_{\sigma'}|g|e_{2\sigma} e_{1\sigma'}\rangle\langle e_{2\sigma} e_{1\sigma'}|g|a_{m\sigma} f_{\sigma'}\rangle}{\Delta(e_2 e_1, a_m f)} a^+_{m'\sigma} a_{m\sigma} +$$

$$+ \; \delta_{\sigma\sigma'} \frac{\langle a_{m'\sigma} f_{\sigma'}|g|e_{2\sigma} e_{1\sigma'}\rangle\langle e_{2\sigma} e_{1\sigma'}|g|f_{\sigma'} a_{m\sigma}\rangle}{\Delta(e_2 e_1, a_m f)} a^+_{m'\sigma} a_{m\sigma} \tag{11.25}$$

where $\Delta(e_2 e_1, a_m f)$ is the energy of the two–electron excitation.

When $\sigma \neq \sigma'$ only the direct repulsion (the first term) occurs, if however $\sigma = \sigma'$ both direct and exchange interactions are effective. This process renormalizes the contribution to the crystal field potential originating from inter–electronic interactions. Physically it represents an additional interaction arising from a distortion of the electron orbitals.

It is worth noting that the second (and higher) order renormalizing corrections, see Eq.11.24 and Eq.11.25, contain the energy denominators which make them dependent on individual levels.

The fact that some of the single site terms produced in second order behave differently from similar ones produced in first order may be associated with so-called correlation crystal field [177, 27].

Analogous intra–atomic processes $a_4^+ a_3^+ e_1 e_2 : e_2^+ e_1^+ a_2 a_1 \rightarrow a_4^+ a_3^+ a_2 a_1$, where e are the excited states beyond A site lead to so-called nepheloauxetic effect being responsible for observed differences between the Slater integrals for ion in free and bound state, respectively.

The higher order corrections the number of which rapidly increases are characterized by more and more complicated structure. However, if the perturbation expansion is properly converged which is ensured by an appropriate choice of the initial functions (construction of H_0) they become quickly negligible.

CHAPTER 12

Specific mechanisms of metallic state contributing to the crystal field potential

By now there is a pretty vast amount of experimental material referring to the crystal field in metals. Its appreciable and best systematized part regarding the lanthanides and lanthanide intermetallics is crowned by review works by Touborg [178], Schmitt et al [179] and Barbara et al [180]. Results of some successful trials of ab initio calculations [181, 77, 182, 183, 184] are also available (see chapters 13–15 and 16). However, the problem is complex enough that even a consensus regarding the hierarchy of various mechanisms combining in the crystal–field effect has not been unambiguously settled. The more so as in the current stage of development of the theory, i.e. the density functional theory the partition of the effect into the individual mechanisms (contributions) is an inessential (and perhaps unattainable) matter.

Before we move on to discuss particular components of the crystal field potential in metals we have to assume an attitude towards two fundamental questions: – is the crystal–field interaction in metals one–electron problem like in ionic compounds? and – does it satisfy the superposition model requirement?

As opposed to ionic compounds, or broadly speaking to non–metallic compounds, interpretation of the crystal–field effect in metallic materials is based on information being deduced from merely the few lowest levels of the ground state experimentally available. Inaccessibility to a larger amount of experimental data makes the examination and parameterization of particular mechanisms of the effect difficult, in particular the correlation effects. Verification of some of them, as e.g. the spin–correlated crystal–field effect is simply precluded. To some extent occurring the eight or higher order terms in the expansion of the crystal–field potential could evidence the correlation effects [177] but so far no trials in this direction have been undertaken. Another possibility of verification of the one–electron character of crystal–field potential in metals consists of checking up the expected regularity of the crystal–field effect with respect to the ion size and lattice distances along e.g. the lanthanide series. Such analysis performed for the lanthanides has not revealed noticeable fluctuations which could be indicative the significance of the correlation effects [185]. So, the one–electron crystal–field model is assumed to be valid in metals [188].

In turn, the superposition model of crystal field potential in metals seems to be false. Although several mechanisms, as in non–metallic compounds, come under the model there is one (at least one) mechanism arising from the presence of conduction electrons and making their condensation in the form of the virtual bound states possible which breaks loose from the model. Moreover, this virtual bound state mechanism affecting the second and fourth order terms only turns out to be the predominant component which determines the sign of the crystal field parameters.

Let us notice that negative values of all three intrinsic parameters are yielded by the point charges contribution itself since these are positively charged ionic cores in metals. In addition, mutual penetration of the central ion and ligands charge densities as well as the exchange interaction contribute in the same direction. From the remaining contributions which have been important previously the polarization of the cores is negligible due to their compactness and the orthogonalizing field is considerably weaker too. Thus, what is needed is a new mechanism providing positive inputs to the intrinsic parameters of the second and fourth order so large as to reverse their signs into plus but leaving the sign of the sixth order parameter unaltered. These requirements are fulfilled by the virtual bound state mechanism [186], e.g. the $5 - d$ vbs in the lanthanide metallic materials. This state becomes split in the crystal–field and its anisotropic charge distribution strongly interacts with the unpaired electrons ($4f$ in the case)) (chapter 14). Fortunately, ignoring the configuration mixing, this mechanism does not contribute to b_6 parameter. The occurrence of the $5 - d$ vbs in the case of rare–earth metals has been independently confirmed by experiments (chapter 14) and theoretical calculations in the frame of the density functional theory [90, 187]. This mechanism will be considered in detail in chapter 14 among other fundamental mechanisms being characteristic for the metallic state: screening by the conduction electrons (chapter 13), hybridization of localized (including virtual bound state) and delocalized states (chapter 15).

Nowadays, the prevailing theoretical approach based on the density functional theory and local density approximation is formulated in terms of the one–electron model leaving the superposition problem, as useless in the approach, aside. Nevertheless, this global calculated potential can be decomposed into on–site (the virtual bond state input), nearest neighbor shell and remaining lattice contributions.

Construction of the complete theoretical model of crystal–field in metals [77] is nicely contained within the frame of the perturbational scheme by

Stevens [76] employing projection operators and the second quantization technics convenient for degenerate states. Presentation of this formalism will precede further through considerations.

CHAPTER 13

Screening the crystal field in metallic materials

Among many factors forming the crystal–field effect in metallic materials [186, 189, 181, 182, 190] the screening by the conduction electrons can not be neglected. Considering the unquestionable freedom and mobility of the conduction electrons this mechanism seems to be obvious and at first glance highly effective. From such point of view the experimental data are rather intriguing. They are not managed to be systematized in a simple way. They seem to be unpredictable and up to now there is a symptomatic uncertainty among crystal–field model users about the screening role and efficiency of the conduction electrons in the crystal–field effect. Some authors ascribe them a minimum shielding role [191, 106], others – a considerable one, up to an antishielding one reversing the sign of the simple point–charge model parameters [186, 192].

In principle, there are two approaches to the screening problem in metals. The first is based on the free–electron gas model and resolves itself into the formalism of the static dielectric function of the medium [127, 168]. Two basic methods are at our disposal. Both of them are strictly defined approximations of general method of finding charge density distribution in electron gas and ionic core charges immersed in it, i.e. in the corresponding self–consistent Hartree field. The first called the Thomas–Fermi method is a long–wave approximation (for small q) suitable rather for slow variable perturbing potentials. The second known as the Lindhard method consists in a more exact calculation of the charge density by summing all Fourier components of the potential including large q too. However, taking into account only the linear response of the system to the perturbation remains an essential simplification of the approach [127, 168, 193].

A clear cut discontinuity of the Fourier spectrum of any crystal lattice potential appears in the fact that the mutual position of the smallest Fourier components of the crystal lattice potential versus the Fermi vector on the q-axis determines the screening efficiency. In practice, the whole variation of the dielectric function $\varepsilon(\mathbf{q})$ from infinity down to unity takes place within the short range $0 < q/k_F < 4$ [190]. On the other hand, there are usually only several or at most a dozen or so allowed reciprocal vectors **q** in this range and often they are greater than k_F. In this fact the reason of the apparent unpredictability or randomness of the screening effect in metals should be sought. The more so as the contributions coming from different **q** components may be either positive or negative. The Fourier expansion of the lattice potential, contrary to the scalar lattice sum, reveals its Fourier spectrum which explicitly depends on both direction and modulus of **r**. Hence, due to the $\varepsilon(\mathbf{q})$ function, the screening effect explicitly depends on **r**. Thus, despite the fact that the $\varepsilon(\mathbf{q})$ is independent on direction of **q** we deal with the anisotropy of the screening. Moreover, the Fourier form of the lattice potential is the optimum one for the screening including.

In the second, so–called dynamic approach (at neglecting the lattice vibrations) the electron system in a metal is treated as the plasma state in which the electrons are involved in a collective oscillations of characteristic plasma frequency ω_p. This frequency for typical metals lies in infrared range and the energy quantum is large enough to take into account only the ground state of the plasma, i.e. the zero–point plasmon $1/2(h\omega_p)$. So the dynamic screening approach consists in the re–description of the motion of the electrons in terms of the plasma oscillations plus short–range (about 1Å) interactions among electrons [194, 195]. The screening calculated in the dynamic approach is less effective than that found in the former approach. This is a consequence of the fact that the electrons involved in the collective motion are not fully free to shield the ionic core potentials.

However, the two above approaches possess serious constitutional short-comings. Saying nothing of other simplifying assumptions they make allowances only for linear response of the system to the perturbation, i.e. to the crystal field potential in this case. Besides, these methods include the screening effect ignoring the self–consistent procedure which turns out to be crucial in the screening mechanism. On the other hand, a direct dealing with the net crystal–field potential is rather a strong point of these approaches.

Being of opinion, which is still open, that detailed knowledge of charge density distribution in crystal is enough to determine the real crystal field potential and the actual parameters there is another possibility to calculate the crystal–field spectrum in the frame of the density functional theory (DFT) and in consequence to estimate the screening effect [196, 197, 198, 199] (see also chapter 16). At first, an accurate charge density is needed. This is so since the non–spherical part of the crystal potential, i.e. the crystal field potential, constitutes only an insignificant part of the whole potential and the crystal–field effect is very sensitive to even tiny changes in the electronic charge distribution.

The density functional theory is based on the ascertainment that any external potential is up to a constant uniquely determined by the ground state charge density distribution through a single–valued functional, or in other words, that the whole information about the ground state is already contained in the single particle density. The calculations of this charge distribution are based on DFT at the local density approximation (LDA) level [90, 187]. The calculation is based on an optimized linear combinations of atomic orbitals scheme and on an "open–core shell" treatment with the self–interaction correction for non–localized and localized electrons, respectively. Fortunately, the LDA calculations are known to yield accurate charge distributions even in strongly correlated systems.

The crystal–field spectrum and eigenfunctions can be obtained by means of mapping DFT to a crystal–field model Hamiltonian. The screening effect can be separated as the difference between the DFT results and those of the simple electrostatic approach.

13.1   The Fourier form of the crystal lattice potential

Since the dielectric function $\varepsilon(\mathbf{q})$ is in a natural way defined in the reciprocal space we will employ the Fourier form of the crystal lattice potential instead of that in the real space. Incidentally, two different methods of separation of the crystal field potential (understood narrowly as the aspherical potential seen by the open–shell electrons of the central ion) from the total crystal lattice potential are sketched.

The electrostatic potential of a single charge Ze in the empty space, i.e. in the system of spherical symmetry, and volume Ω, may be written with the Fourier expansion in the form

$$V(\mathbf{r}) = -\frac{Ze^2}{r} = -\sum_q \left(\frac{4\pi Ze^2}{\Omega q^2}\right) e^{i\mathbf{q}\cdot\mathbf{r}} \qquad (13.1)$$

where the summation is over the whole quasi–continuous wave vector spectrum which is limited only by the boundary conditions of the system. Replacing the summation by integration with the density of states allowed $\Omega/(2\pi)^3$ and averaging for the angles gives

$$\begin{aligned}
V(\mathbf{r}) &= -\int_0^\infty \left(\frac{4\pi Ze^2}{\Omega q^2}\right) \frac{\sin(qr)}{qr} \frac{\Omega}{(2\pi)^3} 4\pi q^2 dq = -\frac{2Ze^2}{\pi r} \int_0^\infty \frac{\sin t}{t} dt = \\
&= -\frac{Ze^2}{r} \qquad (13.2)
\end{aligned}$$

i.e. the return to the initial form in the real space. In principle, there are two methods of calculation of the $V(\mathbf{r})$ in crystal lattice. The first, that seemingly ignores the translational invariance of the potential and treats the lattice as an assembly of individual charges, resolves itself into calculation of so called lattice sum. This method is fully adequate for the ionic crystals for here it directly corresponds to the physical actuality. However, considering the slow convergence of typical sequences of the contributions from the succeeding coordination spheres, calculation of the so–called Madelung constant needs usually a specific clustering the terms. Obviously the method ·can not be applied to the empty lattice of bare atomic cores in metals because the lattice sum is then an extensive quantity proportional to $N^{2/3}$, where N is the number of the cores.

But such a model in the absence of its electrical neutrality does not correspond to any solid. Contrary to the total potential $V(\mathbf{r})$ the crystal–field potential $V_{\mathrm{CF}}(\mathbf{r})$ that consists of higher terms of the power expansion of the $V(\mathbf{r})$ in relation to $\mathbf{r}$, due to their much faster convergence, can be calculated similarly to the lattice sum method for both ionic and mettalic lattices.

The second approach taking into account the translational symmetry of the lattice gives values of the potential at each point in the form of the Fourier expansion over the reciprocal lattice vectors. Now, we have at our disposal separate q-components of the potential including the vast constant part (infinite in the perfect lattice) which is compensated with the constant potential of the uniform negative charge distribution. The result obtained with the second method is more complete, apart from values of the potential it gives its Fourier spectrum in addition. Besides, it allows us to include the screening by the conduction electrons in a more direct and convenient way.

The resultant potential $V(\mathbf{r})$ at any point $\mathbf{r}$ of the crystal lattice is a superposition of single potentials $v(\mathbf{r} - \mathbf{R}_j)$ produced by the cores localized at points $\mathbf{R}_j$, where index j runs over all lattice sites. Thus, it can be presented in the form

$$V(\mathbf{r}) = \sum_j v(\mathbf{r} - \mathbf{R}_j) \tag{13.3}$$

to focus our attention on the merits of the problem a simple Bravais lattice with one atom in the unit cell will be considered. The Fourier expansion each of the local potentials $v(\mathbf{r} - \mathbf{R}_j)$ has the form

$$v(\mathbf{r} - \mathbf{R}_j) = \frac{1}{\Omega} \sum_q V_q e^{i\mathbf{q}(\mathbf{r} - \mathbf{R}_j)} \tag{13.4}$$

where V_q is the Fourier transform of the potential which for the bare Coulomb potential ($\sim r^{-1}$) is equal to $-4\pi Z e^2 / q^2$. Consequently

$$V(\mathbf{r}) = \frac{1}{\Omega} \sum_j \sum_q V_q e^{i\mathbf{q}(\mathbf{r} - \mathbf{R}_j)} \tag{13.5}$$

or in the equivalent form

$$V(\mathbf{r}) = \frac{1}{\Omega} \sum_q e^{i\mathbf{q}\cdot\mathbf{r}} \cdot \sum_j e^{-i\mathbf{q}\cdot\mathbf{R}_j} \tag{13.6}$$

The term $\sum_{j} e^{-i\mathbf{q}\cdot\mathbf{R}_j}$ is the structure factor, which is equal to zero unless $\mathbf{q}$ is a reciprocal lattice vector. If the request is fulfilled it is equal to N. Hence, it follows that

$$V(\mathbf{r}) = \frac{1}{\Omega_0} \sum_{q} V_q e^{i\mathbf{q}\cdot\mathbf{r}} \qquad (13.7)$$

where Ω_0 is the volume allocated to a simple core. This expression is quite similar to Eq.13.1 but this time the summation is over the reciprocal vectors only. Eq.13.7 is a special case, for the simple Bravais lattice, of the more general expression [190, 200] which for the screened potential with the dielectric function $\varepsilon(q)$ takes the form

$$V(\mathbf{r}) = \frac{1}{\Omega_0} \sum_{q} \frac{V_q}{\varepsilon(q)} \, e^{i\mathbf{q}\cdot\mathbf{r}} \cdot S(\mathbf{q}) \qquad (13.8)$$

where the structure factor $S(\mathbf{q}) = (1/n)\sum_{j} e^{-i\cdot\mathbf{q}\cdot\mathbf{R}_j}$ is the sum over all n elements of the basis, i.e. all atoms of the unit cell.

Let us consider the spectrum of the reciprocal vectors. The face centered cubic lattice (or A_1) with the lattice constant a and the sites of which are occupied by ionic cores $+Ze$ charge has been chosen as a typical example. Fro the long–wave side (for small q's) the spectrum is delimited and clearly discrete. The smallest value of q corresponds to the largest interplane distance, i.e. to the distance between the most close–packing planes. For the considered case $q_{min} = 2\pi\sqrt{3}/a$ and corresponds to the set of (111) planes. From the short–wave side (for large q's) the spectrum becomes more and more close–set down to quasi–continuous one. The allowed wave–lengths can be as small as possible provided they are integer dividers of the interplane gaps and obey the boundary conditions. Taking into account the fact that both $\mathbf{q}$ and $-\mathbf{q}$ are the reciprocal vectors, Eq.13.7 can be rewritten in a more convenient form

$$V(\mathbf{r}) = \frac{2}{\Omega_0} \sum_{q} V_q \cos(\mathbf{q}\cdot\mathbf{r}) \qquad (13.9)$$

Due to the discontinuity of the q-spectrum, with respect to both their value and direction, the summation in Eq.13.9 can not be replaced by integration this time. The following features of Eq.13.9 should be noticed:

- The sum is convergent because it describes the intense quantity. All terms from the succeeding spherical shells of π/r thickness have the same sign. Since the reciprocal lattice points are distributed with uniform density in q-space their number within each the shell is proportional to q^2. Thus, at large q the sum of the terms of each shell tends to a constant value, as for its absolute values, and the sequence becomes oscillating. Its sum may be found either by specific grouping the terms or by means of a suitable convergence factor.

- It includes the potential generated by the central ion. Fortunately, this is of no importance from the crystal–field potential point of view for its aspherical part only is taken into account.

- The aspherity of the potential which is mainly derived from the range of the smallest q. The strongest contributions come from the planes parallel to $\mathbf{r}$ since then $\mathbf{q} \cdot \mathbf{r} = 0$.

- The periodic change of sign of the contribution for a fixed $\mathbf{r}$ at q variation.

- The lesser r the more Eq.13.9 becomes similar to the spherical symmetry case, i.e. to the free ion situation. Longer and longer sequence of the initial terms with the positive *cosine* ensures the adequate attraction.

In the considered cubic system the interplane distances are equal to $d = a/(h^2 + k^2 + l^2)^{1/2}$, where (hkl) is the Miller index of a set of parallel planes. Thus, the vector $\mathbf{q}$ is given by

$$q = \frac{2\pi n}{a}(h\hat{\mathbf{x}} + k\hat{\mathbf{y}} + l\hat{\mathbf{z}}) \tag{13.10}$$

where n is any integer which in theory of diffraction is called the interference order, and $\hat{\mathbf{x}}$, $\hat{\mathbf{y}}$, $\hat{\mathbf{z}}$ are the unit vectors. Hence

$$q = \frac{2\pi n}{a}(h^2 + k^2 + l^2)^{1/2} \tag{13.11}$$

Substituting Eqs 13.10 and 13.11, $\mathbf{r} = x\hat{\mathbf{x}} + y\hat{\mathbf{y}} + z\hat{\mathbf{z}}$ and $V_q = -4\pi Z e^2/q^2$ to Eq.13.9 one gets

$$V(x,y,z) = \frac{-2Ze^2a^2}{\Omega_0 \pi n^2} \sum_{hkl} \frac{\cos\left[\frac{2\pi n}{a}(hx + ky + lz)\right]}{h^2 + k^2 + l^2} \tag{13.12}$$

Each component of the above sum corresponds to the input to the total potential coming from the set of parallel planes (hkl). Taking into account the conditions limiting possible reflections for the fcc lattice ($Fm3m$ space group) and the multiplicity of the corresponding planes [201] all the components of the potential can be calculated. Having at our disposal the Fourier form of the crystal lattice potential the screening can be included by the direct and natural way simply by dividing each term by the proper $\varepsilon(\mathbf{q})$ value. Then, the separation of the aspherical part of the potential Eq.13.12 and its parameterization can be achieved in two ways. First, by a suitable differentiation of the potential with respect to particularly defined combination of the differentiation operators $\partial/\partial x$, $\partial/\partial y$, $\partial/\partial z$. The value of the derivative for $\mathbf{r} = 0$ is the corresponding crystal field parameters. For instance, in any cubic system, including the fcc lattice, the coefficient at the $x^4 + y^4 + z^4 - (3/5)r^4$ fourth order uniform polynomial is equal to the B_{40} parameter according to

$$B_{40} = -\frac{1}{4} \cdot \frac{2}{9} \left(\frac{\partial^4}{\partial x^4} - 3\frac{\partial^4}{\partial x^2 \partial y^2} \right) V(x, y, z) \Bigg|_{r=0} \qquad (13.13)$$

Eq.13.13 for the unscreened field $V(x, y, z)$ as well as its simple modification for the screened field could be very useful but for the problem of its convergence. Due to its lattice and non–local origin the sum over hkl (Eq.13.12) is definitely divergent. Fortunately, a convergence factor can be introduced [190] which besides possesses a simple physical meaning. This factor defined as $\exp[-(q^2 b^2)/4]$ can be considered as the Fourier transform of the positive ionic charge density smeared out into a Gaussian blob, $\varrho(r) \sim \sum_j \exp[-(\mathbf{r} - \mathbf{R}_j)^2/b^2]$, where b is the diffusion parameter [190].

The second method exploits the expansion of plane wave $e^{i\mathbf{q}\cdot\mathbf{r}}$ or $\cos(\mathbf{q}\cdot\mathbf{r})$ directly into the spherical waves [28]. One gets

$$\cos(\mathbf{q} \cdot \mathbf{r}) = \frac{1}{2} \sum_{l \text{ even}} (2l + 1)i^l \, j_l(qr) \, P_l(\cos\theta) \qquad (13.14)$$

where θ is the angle between the $\mathbf{q}$ and $\mathbf{r}$ vectors, and the summation due to the oddness of the Legendre polynomials $P_l(x)$ for l odd runs over l even only, $j_l(qr)$ is the usual spherical Bessel function of l order.

Averaging $j_l(qr)$ over the radial wave function of the open nl shell, R_{nl}, the specific intrinsic parameter $b_l(\mathbf{q})$ for each set of planes characterized by vector $\mathbf{q}$ is obtained

$$b_l(\mathbf{q}) = \int_0^\infty R_{nl}^2(r) j_l(qr) r^2 dr \qquad (13.15)$$

with the normalization condition $\int_0^\infty R_{nl}^2(r) r^2 dr = 1$. The set of planes (hkl) plays here the role of a ligand. To obtain the total crystal field parameter all the local systems z axes of which coincide with the $\mathbf{q}$ vectors have to be one after the other transformed into the global system z-axis of which is in general one of the main crystallographic axes.

13.2 The dielectric static screening function $\varepsilon(\mathbf{q})$

The dielectric screening function is defined as the quotient of the unscreened to screened potentials (see Eq.13.8). Assumptions of slow–variability of the perturbing potential along distances being comparable with the lattice constants which allows us to specify the momentum and position of single electrons simultaneously (with regard to the uncertainty principle) leads to the Thomas–Fermi approximation (TF) for the dielectric function of the semi–classic free–electron gas [127, 168]

$$\varepsilon(q) = 1 + \frac{\lambda^2}{q^2} \qquad (13.16)$$

where $\lambda^2 = 4\pi e^2 N(E_F)$, and the density of states at the Fermi level $N(E_F) = mk_F/(\hbar^2\pi^2)$ where k_F is the Fermi vector and m is the electronic mass. λ is called the Thomas–Fermi vector and its reciprocal $1/\lambda$ – the screening radius. In metals the screening radius is of order of the interatomic distances, in semiconductors much more larger. The bigger λ the more efficiency of the screening. This results mean that for $q \to 0$, $\varepsilon \to \infty$, i.e. a long–wave external field is almost entirely screened through redistribution of the electron density. For instance, for aluminum $\lambda \approx 2.06\text{Å}^{-1}$ and $1/\lambda \approx 0.49\text{Å}$, whereas $k_F = 1.75\text{Å}$ [202]. This small value of the screening radius is a well illustrative of high efficiency of the screening in this approach. Starting from the Fourier expansion form of the screened Coulomb potential of a single ion of a charge Ze situated inside the free–electron gas of volume Ω we directly obtain

$$
\begin{aligned}
V(\mathbf{r}) &= \int\limits_0^\infty \left[\frac{-4\pi Ze^2}{\Omega q^2(1 + \lambda^2/q^2)} \right] \frac{\sin(qr)}{qr} \frac{\Omega}{(2\pi)^3} 4\pi q^2 dq = \\
&= -\frac{2Ze^2}{\pi r} \int\limits_0^\infty \frac{\sin(qr)}{q^2 + \lambda^2} q dq = -\frac{2Ze^2}{\pi r} \left(\frac{\pi}{2} e^{-\lambda r} \right) = \\
&= -\frac{Ze^2}{r} e^{-\lambda r}
\end{aligned}
\qquad (13.17)
$$

This is the expression for the screened $V(\mathbf{r})$ in the real space. It should be underlined that all limitations arising from existence of a crystal lattice are ignored.

Due to the commonness of the TF approximation this is the well known exponential dependence of the screened potential on the distance from its source r or otherwise this is the integral form of the approximation. However, in crystal lattice the reciprocal vector spectrum is discontinuous and the integration as above in Eq.13.17 is not permitted. The $\varepsilon(q)$ plots according to the TF approach (Eq.13.16) for three different values of k_F are shown in Fig.13.1. As it is well known the basic limitation of the TF approximation is the requirement of slow–variability of the potential. Unfortunately, this condition is not usually obeyed by ionic core potentials in the lattice, particularly in the vicinity of the cores.

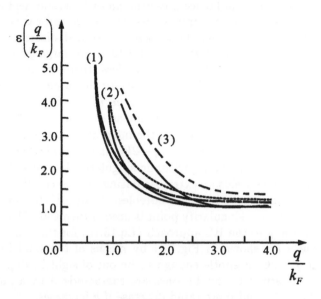

Fig. 13.1 Dielectric screening function $\varepsilon(q/k_F)$; solid lines – acc. to the Thomas-Fermi formula (Eq.10.16); dashed lines – acc. to the Hartree-Lindhard formula (Eq.10.18) 1) $k_F = 1.75$ Å^{-1}, 2) $k_F = 1.25$ Å^{-1}, 3) $k_F = 0.50$ Å^{-1}

The other approach not limited to the long–wave perturbing fields is the quantum approach by Lindhard based on the Hartree approximation (HL). Now, the formula obtained for the dielectric function

$$\varepsilon(q) = 1 + \frac{\lambda^2}{q^2}X(q) \qquad (13.18)$$

differs from Eq.13.1 in the dimensionless factor $X(q)$ of the form

$$X(q) = \frac{1}{2}\left(\frac{4 - (q/k_F)^2}{4(q/k_F)}\ln\left|\frac{2 + q/k_F}{2 - q/k_F}\right| + 1\right) \qquad (13.19)$$

For the long waves (small q's), $X(q) \to 1$. For the short waves (large q's), $X(q) \to 0$ as q^{-2}. The function $X(q)$ has a singularity at the point $q = 2k_F$, in which its first derivative tends to $-\infty$ and the second derivative suffers for discontinuity. The singularity, which is the consequence of the abrupt decay of population of the states at the Fermi level, is the source of numerous important outcomes. However, the singularity is substantial only when we deal with the continuous, or more precisely, quasi–continuous spectrum of **q** in this range of the wave numbers.

In the case of the crystal lattice potential and the crystal field effect the condition is not fulfilled. It so happens because the reciprocal vector spectrum of the lattice in the vicinity of $q = 2k_F$ is most of all discrete. For small q's, if $X(q) \rightarrow 1$, Eq.13.3 transforms into Eq.13.1. For large q's in both TF and HL cases $\varepsilon(q) \rightarrow 1$ (Fig.13.1). Nevertheless, there are two important differences in both the plots [127, 202]. First, for the large q's, $\varepsilon(q)$ according to Eq.13.18 and Eq.13.19 tends to $1 + 16me^2 k_F^3/(3\pi\hbar^2 q^4)$ and therefore $\varepsilon(q)$ function nears asymptotically to 1 not as q^{-2} but as q^{-4}. This fact changes the character of convergence of the integral over q. Secondly, in the vicinity of the point $q = 2k_F$ the term $(q - 2k_F) \ln |q - 2k_F|$ which is responsible for the mentioned above logarithmic singularity becomes dominating in Eq.13.19 [127]. This yields an extra oscillating term $\sim \cos 2k_F r/r^3$ in the potential producing so called Friedel aureoles. However, the continuity of the q–spectrum near the singularity point is needed then too. The $\varepsilon(q)$ plots calculated according to the HL approach (Eq.13.18) for the same values of k_F as previously are shown in Fig.13.1 by means of the solid lines. The singularity at $q = 2k_F$ is subtle enough to be out of sight in Fig.13.1. The HL plots in comparison to the TF ones are characterized by a more rapid rise of $\varepsilon(q)$ if $q \rightarrow 1_+$ and more rapid decrease if q increases.

Considering the almost complete drop of the $\varepsilon(q)$ function to unity for $q > 4k_F$ and on the other hand the distinct discontinuity of the crystal field spectrum below $q = 4k_F$, the screening, in fact, affects only few of the initial Fourier components of the lattice potential. On the other hand these are the most important terms in the potential mainly responsible for its aspherical part. So, the mutual position of the smallest q with respect to k_F on the q-axis determines efficiency of the screening. A relative weakness of the screening of the crystal field potential in metals results predominantly from the fact that the lower threshold of summation over $\mathbf{q}$, corresponding to the largest interplane distance, is a finite quantity relatively large for which $\varepsilon(q)$ is not far from 1 and regularly decreases with further increase of q. In other words, the most effective, from the screening point of view, range of small q lies beyond the summation range. This is also the reason of inadequacy of the TF approximation.

For the considered example of fcc aluminum matrix there are only eight allowed $\mathbf{q}$ vectors within the range $0 < q < 4k_F$ [190], and all of them are greater than k_F. Therefore the singularity of the dielectric function at $q = 2k_F$ is of no importance in this case. Thus, using the screening radius $1/\lambda$ which is the integral parameter of the Thomas–Fermi model (Eq.13.17) whether taking into account the oscillating long–range term in the potential, $\sim \cos(2k_F)/r^3$ which is a consequence of the continuous perturbation in the Hartree–Lindhard approach is for the perfect metallic lattice altogether groundless. However, we deal with the consequences of the singularity in the case of an admixture to the perfect metallic matrix. Then the $2k_F$ oscillations of the potential and the charge aureoles associated with them can be observed.

From the reasons given above also in the superconducting materials the characteristic changes of the density of states distribution near the Fermi level as well as the energy gap existence should not affect the screening of the crystal–field potential appreciably.

Contrary to the electrostatic field of crystal lattice its phonon spectrum is quasi–continuous starting from $q = 0$ up to q_{max} which is related to the Debye frequency ω_D by the dispersion law. If $k_F < (1/2)q_{max}$ then the so called Kohn effect [168] can be observed. It manifests itself in the form of anomalies in the phonon spectrum influenced by a sudden decrease of the screening efficiency when q passes $2k_F$.

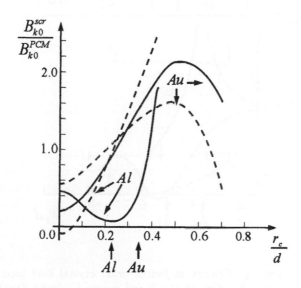

Fig. 13.2 The screening efficiency of the fourth (solid lines) and sixth order (dashed lines) crystal field terms in cubic Al and Au matrices as a function of the pseudopotential parameter r_c; d is the lattice constant, the actual values of the abscissa for Al and Au are marked with arrows (acc. to Buzukin and Khaliullin [182])

As it was mentioned earlier the expanding the screened lattice potential expressed in the reciprocal lattice terms into the spherical harmonics series leads directly to the crystal–field parameters with the screening effect included [182]. Then, dependence of the screening efficiency of the three following parameters clearly reveals:

- the orbital size of the unpaired electrons via its radial distribution,

158

- the Fermi vector k_F via the dielectric function $\varepsilon(q)$,

- the parameter characterizing the potential (or pseudopotential) of ionic core via the $V(q)$, e.g. in the empty–cores approximation [202]

$$V(q) = \frac{4\pi e^2 Z}{\Omega_0 q^2} \cos(q r_c) \tag{13.20}$$

this is the pseudopotential parameter r_c.

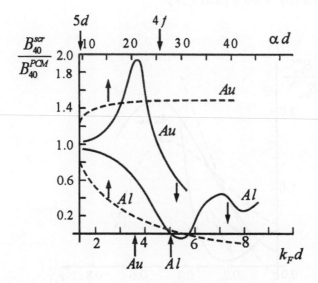

Fig. 13.3 The screening efficiency of fourth order crystal field term in cubic Al and Au matrices as a function of the Fermi vector k_F (solid lines) and the open-shell orbital size parameter a (dashed lines) (acc. to Buzukin and Khaliullin [182])

The above dependencies are rather complicated. An examplary screening efficiency defined as the ratios of $B_{k0}^{scr}/B_{k0}^{PCM}$, where $k = 4, 6$, in cubic Al and Au matrices as function of the pseudopotential parameter r_c (Eq.13.20), the Fermi vector k_F (Eqs 13.16 or 13.18) and the size parameter α of unpaired electron state (e.g. for the lanthanide ions $R_{4f} \sim r^3 \exp(-\alpha r)$) are shown after Khaliullin and Buzukin [182] in Figs 13.2 and 13.3.

In general, in simple metals like aluminum the true screening is observed whereas in the transition metals, e.g. in Au, the antiscreening takes place. It happens so due to occurring two competitive mechanism – on the one hand the conduction electrons influenced by the lattice potential tend to concentrate in the core areas (screening), on the other – on account of orthogonality requirement (exclusion principle) these electrons are pushed out from the areas (antiscreening). In the simple metals the first tendency prevails, in the transition metals with their electron abundance – the second one.

13.3 The dynamic mechanism of the screening – zero–point plasmon

The second approach resolves itself, in principle, into an intimate reorganization of the whole Hamiltonian of interacting electrons and nuclei. A part of energy can be separated in the form of zero–oscillations of simple harmonic oscillators, i.e. the zero–point plasmons for all wave vectors k of the range $k < k_c$, where k_c is the critical k value for the harmonicity of the plasma oscillations. The magnitude of the quantum $\hbar\omega_p$ (Eq.13.23), i.e. the plasmon energy, is crucial to the success of the dynamic description of the screening. For electron densities occurring in all metals $\hbar\omega_p$ is found to lie between about 3 and 25 eV [194]. Such a separation of general Hamiltonian can be attained by means of a certain unitary transformation or succession of the transformations construction of which has been given by Bohm and Pines [203, 204] on the strength of the random phase approximation (RPA). In consequence of the above transformation a radial modification of all electrostatic terms in the Hamiltonian takes place, and more precisely, a considerable change of their range. Additionally, some secondary modifications of electron effective mass (the kinetic energy), the spin–orbit coupling as well as arising other terms involving electron–plasmon and nuclei–plasmon interactions are included. Ignoring these last rather subtle modifications we have at our disposal the Hamiltonian which compared to the initial one contains the sum of zero–point energy of the plasmons and all terms of electrostatic interactions being multiplied by the so called screening function $F(k_c r_{ij})$ in result of which the range of the interactions is considerably reduced. For example, the interelectronic repulsion $\frac{e^2}{|\mathbf{r}_i - \mathbf{r}_j|}$ is replaced by the function $\frac{e^2}{|\mathbf{r}_i - \mathbf{r}_j|} F(k_c r_{ij})$. The function $F(k_c r)$ is defined [194, 204] by

$$F(k_c r) = 1 - \frac{2}{\pi} \int\limits_0^{k_c r} \frac{\sin x}{x} dx \qquad (13.21)$$

and its graph is presented in Fig.13.4.

Using the correct value of k_c is the crucial problem in this approach. The more its value the shorter range of the residual part of Coulomb interactions (after substracting the oscillation of the electron density energies). A very rough upper bound to the value of k_c may be estimated based on the following argumentation. The dispersion dependence of plasma oscillations $\omega(k)$ in the frame of classical electrodynamics of free–electron gas is given approximately by

$$\omega^2(k) = \omega_p^2 + k^2 \langle v_i^2 \rangle \qquad (13.22)$$

where $\omega(k)$ is the angular frequency corresponding to the wave number k (i.e. to the wave–length $2\pi/k$), $\langle v_i^2 \rangle$ is the average of the squared velocities of the electrons, $\langle v_i^2 \rangle = 2E_F/m$, where E_F is the Fermi energy, and ω_p is the basic angular frequency of plasma for $k = 0$ being equal to

$$\omega_p = \left(\frac{3e^2}{mr_s} \right)^{1/2} \qquad (13.23)$$

where r_s being defined by $\frac{4\pi}{3}r_s^3 = \Omega/N$ is the radius of the sphere allocated to individual conduction electron.

Converting Eq.13.23 to the atomic length unit (Bohr unit = $\hbar^2/me^2 = 0.53$ Å) and expressing energy in rydbergs (1 $ry = me^4/2\hbar^2 \approx 13.60$ eV) one gets

$$\omega_p = \frac{2\sqrt{3}}{r^{3/2}} \, [ry] \qquad (13.24)$$

The plasma oscillations may be represented approximately by a finite set of harmonic oscillators with angular frequency ω_p one for each density fluctuation with $k < k_c$. Taking into account the dispersion law, Eq.13.22, otherwise weak or flat, the harmonic description is valid for

$$\omega_p \gg k v_0 \qquad (13.25)$$

where v_0 is the speed of an electron at the Fermi surface. For larger k, however, these oscillations will be damped by the random thermal motion of the electrons which breaks its harmonicity. Hence as a very rough upper bound of the value k_c we may take

$$k_c \approx \omega_p/v_0 \qquad (13.26)$$

But since

$$v_0 = \frac{\hbar k_F}{m} = \left(\frac{9\pi}{4} \right)^{1/3} \frac{\hbar}{mr_s} \qquad (13.27)$$

we have

$$k_c \approx \sqrt{3} \left(\frac{4}{9\pi} \right)^{1/3} \frac{em^{1/2}}{\hbar} r_s^{-1/2} \qquad (13.28)$$

and after converting to the atomic units

$$k_c \approx \sqrt{3} \left(\frac{4}{9\pi} \right)^{1/3} r_s^{-1/2} \approx \frac{0.902}{r_s^{1/2}} \qquad (13.29)$$

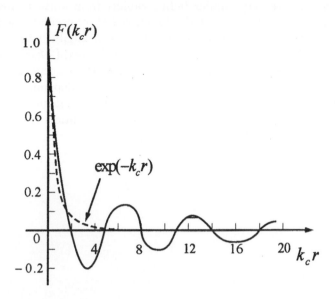

Fig. 13.4 The dynamic screening function $F(k_c r)$ acc. to Eq.13.21 – solid line, and $\exp(-k_c r)$ function – dashed line

For instance, in the case of silver $r_s = 3.02$ *a.u.* and $k_c \approx 0.52 \ a_0^{-1} \approx 1.00 \ \mathring{A}^{-1}$ which is a typical value of this screening parameter. Its physical significance is more complicated but qualitatively it represents the magnitude of the maximum wave vector for which it is useful to introduce plasmons.

In the Debye and Huckel theory [195, 205] k_c corresponds to so called Debye length λ_c or Debye sphere radius. From the probabilistic interpretation of wave function results that if the interacting electrons can not be localized inside a sphere of radius k_c^{-1} the collective character prevails in their behavior. On the other hand if kinetic energy of the electrons is large enough so that many wave lengths of the electrons are contained along the k_c^{-1} radius, the collective character decays in favour of independent individual behaviors. A high concentration of the conduction electrons and low temperature favour the collective behavior.

Unfortunately, later measurements have shown that plasmons exist at values of the wave vector k far beyond the k_c^{RPA} (Eqs 13.26 and 13.29) and that the corrected value of k_c may be even 1.5 times larger than the RPA value, as e.g. for silver [204, 70]. So, we should treat the value of k_c with some caution and the results have to be discuss for a range of values of k_c.

As seen in Fig.13.4 both functions $F(k_c r)$ and $\exp(-k_c r)$ for values of

$k_c r$ less than 2, are very similar failing rapidly from unity to nearly zero (~ 0.1). At larger values of $k_c r$ the function $F(k_c r)$ oscillates with slowly decreasing amplitude. These oscillations become negligible when multiplied by e^2/r. Both the plots $(e^2/r)F(k_c r)$ and $(e^2/r)\exp(-k_c r)$ coincide in practice and therefore k_c may be recognized as the screening parameter. It corresponds, in fact, to the Thomas–Fermi type dependence but with a modified screening constant λ. This type of the dependence seems to be the best for describing screened electrostatic potentials provided the k_c parameter is properly chosen.

CHAPTER 14

Virtual bound state contribution to the crystal field potential

The contribution of the so–called virtual bound states or semilocalized states within atomic core area is one of the three main mechanisms specific for the metallic state being responsible for the crystal field potential in metallic phases. The two remaining are the electrostatic contribution of atomic cores screened by the conduction electrons and hybridization effect of the localized and conduction band states. Two features distinguish this mechanism from the others. First is its two–stage, indirect nature leading in consequence to the effect being opposite to that produced by the immediate interaction and the second is a collective character of this effect hindering its decomposition into the intrinsic contributions. The contribution of the $5d$ virtual bound state into the crystal field potential of the $4f$ open–shell electrons in the lanthanide compounds is a model example of this mechanism and this example will be analyzed in detail below. This may be unhamperedly expanded for the $6d - 5f$ systems, i.e. for the actinide ions.

14.1 The resonance scattering of conduction electrons by a central potential

The phenomenon of the resonance scattering of conduction bands electrons by atomic core potentials is the source of specific condensation of the electrons on the virtual bound states in metals. Since this is from the natural reasons a spherically symmetric problem it is advisable to begin the consideration from reminder of the solution of the Schrödinger equation for a free–electron in the spherical coordinate system. This equation has the most elementary form

$$\frac{-\hbar^2}{2m}\nabla^2\Psi(\mathbf{r}) = E\Psi(\mathbf{r}) \qquad (14.1)$$

and its solutions in the factorized form

$$\Psi(\mathbf{r}) = \phi_l(r)Y_l^m(\mathbf{r}/r) \tag{14.2}$$

are in request. The spherical harmonics $Y_l^m(\mathbf{r}/r)$ are the eigenfunctions of the angle part of the ∇^2 operator with the eigenvalues $l(l+1)$, and hence the associated radial functions $\phi_l(r)$ have to fulfil the equation

$$\frac{d^2\phi_l}{dr^2} + \frac{2}{r}\frac{d\phi_l}{dr} + \left[\frac{2mE}{\hbar^2} - \frac{l(l+1)}{r^2}\right]\phi_l(r) = 0 \tag{14.3}$$

After substituting the relation

$$E = \frac{\hbar^2 k^2}{2m} \tag{14.4}$$

Eq.14.3 assumes the more convenient form

$$\frac{d^2\phi_l}{d(kr)^2} + \frac{2}{kr}\frac{d\phi_l}{d(kr)} + \left[1 - \frac{l(l+1)}{(kr)^2}\right]\phi_l(kr) = 0 \tag{14.5}$$

In general, there are two types of solutions of the above equation. The first are the spherical Bessel functions of index l

$$j_l(kr) = \sum_{m=0}^{l} \frac{(l+m)!}{m!(l-m)!} \frac{1}{2^m} \frac{\sin\left(kr - \frac{l-m}{2}\pi\right)}{(kr)^{m+1}} \tag{14.6}$$

This is the regular solution tending to zero for $kr \rightarrow 0$ and having the following asymptotic dependence for $kr \rightarrow \infty$

$$j_l(kr) \approx \frac{\sin\left(kr - \frac{l\pi}{2}\right)}{kr} \tag{14.7}$$

To the second class belong the spherical Neumann functions of index l

$$n_l(kr) = -\sum_{m=0}^{l} \frac{(l+m)!}{m!(l-m)!} \frac{1}{2^m} \frac{\cos\left(kr - \frac{l-m}{2}\pi\right)}{(kr)^{m+1}} \tag{14.8}$$

This solution possesses, in turn, a sigularity at the point $kr = 0$, if $kr \rightarrow 0$, $n_l(kr) \rightarrow \infty$. Its asymptotic approximation for $kr \rightarrow \infty$ is given by

$$n_l(kr) \approx \frac{\cos\left(kr - \frac{l\pi}{2}\right)}{kr} \tag{14.9}$$

For any regular solution of Eq.14.5 within a large sphere of radius R the allowed k values result from the natural boundary condition – vanishing the wave function for $r = R$ which occurs for

$$k = \frac{\left(n + \frac{1}{2}l\right)\pi}{R} \tag{14.10}$$

When at the origin of the coordinate system a perturbing potential $V(r)$ or pseudopotential $W(r)$ produced by an atomic core localized at this place exists the extended Schrödinger problem may be solved by partitioning the space about the potential source into two areas:

- the vicinity of the core in which the Schrödinger equation is solved by the direct integrating it from the centre of the system to the surface of the demarcating sphere with ignoring possible singular, i.e. non–physical, solutions, and

- the outer area in which $V(r) \equiv 0$ is assumed and where the general solution can always be presented in the form of linear combination of the appropriate spherical Bessel function $j_l(kr)$ and spherical Neumann function $n_l(kr)$:

$$\phi_l = A_l[\cos \delta_l \; j_l(kr) - \sin \delta_l \; n_l(kr)] \qquad (14.11)$$

where the meaning of δ_l will become clear further (Eq.14.12).

Subsequently, the two solutions being valid within the two separate areas ougth to be sewed smoothly together at the partition border throuh equating their values and values of their derivatives. The parameters which should be reconciled during the sewing procedure are the amplitude of the wave function A_l and the phase shift δ_l (Eq.14.11, 14.12). The combined function being the global solution behaves far from the centre, i.e. for large r, as

$$\phi_l \approx A_l \frac{\sin \left(kr - \frac{l\pi}{2} + \delta_l \right)}{kr} \qquad (14.12)$$

where δ_l is the phase shift as it results from comparison of Eq.14.12 with the asymptotic form of the solution for the free electron (Eq.14.7). Let us estimate quantitatively the extent of the modifications of the eigenfunctions caused by the potential centred at the origin. This modification can be perceived as absorption of a part of the wave function into the core region and in consequence as a phase–shift of the function far from the centre. The absorbtion of the wave function is followed by an increase of electron density in this area which by means of the Friedel sum rule [168] can be recasted to the degree of localization of the electrons with given quantum number l according to

$$\Delta n = \frac{2}{\pi} \; (2l + 1) \; \delta_l(E_F) \qquad (14.13)$$

where Δn is the rise of the number of electrons localized within the core area and $\delta_l(E_F)$ the phase shift determined at the Fermi level. Factor $2(2l + 1)$ is simply the degeneracy of the state ϕ_l.

According to the phase shift $\delta_l(E_F)$ the degree of localization Δn can vary continuosly but all the time the virtual bound state remains the resonance state. The situation changes qualitatively when the phase shift passes π value. Then, the number of allowed values of k, where

$$k = \frac{\left(n + \frac{1}{2}l\right)\pi - \delta_l}{R} \qquad (14.14)$$

increases by 1 and localization of the electron takes place. Now, the resonance state becomes localized and exponentially decays with the distance. This corresponds to imaginary lengths of the wave vector k. Considering the interelectronic repulsion $\Delta n \leq 1$ for virtual bound states is commonly assumed. The existence of virtual bound states in metals, alloys and especially in impurity systems were anticipated and postulated by Coles and Orbach and Williams and Hirst [186].

14.2 The nature of the virtual bound state

An appreciable overlap of broad, in general, bands originating from external electronic states of the s and p types with other discrete levels or rather with corresponding to the $d-$ and $f-$ electronic states narrow bands is an ordinary effect occurring in the $d-$ and $f-$ electron metallic systems. In such a situation no stationary one–electron wave functions could be attributed to the actual states of these types. They would have to decay in time becoming converted into the propagating plane waves. However, in such a system the phenomenon of resonance scattering of conduction electrons by the core potentials takes place. The arising resonance states are the effect of superposition of many one–electron states from the conduction band somewhat condensed within the scattering core area due to the phase shift. These states are not fully localized, their wave functions oscilate with a finite amplitude in the whole volume of the crystal. The closer the energies of the scattered conduction electrons and the unoccupied core level the greater the phase shift and in consequence a reinforced localization of the free electrons. Therefore these of the localized core states which have the highest energy but lower than the Fermi energy pretend first of all to those performing the role of virtual bound states. Thus, these are the $d-$ and $f-$ states most often. Such a space distribution of the increased electron density in the vicinity of the core is forced by the requirements of minimum of energy and orthogonality of the states. The many–electron character of the virtual bound states is just their fundamental feature. These are dynamic states being characterized by a certain diffusion of their energy (diversity of k values) and finite life time.

14.3 Spin–polarization of the virtual bound state

In general it is assumed that virtual bound states are non–magnetic in the spin sense. It means that in such states of many–electron nature both spin orientations are represented with the same probability. The most direct experimental confirmation of the assumption is the lack of EPR effect coresponding to these states in the examined systems.

However, the problem of spin–polarization of virtually bound electrons is not finally prejudged and has been treated variously in different approaches [206, 207]. There is a qualitatively different situation when the semilocalization refers to states of completely unoccupied shell than in the case of a partly populated shell. According to the energy situation of the virtual state within the core electron configuration spectrum the factor determining the spin correlation can be either the Coulomb interelectron repulsion as in the Anderson model [208] or the interelectron exchange interaction as in the Hund rule [206]. The primary role plays of course the Pauli exclusion principle. Thus, there are such potentially possible circumstances in which the virtual bound states can be spin polarized [208]. The spin polarization affects vitally the exchange interaction between the virtually bound electrons $(5d)$ and those of the open–shell of the core $(4f)$ and as a result of this has its contribution to the crystal field potential.

14.4 Experimental manifestations of existing the virtual bound states and methods of estimating their localization degree

The degree of localization of elecrons in the $5d$ virtual bound state, denoted by ε_d, enters the crystal field model in metals as an independent parameter. Theoretically it can be calculated by means of the Green function of the state [182]. Its magnitude can be independently estimated based on analysis of data of the following measurements:

- the crystal effect [209, 210, 211],

- the excess saturation magnetic moment in metals [212, 213], the extra contribution originates from the orbital magnetic moment of the virtual state,

- the Knight shift [212],

- the additional contribution to the electrical resistivity [219],

- the electric field gradient within the core area [214]

- the hyperfine field [213]

However, interpretation of the results is usually restricted by rather strong and poorly justified assumptions and unfortunately the values of ε_d obtained not always are reproducible and reliable. For the rare earth metals they are of order of $0.001 - 0.01$ only. In the impurity systems of the metals in the noble metal matrices Ag or Au, ε_d varies from 0.01 to 0.1 [210, 212], and even a model assumption $\varepsilon_d = 1$ has been postualted [206].

14.5 The crystal–field splitting of the virtual bound state

The virtual bound state $5d$ bearing a resemblance to the external atomic state of the same energy and angular momentum which is wide–spread in space suffers in the crystal field with a splitting according to the symmetry of the field and analogously as other open–shell states. In the cubic fields commonly encountered in metals, intermetallic compounds and alloys the five d states split into three t_{2g} and two e_g states. Considering the position of the $5d$ wave functions with respect to the surroundings of the central ion this effect should be much stronger than that in which we are mainly interested, i.e. the crystal field effect within the unfilled $4f$ shell. The most important characteristic of the splitting from the considered mechanism point of view is its angular symmetry or more precisely the symmetry, i.e. the irreducible representation, of the lowest level of the $5d$ shell. This is just the level which is virtually bound. It possesses a strictly specified angular symmetry. This unspherical distribution of the $5d$ electron density yields this specific intraatomic crystal field effect. Due to the space nearness and mutual penetration of the interacting electron densities $4f$ and $5d$ this is a strong effect. The external crystal field potential is not the only one factor determining the space distribution of the $5d$ electron density. The spin–orbit coupling is the second factor. Simultaneous taking into account the crystal field potential and the spin–orbit coupling [215] leads to certain modifications which can manifest in:

- a distinct space distribution of the $5d$ electron density,

- spin polarization of the ground (populated) $5d$ level and hence in a change of the $4f - 5d$ exchange interaction,

- possible appearance within the Boltzmann population range more than one level of the $5d$ manifold and in all consequences of the fact, especially in dependence of the crystal field potential both its symmetry and intensity on temperature.

The status of the $5d$ virtual bound state indicates that the crystal–field effect relative to it should be classified as the strong field case althoug according to the degree of localization ε_d its effective strength may be weakened.

14.6   The primary crystal field effect relative to the open–shell states $(4f)$

The causative factor of the considered effect is the Coulomb interaction of the unspherical part of the $5d$ electron density with the $4f$ electrons. In general, this interaction may be in the simplest way described in the second quantization form as

$$\mathcal{H}_{4f/5d} = \sum_{m_1,m_2,m_3,m_4} \sum_{\sigma_1\sigma_2} V^{\sigma_1\sigma_2}_{m_1,m_2,m_3,m_4} \, f^+_{m_1\sigma_1} d^+_{m_2\sigma_2} d_{m_4\sigma_2} f_{m_3\sigma_1} \quad (14.15)$$

where f^+, f and d^+, d are the creation and annihilation operators which create and destroy f and d electrons in generalized Wannier states of the central ion, respectively, m_i are the magnetic quantum numbers corresponding to the interacting states and the specified spin regulation results from the well known independence of the Coulomb interaction on the electron spin.

$$\begin{aligned}V^{\sigma_1\sigma_2}_{m_1,m_2,m_3,m_4} &= \langle m_1\sigma_1, m_2\sigma_2 | e^2/r_{ij} | m_3\sigma_1, m_4\sigma_2 \rangle \\ &- \langle m_1\sigma_1, m_2\sigma_2 | e^2/r_{ij} | m_4\sigma_2, m_3\sigma_1 \rangle \delta_{\sigma_1\sigma_2} \quad (14.16)\end{aligned}$$

is the total matrix element of the interaction – the sum of the direct and exchange terms, respectively.

Knowledge of the matrix elements of the $\mathcal{H}_{4f/5d}$ interaction affords possibilities for immediate conversion of the effect to the crystal–field theory or in other words for expressing the matrix elements by the crystal–field parameters and vice versa. Formally, the conversion can be carried out with several equivalent methods. Their common plot is the expression given by Stevens [76] for pairs of second quantization operators $a^+_{m\sigma} a_{m'\sigma'}$ in terms of combinations of appropriate products of orbital and spin angular momentum operators. Only the spin diagonal one–electron operators $(\sigma = \sigma')$ are effective in the crystal field Hamiltonian. Taking into account the above spin restriction, i.e. the condition $\sigma_1 = \sigma_2$, Eq.14.15 can be reduced to the form being the sum of one–electron operators

$$\mathcal{H}_{4f/5d} = \sum_i \sum_n \sum_q A^n_q \, O^{(n)}_q(\mathbf{l}_i) \quad (14.17)$$

where $O^{(n)}_q(\mathbf{l}_i)$ are the Smith–Thornley operators [33] and A^n_q the crystal field parameters in this parameterization. The general transformation from Eq.14.15 to Eq.14.17 is sketched in chapter 11. In this chapter we will use an intuitively simpler relation between the matrix elements of any interaction involving the open–shell electrons and the corresponding partial crystal–field parameters (see chapter 11). This relation for the considered interaction has the form

$$(B_{kq})_{5d} = \frac{2k+1}{2l+1} \left[\begin{pmatrix} l & k & l \\ 0 & 0 & 0 \end{pmatrix} \right]^{-1} \sum_m \sum_{m'} (-1)^m \langle m|\mathcal{H}_{4f/5d}|m'\rangle \times$$
$$\times \begin{pmatrix} l & k & l \\ -m & q & m' \end{pmatrix} \tag{14.18}$$

where $(B_{kq})_{5d}$ is the partial crystal–field parameter being derived from the $4f - 5d$ effect only. By way of example the calculations of $(B_{40})_{5d}$ for the $Er^{3+}(4f^{11})$ ion in Au matrix separately for the direct $(B_{40})^J_{5d}$ and exchange $(B_{40})^K_{5d}$ contributions for a rather academic model of $5d$ singlet $\frac{1}{\sqrt{2}}|m_l = 2\rangle + \frac{1}{\sqrt{2}}|m_l = -2\rangle$ as the virtual bound state is sketched below. To simplify the problem the total condensation of the $5d$ electron on this singlet, $\varepsilon_d = 1$, is assumed and all corrections including the screening with the conduction electrons are ignored. Owing to the symmetry properties of the $3\text{-}j$ symbols it can be easily demonstrated that the result obtained for the above singlet is identical with that for the state $|m_l = 2\rangle$. Thus, for the direct contribution we have

$$(B_{40})^J_{5d} = \frac{9}{7} \left[\begin{pmatrix} 3 & 4 & 3 \\ 0 & 0 & 0 \end{pmatrix} \right]^{-1} \sum_m (-1)^m \times$$
$$\times \langle 3, m; 2, 2|e^2/r_{12}|3, m; 2, 2\rangle \begin{pmatrix} 3 & 4 & 3 \\ -m & 0 & m \end{pmatrix} \tag{14.19}$$

It so happens that the only one $C_q^{(k)}$ operator in the expansion of e^2/r_{12} interaction giving non–zero matrix element in Eq.14.19 is the $C_0^{(4)}$ one associated with the $F^4(4f, 5d)$ Slater integral. The $C_0^{(6)}$ operator does not obey the triangle rule and is ineffective and, in turn, $C_0^{(2)}$ operator does not contribute due to the relation [31]

$$\sum_{m=-3}^{3} \begin{pmatrix} 3 & 2 & 3 \\ -m & 0 & m \end{pmatrix} \begin{pmatrix} 3 & 4 & 3 \\ -m & 0 & m \end{pmatrix} = 0 \tag{14.20}$$

Hence

$$(B_{40})^J_{5d} = \frac{9}{7} \left[\begin{pmatrix} 3 & 4 & 3 \\ 0 & 0 & 0 \end{pmatrix} \right]^{-1} \sum_m (-1)^m (-1)^{3-m} \times$$
$$\times \begin{pmatrix} 3 & 4 & 3 \\ -m & 0 & m \end{pmatrix} (-1)^3 7 \times$$
$$\times \begin{pmatrix} 3 & 4 & 3 \\ 0 & 0 & 0 \end{pmatrix} \begin{pmatrix} 2 & 4 & 2 \\ -2 & 0 & 2 \end{pmatrix} \times$$

$$\times \ 5 \begin{pmatrix} 2 & 4 & 2 \\ 0 & 0 & 0 \end{pmatrix} \begin{pmatrix} 3 & 4 & 3 \\ -m & 0 & m \end{pmatrix} \times$$

$$\times \ F^4(4f, 5d) \qquad (14.21)$$

where

$$F^4(4f, 5d) = e^2 \int_0^\infty \int_0^\infty R_{4f}^2(r_1) R_{5d}^2(r_2) \frac{r_<^4}{r_>^5} r_1^2 r_2^2 \, dr_1 dr_2$$

and R_{4f}, R_{5d} are the radial functions of the $4f$ and $5d$ electron, respectively. Using the tables of the 3-j symbols [31] one gets

$$(B_{40})_{5d}^J = \frac{1}{21} F^4(4f, 5d) \qquad (14.22)$$

For the Er^{3+} ion in dilute Er-Au alloy $F^4(4f, 5d) = 11750 \ cm^{-1}$ [216] which gives $(B_{40})_{5d}^J = 560 \ cm^{-1}$ (or $70 \ cm^{-1}$ in the Stevens parameterization).

In the more realistic case of the cubic crystal field the virtually bound electron occupies, with the same probability, the three levels: $\frac{1}{\sqrt{2}}|2\rangle - \frac{1}{\sqrt{2}}|-2\rangle$, $\frac{1}{\sqrt{2}}|1\rangle \pm \frac{1}{\sqrt{2}}|-1\rangle$ of the t_{2g} triplet.

Apart from the calculated matrix element the two others contribute to the $(B_{40})_{5d}^J$. Since $\begin{pmatrix} 2 & 4 & 2 \\ -1 & 0 & 1 \end{pmatrix} / \begin{pmatrix} -2 & 0 & 2 \\ 0 & 0 & 0 \end{pmatrix} = 4$ they are, regarding their absolute values, four times each greater than the previous one but negative. Averaging the three values according to their equal population we get

$$(B_{40})_{5d}^J = -\frac{1}{9} F^4(4f, 5d) \qquad (14.23)$$

which for the previously given value of $F^4(4f, 5d)$ yields $(B_{40})_{5d}^J = -1306 \ cm^{-1}$.

As seen, the direct contribution of the $5d$ singlet $\frac{1}{\sqrt{2}}|2\rangle + \frac{1}{\sqrt{2}}|-2\rangle$ descending from the e_g doublet under the influence of a tetragonal distortion is positive (!) whereas that produced by the whole triplet t_{2g} is stronger and negative. This fact is worthy of noticing since it reveals the true capacity of the model.

Let us consider, in turn, the contribution of the $4f - 5d$ exchange interaction to the B_{40} parameter, i.e. $(B_{40})_{5d}^K$. The input can be calculated from the expression

$$(B_{40})_{5d}^J = \frac{1}{2} \cdot \frac{9}{7} \left[\begin{pmatrix} 3 & 4 & 3 \\ 0 & 0 & 0 \end{pmatrix} \right]^{-1} \sum_m (-1)^m \times$$

$$\times \ \langle 3, m; 2, 2 | e^2/r_{12} | 2, 2; 3, m \rangle \begin{pmatrix} 3 & 4 & 3 \\ -m & 0 & m \end{pmatrix} \qquad (14.24)$$

The coefficient $1/2$ in front of Eq.14.24 restricts the number of states combining to the $5d$ virtual bound state to those only which possess the same spin as the $4f$ electron. It is easy to show that in this case on account of the difference in parity of the angular momentum quantum numbers l of the interacting electrons only the odd order operators: $C_{-1}^{(1)}$, $C_0^{(1)}$, $C_1^{(1)}$; $C_{-3}^{(3)}$, $C_{-2}^{(3)}$, $C_{-1}^{(3)}$, $C_0^{(3)}$, $C_1^{(3)}$ and $C_{-5}^{(5)}$, $C_{-4}^{(5)}$, $C_{-3}^{(5)}$, $C_{-2}^{(5)}$, $C_{-1}^{(5)}$, $C_0^{(5)}$, $C_1^{(5)}$, are effective. They are coupled with the following Slater integrals $G^{(1)}(4f, 5d)$, $G^{(3)}(4f, 5d)$ and $G^{(5)}(4f, 5d)$, respectively. Thus

$$
(B_{40})_{5d}^{\mathrm{K}} = \frac{9}{14} \left[\begin{pmatrix} 3 & 4 & 3 \\ 0 & 0 & 0 \end{pmatrix} \right]^{-1} 35 \sum_m (-1)^m \times
$$

$$
\times \left\{ \left[\begin{pmatrix} 3 & 1 & 2 \\ 0 & 0 & 0 \end{pmatrix} \right]^2 \left[\begin{pmatrix} 3 & 1 & 2 \\ -m & m-2 & 2 \end{pmatrix} \right]^2 \times \right.
$$

$$
\times \begin{pmatrix} 3 & 4 & 3 \\ -m & 0 & m \end{pmatrix} G^{(1)}(4f, 5d) +
$$

$$
+ \left[\begin{pmatrix} 3 & 3 & 2 \\ 0 & 0 & 0 \end{pmatrix} \right]^2 \left[\begin{pmatrix} 3 & 3 & 2 \\ -m & m-2 & 2 \end{pmatrix} \right]^2 \times
$$

$$
\times \begin{pmatrix} 3 & 4 & 3 \\ -m & 0 & m \end{pmatrix} G^{(3)}(4f, 5d) +
$$

$$
+ \left[\begin{pmatrix} 3 & 5 & 2 \\ 0 & 0 & 0 \end{pmatrix} \right]^2 \left[\begin{pmatrix} 3 & 5 & 2 \\ -m & m-2 & 2 \end{pmatrix} \right]^2 \times
$$

$$
\times \begin{pmatrix} 3 & 4 & 3 \\ -m & 0 & m \end{pmatrix} G^{(5)}(4f, 5d) \right\} \tag{14.25}
$$

where

$$
G^{(n)}(4f, 5d) = e^2 \int_0^\infty \int_0^\infty R_{4f}(r_1) R_{5d}(r_1) R_{4f}(r_2) R_{5d}(r_2) \frac{r_<^n}{r_>^{n+1}} r_1^2 r_2^2 \, dr_1 \, dr_2
$$

Using the 3-j symbol tables again we get

$$
(B_{40})_{5d}^{\mathrm{K}} = \frac{33}{980} G^{(1)} + \frac{11}{735} G^{(3)} + \frac{5}{6468} G^{(5)} \tag{14.26}
$$

For the Er^{3+}:Au system considered the Slater integrals are equal to $G^{(1)} = 10331 \ cm^{-1}$, $G^{(3)} = 8586 \ cm^{-1}$ and $G^{(5)} = 6585 \ cm^{-1}$ [216]. After substituting these values to Eq.14.26 we have $(B_{40})_{5d}^{\mathrm{K}} = 481 \ cm^{-1}$ which makes as much as 0.86 of the direct contribution and both effects reinforce each another which is rather an untypical behaviour. The presented example pointedly illustrates the weight of this specific mechanism against the background of other components of the crystal field potential. This is one of the strongest contributions, frequently dominating [182], and which is particularly important of opposite sign, in general, than the main remaining terms. It results from the "offset" character of the mechanism, i.e. the electrostatic interaction of the external potential with the $4f$ electrons through the mediation of the $5d$ virtual bound state electron. The scheme of the mechanism is shown in Fig.14.1.

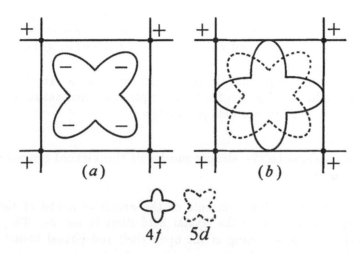

Fig. 14.1 The offset character of the crystal field mechanism via the virtual bound state – schematic distribution of the electron density : a) without participation of the $5d$ electron, b) through the mediation of the $5d$ virtual bound state

This distinguishing mark of the virtual bound state mechanism gave probably inspiration to introducing it to the crystal–field theory. Troublesome disagreement in signs of calculated and measured crystal field parameters in metals, particularly those of the fourth order [186], could not be explained otherwise.

In the $4f - 5d$ Coulomb interaction an exeptionally large proportion of the exchange term to the direct one, from 0.7 to 0.9 [182] is observed. Perhaps it is connected with a greater numbers of the effective operators in the case of exchange interaction (compare Eq.14.19 versus Eq.14.24).

In typical cases the direct and exchange contributions have opposite signs, the first are negative and the second positive and their mutual partial concellation takes place although as it results from the above example both the terms can also amplify the effect together.

It obliges us to a high accuracy in the calculations, to avoid coarse approximations particularly those concerning the radial distribution of the virtual bound state and the Slater integrals in metals [182] and to including the screening effect and hybridization of the virtual bound state.

The second characteristic feature of the mechanism is the independence, in principle, of its effect on the strength of the external crystal field. In other words, neglecting the differences in ε_d, the effect of the intraatomic $4f - 5d$ interaction is constant for a given symmetry and therefore the notion of the spectrochemical series in relation to the ligands has no application in this case.

This effect can not be separated into contributions originating from individual neighbouring atomic cores in the crystal lattice and in consequence of it the effect does not come under the superposition model as well as angular overlap model (AOM).

One should also notice that the sixth order terms of the external crystal field do not split the d–shell. So, in principle, this mechanism does not affect the sixth order crystal–field parameters.

14.7 Corrections to the simple model of the virtual bound state mechanism

There are three main improvements of the primitive model of the virtual bound state impact on the crystal field effect in metals. They consist in including the screening of the open–shell and virtual bound state electron Coulomb interaction through the conduction electrons, the non-orthogonality of the virtual bound state and the hybridization of the state with the conduction electrons.

One of the conventional methods of including the screening of electrostatic interactions particularly for the short distance interactions is the dynamic approximation (chapter 13). The modern approaches are much more realistic and subtle [196].

The dynamic approximation reduces itself to multiplying all operators of any electrostatic interaction by the screening function $F(k_c|\mathbf{r}_i - \mathbf{r}_j|)$ [217] (see chapter 13). For the interelectron repulsion it corresponds to replacing the expansion

$$\frac{e^2}{|\mathbf{r}_1 - \mathbf{r}_2|} = e^2 \sum_{l=0}^{\infty} \frac{r_<^l}{r_>^{l+1}} \, P_l \left(\frac{\mathbf{r}_1 \cdot \mathbf{r}_2}{r_1 \cdot r_2} \right) \tag{14.27}$$

where $r_<$ and $r_>$ are the smaller and larger values from r_1 and r_2, respectively, and P_l is the Legendre polynomial of l degree, by the modified expansion

$$\frac{e^2}{|\mathbf{r}_1 - \mathbf{r}_2|} F(k_c|\mathbf{r}_1 - \mathbf{r}_2|) = e^2 \sum_{l=0}^{\infty} a_l(r_1, r_2) \, P_l \left(\frac{\mathbf{r}_1 \cdot \mathbf{r}_2}{r_1 \cdot r_2} \right) \tag{14.28}$$

in which $a(r_1, r_2)$ may be found based on the orthogonality of the Legendre polynomials

$$a_l(r_1, a_2) = \frac{2l+1}{2} \int_{-1}^{1} \frac{F(k_c|\mathbf{r}_1 - \mathbf{r}_2|)}{|\mathbf{r}_1 - \mathbf{r}_2|} P_l\left(\frac{\mathbf{r}_1 \cdot \mathbf{r}_2}{r_1 \cdot r_2}\right) d\left(\frac{\mathbf{r}_1 \cdot \mathbf{r}_2}{r_1 \cdot r_2}\right) \quad (14.29)$$

The screening function $F(k_c|\mathbf{r}_1 - \mathbf{r}_2|)$ may be with a good accuracy replaced by exponentional function $\exp(-k_c|\mathbf{r}_1 - \mathbf{r}_2|)$ especially for small $|\mathbf{r}_1 - \mathbf{r}_2|$. Roughly speaking k_c is a screening parameter. Its physical significance is more complicated but quantitatively it represents the magnitude of the maximum wave vector or its critical value for which it is useful to introduce plasmons in electron gas of metal. Unfortunately this quanity is not well recognized so far. Its value resulting from the random phase approximation (RPA) seems to be in the light of several experimental data considerably lowered, even $1.5 - 2$ times [217, 218].

In the case of the considered intraatomic interaction between the $4f$ and $5d$ electrons the screening through the conduction electrons is not too effective but plays a much more important role than it could seem since its exchange part is screened much more efficiently than the direct term (about ten times) [218].

Due to the competitive character of both the terms the net effect can be noticeable. The parameter k_c which governs the effect should not be treated as a free parameter of the model but ought to be estimated independently.

- In a similar way as in ionic systems the renormalization effects arising from the non–orthogonality of the wave functions involved can be significant. Starting from the correct wave function of the virtual bound state is the key–point of estimating the effect. On account of characteristic space expanse and diffusion of energy for such states which arise from their incomplete localization most of the conventional approximations, including e.g. the simple Slater orbitals, turn out to be inadequate. The main non–orthogonality in the problem is that between the virtual bound state and outer localized states of neighbouring atomic cores. Rough estimate based on the Löwdin orthogonalization method [218] and the virtual $5d$ state in the form of a single Slater orbital suggests however that the renormalization effects in the case of Er^{3+}:Au system are of secondary meaning, of order of few percent [70]

- The virtual bound states undergo the hybridization mixing as the open–shell states. The hybridization admixtures complicate somewhat the virtual bound state effect. In the cubic fields the admixtures of the s–, p– and d–type are ineffective but that of f–type is the source of contribution to the sixth–order crystal field parameters which is impossible for the pure $5d$ state. Some troubles and discrepancies in quantitative estimation of the effect result mainly from diversity of the virtual bound state specification. In the perturbational first order approach by Dixon and Wardlaw [77] the effect is neglected. On the other hand according to the Buzukin and Khaliullin results [182] the effect produces contribution to the crystal field parameters being comparable to the main hybridization effect of the $4f$ electrons (chapter 15). In the cubic crystal field of typical metals such as Al, Cu, Ag, Au of the fcc type structure the hybridization corrections to both the fourth– and sixth– order parameters are positive and increase when the atomic weight of the metal rises.

CHAPTER 15

Hybridization or covalent mixing between localized states and conduction band states in metallic crystals

15.1 The essence of the hybridization

The fundamental difficulty in description of the band structure of the transition d–electron and f–electron metals and their intermetallic compounds and alloys arises from the specific situation of the open–shell electrons in the metals compared to that in the simple metals. In the metallic crystals the d– and f– electrons are neither completely localized like the true inner states of metal core, nor fully free as the conduction electrons are. In consequence, the wave functions of these states are not managed to be effectively defined in the form of their expansion into the conventional series of the orthogonalized plane waves (OPW) as in the case of simple metals pseudopotential theory [127]. However, they can be effectively expanded into the OPW series within a supracomplete system comprising apart from the plane waves and core localized state wave functions explicitly the open–shell states of the free ion (or atom) according to the pseudopotential theory of the transition metals [221, 222]. Thus, the actual d– or f– states are the covalence mixings (hybrids) of all initial states employing to the description. This mixing effect is called hybridization. So, from the very assumption the hybridization is a purely one–electron effect. Some modifications of the central potential of each lattice site resulting from differences between the bound in crystal and free ions are the source of the effect. These are the charge density distribution corresponding to valence states in metals, deformation of the open–shell states as well as overlap of the potentials coming from the neighbouring ions. The one–electron hybridization effect should be evidently distinguished from the two–electron interactions between the localized and delocalized electrons, as e.g. the exchange $s - d$ interaction which enters the Zener theory. The hybridization effect is determined, apart

177

from energy difference of the mixing states, by matrix elements of all possible interelectron interactions which in the frame of the second quantization formalism can be expressed in the form

$$\mathcal{H}_1 = \sum_m \sum_\mathbf{k} \sum_\sigma (V_{m\mathbf{k}} a^+_{m\sigma} c_{\mathbf{k}\sigma} + V_{\mathbf{k}m} c^+_{\mathbf{k}\sigma} a_{m\sigma}) \tag{15.1}$$

where $a^+_{m\sigma}$, $a_{m\sigma}$ are the creation and annihilation operators of the localized open–shell electron of the magnetic and spin quantum numbers m and σ, respectively, $c^+_{\mathbf{k}\sigma}$, $c_{\mathbf{k}\sigma}$ — the analogous operators for the conduction electron with the wave vector $\mathbf{k}$ and $V_{m\mathbf{k}}$ and $V_{\mathbf{k}m}$ are the corresponding matrix elements.

Besides the simple one–electron first–order terms in $\mathcal{H}_1$ several others, the operator part of which may be by means of contraction of creation and annihilation operators cast into the one–electron form of the type of a^+c or c^+a as in Eq.15.1 contribute to $\mathcal{H}_1$. The physical meaning of the interactions is inherent in the matrix element associated with the pair of operators. In crystal lattice, considering the obvious disagreement in symmetry of localized $a_{m\sigma}$ state and delocalized $c_{\mathbf{k}\sigma}$ Wannier function centered at the same site, non–zero matrix elements come primarily from the Wannier functions of conduction electrons centered at the neighbouring sites [208]. If the energy of interacting conduction electron E_k differs noticeably from the localized electron energy E_a the hybridization term is small enough to use the perturbation calculations for including the effect. However, this term becomes divergent if the energy reaches its resonance value ($E_k \approx E_a$) but then we enter the scope of qualitatively new phenomenon — the resonance scattering of the conduction electrons (chapter 14). In this chapter the discussion is limited to conditions being far from the resonance regime.

15.2 Hybridization contribution to the crystal field parameters

Quantitative analysis of the hybridization effect and its input to the crystal field potential may be clearly performed within the frame of the Anderson model [208, 181]. In this model the Hamiltonian consists of the three following terms

$$\mathcal{H} = \mathcal{H}_{\text{ion}} + \mathcal{H}_{\text{cond}} + \mathcal{H}_1 \tag{15.2}$$

where

$$\mathcal{H}_{\text{ion}} = \sum_m \sum_\sigma E_a \hat{n}_{m\sigma} + U \sum_m \sum_{m'} \hat{n}_{m\uparrow} \hat{n}_{m'\downarrow} +$$

$$+ (U - J) \sum_{m>m'} \sum_\sigma \hat{n}_{m\sigma} \hat{n}_{m'\sigma} + \mathcal{H}_{\text{CF}} + \zeta \sum_i \mathbf{l}_i \cdot \mathbf{s}_i \tag{15.3}$$

is the Hamiltonian of electrons of the atomic core open–shell,

$$\mathcal{H}_{\text{cond}} = \sum_{\mathbf{k}} \sum_{\sigma} E_{\mathbf{k}} \hat{n}_{\mathbf{k}\sigma} \tag{15.4}$$

is the Hamiltonian of the conduction band electrons, and $\mathcal{H}_1$ is the Hamiltonian of mutual interaction between the open–shell and conduction electrons defined previously (Eq.15.1), E_a and E_k denote energies of the states, $\hat{n}_{m\sigma}$, $\hat{n}_{m\uparrow}$, $\hat{n}_{m'\downarrow}$, $\hat{n}_{\mathbf{k}\sigma}$ are the number operators of states specified by the subscripts, ζ is the spin–orbit coupling constant for the open–shell electron, U and J stand for the intraatomic Coulomb integrals – direct and exchange, respectively, $\mathcal{H}_{\text{CF}}$ is the conventional crystal–field Hamiltonian; the meaning of the remaining symbols has been given previously (Eq.15.1).

The appropriate eigenfunctions and eigenvalues are obtained in this model through diagonalization of the Anderson Hamiltonian (Eq.15.2) matrix within the whole set of wave functions comprising both localized ($a_{m\sigma}$) and free ($c_{\mathbf{k}\sigma}$) electron states. The diagonalization of the Anderson Hamiltonian by using a suitable canonical transformation, e.g. the Schrieffer–Wolf transformation [223, 224] leads to the appearance in the effective Hamiltonian an extra term $\mathcal{H}_{\text{hbr}}$ acting within the open–shell electrons only defined as

$$\mathcal{H}_{\text{hbr}} = \sum_{m} \sum_{m'} \sum_{\mathbf{k}} \sum_{\sigma} V_{m\mathbf{k}} V_{\mathbf{k}m'} D(E_k) a^+_{m\sigma} a_{m'\sigma} \tag{15.5}$$

in which $V_{m\mathbf{k}}$ and $V_{\mathbf{k}m'}$ are the hybridization matrix elements and $D(E_k)$ reads

$$D(E_k) = \frac{1 - \hat{n}_{\mathbf{k}\sigma}}{E_- - E_{\mathbf{k}}} + \frac{\hat{n}_{\mathbf{k}\sigma}}{E_+ - E_{\mathbf{k}}} \tag{15.6}$$

where $\hat{n}_{\mathbf{k}\sigma}$ are, as previously, the occupation numbers of the conduction band states, E_- is the energy needed to promote one open–shell electron up to the Fermi level and E_+ the energy necessary to remove it from the Fermi level and place on the open–shell.

This additional potential (Eq.15.5) can be expressed in the standard form by means of components of the angular momentum operators of the open–shell electrons. To this end the following expression can be exploited [76]

$$\sum_{\sigma} a^+_{m\sigma} a_{m'\sigma} = \sum_{n} A_{n1} \langle lm | O^{(n)}_{m-m'} | lm' \rangle \sum_{i} O^{(n)}_{m-m'}(l_i) \tag{15.7}$$

where

$$A_{n1} = \frac{3(2n+1)}{[\langle l || O^{(n)} || l \rangle \langle \frac{1}{2} || O^{(1)} || \frac{1}{2} \rangle]^2} \tag{15.8}$$

The second subscript in A_{n1} coefficient being equal to 1 arises from required identity of spins for both states $|m\sigma\rangle$ and $|m'\sigma\rangle$, and $O^{(n)}_{m-m'}(l_i)$ are the

Smith–Thornley operators [33] equivalent to $C_q^{(k)}(\mathbf{r}_i/r_i)$ operators (chapter 2). They can be obtained from $C_q^{(k)}(\mathbf{r}_i/r_i)$ by replacing $\frac{x}{r}$, $\frac{y}{r}$ and $\frac{z}{r}$ by l_x, l_y and l_z angular momentum components with regard to, however, the lack of their commutation. The Smith–Thornley operators are closely related to the Stevens operator eigenvalues [35] but contrary to them they possess the transformational properties of the spherical tensors: $O_0^{(n)}$ transform as D_0, D_1, $D_2 \ldots$, for $n = 1, 2, \ldots$, respectively. Consequently, since

$$\langle lm'|O_{m-m'}^{(n)}|lm\rangle = (-1)^{l-m'} \begin{pmatrix} l & n & l \\ -m' & m-m' & m \end{pmatrix} \langle l||O^{(n)}||l\rangle \quad (15.9)$$

where

$$\langle l||O^{(n)}||l\rangle = \frac{1}{2^n}\left[\frac{(2l+n+1)!}{(2l-n)!}\right]^{1/2} \quad (15.10)$$

and [31]

$$\langle l||C^{(n)}||l\rangle = (-1)^l(2l+1)\begin{pmatrix} l & n & l \\ 0 & 0 & 0 \end{pmatrix} \quad (15.11)$$

we easily get the expression

$$\sum_\sigma a_{m\sigma}^+ a_{m\sigma} = \sum_n (-1)^m \begin{pmatrix} l & n & l \\ -m & m-m' & m' \end{pmatrix} \times$$
$$\times \frac{2n+1}{(2l+1)\begin{pmatrix} l & n & l \\ 0 & 0 & 0 \end{pmatrix}} \sum_i C_{m-m'}^{(n)}(\mathbf{r}_i/r_i) \quad (15.12)$$

Hence

$$\mathcal{H}_{\mathrm{hbr}} = \sum_m \sum_{m'} \sum_{\mathbf{k}} V_{m\mathbf{k}} V_{\mathbf{k}m'} D(E_k) \sum_n (-1)^m \times$$
$$\times \frac{2n+1}{2l+1} \frac{\begin{pmatrix} l & n & l \\ -m & m-m' & m' \end{pmatrix}}{\begin{pmatrix} l & n & l \\ 0 & 0 & 0 \end{pmatrix}} \times$$
$$\times \sum_i C_{m-m'}^{(n)}(\mathbf{r}_i/r_i) \quad (15.13)$$

The coefficients preceding the $\sum_i C_{m-m'}^{(n)}(\mathbf{r}_i/r_i)$ operators are the hybridization contribution to the crystal parameters in the Wybourne notation, $(B_{n,m-m'})_{\mathrm{hbr}}$.

Thus

$$(B_{n,m-m'})_{\text{hbr}} = \sum_m \sum_{m'} \sum_{\mathbf{k}} V_{m\mathbf{k}} V_{\mathbf{k}m'} D(E_k)(-1)^m \times$$

$$\times \quad \frac{2n+1}{2l+1} \frac{\begin{pmatrix} l & n & l \\ -m & m-m' & m' \end{pmatrix}}{\begin{pmatrix} l & n & l \\ 0 & 0 & 0 \end{pmatrix}} \tag{15.14}$$

The summation over m and m' runs over all their values, the required difference $m - m'$ is automatically ensured by the presence of the $3 - j$ symbol in the nominator of the expression.

A precise evaluation of the $D(E_k)$ and $V_{m\mathbf{k}}$ quantities would require advanced band structure calculations. In general, the matrix elements of hybridization can be found based on the band structure calculations by the local density approximation (LDA) method. In many cases, especially when the hybridization effects are weaker the tight binding approach proposed by Harrison [225, 226] may be used. The hybridization matrix elements or in other words the intersitial transfer integrals has within the tight binding model and in the local coordinate system for the central ion and a chosen ligand the following form

$$V_{ll'm} = \eta_{ll'm} \frac{\hbar^2}{m_e} \frac{(r_l^{2l+1} r_{l'}^{2l'+1})^{1/2}}{d^{l+l'+1}} \tag{15.15}$$

in which the conduction electrons are classified from their angular momentum quantum number l' according to expansion of a plane wave in spherical waves, and m reveals the symmetry of the bond with $m = 0, 1, 2, 3$ for σ, π, δ and φ bonds, respectively.

The input parameters are the atomic radii of the respective atoms r_l and $r_{l'}$ and the interatomic distance d, $\hbar$ is the Planck constant divided by 2π and m_e the electron mass.

The angular coefficient $\eta_{ll'm}$ is given by the expression

$$\eta_{ll'm} = \frac{(-1)^{l+1}}{6\pi} \frac{(l+l')!(2l)!(2l')!}{2^{l+l'} l! l'!} \times$$

$$\times \quad (-1)^m \left[\frac{(2l+1)(2l'+1)}{(l+m)!(l-m)!(l'+m)!(l'-m)!} \right]^{1/2} \tag{15.16}$$

Fortunately, different hybridization energies of the same type (ll') as for example $f - d\sigma$, $f - d\pi$, $4f - d\delta$ can be in the case of sufficient localization of the considered functions approximately correlated by means of relationship descending from Eqs 15.15 and 15.16 [182]

$$V_{ll'm} \sim S_{ll'm} \sim (-1)^{l'-m}[(l+m)!(l-m)!(l'+m)!(l'-m)!]^{1/2} \tag{15.17}$$

where $S_{ll'm}$ is the corresponding overlap integral $\langle lm|l'm \rangle$, l represents the open–shell electron wave function whereas l' that of the hybridizing band electron.

There is another correlation between the hybridization integrals and the width of the interacting bands [227], an example of which is postulated proportionality of the $f-d$ covalence contribution to the geometric mean of the d–band and f–state width, Δ_d and Δ_f, respectively

$$V_{fd} \sim (\Delta_f \Delta_d)^{1/2} \tag{15.18}$$

The above correlations remarkably facilitate theoretical estimations of hybridization matrix elements which otherwise are difficult of attainment.

The total hybridization effect is the sum of the intrinsic effects over all neighbours with regard to their coordination factors in the main coordinate system. $\sum_{\mathbf{k}} D(E_k)$ can be calculated by using the well known relationship

$$
\begin{aligned}
\sum_{\mathbf{k}} D(E_k) &= \sum_{\mathbf{k}} \left(\frac{1-\hat{n}_k}{E_- - E_k} + \frac{\hat{n}_k}{E_+ - E_k} \right) \times \\
&= \int \varrho(E) \left(\frac{1-\hat{n}(E)}{E_- - E} + \frac{\hat{n}(E)}{E_+ - E} \right) dE
\end{aligned}
\tag{15.19}
$$

where $\varrho(E)$ is the density of states of the conduction electrons and the integration concerns the energy range corresponding to the $\mathbf{k}$ space in the sum on the left side. Sometimes $D(E_k)$ are replaced by the adjustable parameters D_σ, D_π, etc. for the $s-$ and p–electrons, respectively.

15.3 The scale of the hybridization effect

The hybridization mechanism is one of three basic sources of the crystal field potential in metals. The contribution of this effect is comparable with the screened electrostatic contribution of the crystal lattice and virtual bound state input (chapter 14). The weight of the effect will be illustrated for example of crystal field parameters of Er^{3+} ion ($4f^{11}$ configuration) embedded into the cubic crystal matrices of Al, Cu and Au of the face centered structure A_1.

The experimental values of the parameters B_{40} and B_{60} and the hybridization contribution to them are presented in Table 15.1 [182]

Table 15.1 Hybridization contribution to the crystal field parameters of Er^{3+} ion in cubic matrices of Al, Cu and Au, all values in cm^{-1}

Crystal field parameter	Metallic matrix	Experimental value	Ref.	Hybridization contribution
B_{40}	Al	−380	[228]	0
	Cu	−481	[229]	−117
	Au	−185	[186]	−185
B_{60}	Al	163	[228]	19
	Cu	134	[229]	68
	Au	72	[186]	44

The representative hybridization matrix elements used in the calculation for Cu are 3390 and 1130 cm^{-1} for V_{fs} and $V_{fd\sigma}$, respectively.

The hybridization contribution to the crystal field splitting of the s-band which usually dominates in the vicinity of the Fermi level turns out to be important in the case of the sixth order crystal field terms and negligible for the fourth order ones (see Al matrix). In turn, the covalent mixing with the d-band predominantly contributes to the fourth order terms. Let us note that the contribution to the fourth order parameters is negative and to the sixth order ones – positive.

The hybridization effect of virtual bound states, if any, with the conduction band states is an additional significant mechanism affecting the actual crystal field potential. If it is the $5d$ state, as for the lanthanide ions, its contribution to the fourth–order crystal field parameters is large indeed (chapter 14), frequently dominating, but negative one. Its hybridization correction as positive quantity cancels partly the enormous direct input of the $5d$ virtual bound state.

The covalent mixing as the notion of wider range than the hybridization is not confined to the conduction band states only and can also occur with the valence bands. This is another important mechanism particularly in systems with empty conduction band. This problem was considered in chapter 8. Based on this mechanism, and more strictly on the $p - f$ mixing mechanism Takahashi and Kasuya [230, 231] have explained anomalous magnetic properties in Ce monopnictides. To end with it should be once more emphasized that any two–electron interaction as e.g. the so-called $s - d$ Zener interaction [127] or Coqblin–Schrieffer interaction [232, 233] has no connection with the hybridization which is ex definitione one–electron effect.

15.4 Contribution to the crystal field potential from a split–off state from the conduction band in impurity systems

In impurity systems in which the substitutional admixture ion carrying a net difference in charge from the host ion it substitutes in perfectly periodic host lattice another one mechanism of interaction of electrons localized on the admixture and conduction electrons occurs. In such an impurity system at least one discrete energy level separates from the host band. The split–off discrete state has A_1–like symmetry, i.e. it is invariant under all the operations of the point group of the lattice site being occupied by the impurity [234, 70] but, which is essential, not rotationally invariant. This split–off state in its zeroth order of approximation is a generalized Wannier function originating from conduction band of the crystal matrix which would be centered at the admixture site in the perfect host (without the admixture). The lack of spherical symmetry of the split–off state is a consequence of contributions to the zero approximation function, e.g. the s–like Wannier function, other states of A_1 symmetry coming from the conduction band as well as from non–orthogonality of the specified state with respect to the states localized at the neighbouring atomic cores. This non–orthogonality can be included with a sufficient accuracy by using the orthogonalizing procedure given by Löwdin [218].

The electron density connected with the discrete split–off state interacts in both direct and exchange ways with electrons of the open–shell of the admixture atom producing an extra contribution to the crystal field potential. The calculation formalism is similar to that used in the case of virtual bound state interaction. Due to a rather large values of overlap integrals of the generalized Wannier function and the wave functions of the open–shell electrons of the admixture the exchange part of the Coulomb interaction is relatively important.

CHAPTER 16

Density functional theory approach

Density functional theory (DFT) is one of the most powerful, dynamically developed theory that makes possible a reliable description of complex many-electron systems in a relatively simple and elegant form. It provides a general framework for many different applications in all kind of electronic systems.

Although main interest in DFT research seems to concern the itinerant systems an increasing number of papers regarding the localized 4f and 5f states subjected to the CF interaction points to its growing significance also in this area. The literature connected with both development and application of the theory is very reach. For limited needs of the present sketch, it is neither possible nor advisable to trace whole variety of investigations and their rigorous mathematical foundations. We concentrate rather on these aspects of the theory, which are directly related to the CF effect. A more comprehensive description of the theory and all relevant topics can be found in many review papers and textbooks - see for example [235, 236, 237]. To start with, we recall the main ideas and definitions of DFT. These enable presentation of two main problems developed thereafter: various approximations and improvements, which make DFT applicable to real systems and its connection with the effective Hamiltonian approach and experimental data. The effectiveness of the theory is illustrated by some examples of recent numerical results reported in the literature and, finally, some shortcomings and open questions are outlined. Apart from the mentioned above monographs and numerous original papers, we have found very helpful during preparation of this chapter theses by Richter [238] and Trygg [239].

16.1 Electron density as a key variable

To start with we rewrite the general Hamiltonian of a system of N interacting electrons moving in electrostatic potential of nuclei in a solid in the Born–Oppenheimer approximation (atomic units are used throughout this chapter).

$$\hat{H} = \hat{T} + \hat{U} + \hat{V}, \tag{16.1}$$

where

$$\hat{T} = \sum_i^N \hat{t}_i = \sum_i^N -\frac{\nabla_i^2}{2}, \tag{16.2}$$

$$\hat{U} = \frac{1}{2} \sum_{i,j \neq i}^N \frac{1}{|\mathbf{r}_i - \mathbf{r}_j|}, \tag{16.3}$$

$$\hat{V} = \sum_i^N v(\mathbf{r}_i) \left(= \sum_i^N \sum_t \frac{-Z_t}{|\mathbf{r}_i - \mathbf{R}_t|} \right). \tag{16.4}$$

The potential $\hat{V}$ in Eq.16.4 may be specified to be the Coulomb potential due to the point nuclei but the general theory admits a wide class of arbitrary multiplicative single-particle operators. Leaving imprecise $\hat{V}$ facilitates presentation of the theory.

Thus we stay with the general Hamiltonian Eqs 16.1-16.4 describing a large variety of the electronic systems. It is noteworthy that only the number of electrons N and $v(\mathbf{r})$ are needed to fix it completely and to determine all the electronic properties related to the ground and excited states. In the traditional quantum mechanics the many-electron wave function is the quantity that describes the state of the system. The density functional theory provides an alternative, and complementary, approach. In this reformulation of wave mechanics, the density of the electrons is a fundamental quantity. The complicated N-electron wave function $\Psi(\mathbf{x}_1, \mathbf{x}_2, \ldots, \mathbf{x}_N)$ (where $\mathbf{x}_i$ comprises space coordinate $\mathbf{r}_i$ and spin coordinate σ_i) defined in the $3N$-dimensional real space and the N-dimensional spin space is replaced with much simpler quantity $n(\mathbf{r})$ defined in merely the three–dimensional real space.

Density functional theory is based on two rudimental theorems formulated and proved by Hohenberg and Kohn (HK) [240], both disarmingly simple. The first theorem justifies the use of the electron density $n(\mathbf{r})$ as a basic variable. It is obvious that $v(\mathbf{r})$ determines $n(\mathbf{r})$:

$$v(\mathbf{r}) \to \hat{H} \to \Psi(\mathbf{x}_1, \mathbf{x}_2, \ldots, \mathbf{x}_N) \to n(\mathbf{r}) \tag{16.5}$$

as

$$n(\mathbf{r}) \equiv N \int \ldots \int \Psi(\mathbf{r}_1'\sigma_1, \mathbf{x}_2, \ldots, \mathbf{x}_N) \Psi^*(\mathbf{r}_1\sigma_1, \mathbf{x}_2, \ldots, \mathbf{x}_N) d\sigma_1 d\mathbf{x}_2 \ldots d\mathbf{x}_N \tag{16.6}$$

Here $d\mathbf{x}$ stands for $d\mathbf{r}d\sigma$ that symbolizes integration over $\mathbf{r}$ and summation over spin variable σ. The HK theorem states that, conversely, $v(\mathbf{r})$ *is a unique functional of* $n(\mathbf{r})$ within a trivial additive constant. In the other words, the ground state density is *v-representable*: the specification of the ground state density, $n(\mathbf{r})$, determines the external potential $v(\mathbf{r})$ uniquely. This is very strong theorem. Thus, there is one-to-one correspondence between n and v. Since $n(\mathbf{r})$, as it follows from Eq.16.6, fixes also the number of electrons,

$$N = N[n(\mathbf{r})] \equiv \int n(\mathbf{r}) d\mathbf{r} \qquad (16.7)$$

it determines the ground state wave function Ψ and the energy of the system:

$$E = E[n(\mathbf{r})] \equiv \left\langle \Psi[n(\mathbf{r})] \left| \hat{H} \right| \Psi[n(\mathbf{r})] \right\rangle = \int v(\mathbf{r}) n(\mathbf{r}) d\mathbf{r} + F[n(\mathbf{r})], \quad (16.8)$$

where all beyond the first explicit term is formally included in $F[n(\mathbf{r})]$:

$$F[n(\mathbf{r})] \equiv \left\langle \Psi[n(\mathbf{r})] \left| \hat{T} + \hat{U} \right| \Psi[n(\mathbf{r})] \right\rangle = T[n(\mathbf{r})] + U[n(\mathbf{r})] \qquad (16.9)$$

$F[n(\mathbf{r})]$ is a *universal functional* of $n(\mathbf{r})$ in the sense that it does not depend on $v(\mathbf{r})$ directly. Having the functional $F[n(\mathbf{r})]$ established in an explicit form, its value for any system can be derived by considering various trial densities.

The second HK theorem corresponds to the variational principle for the wave functions. In the traditional approach such a variational principle leads to the usual Schrödinger equation. The theorem states that *the energy functional $E[n(\mathbf{r})]$ in Eq.16.8 attains its minimum for the true ground state electron density*. Hence, the density must obey the Euler–Lagrange equation associated with the stationarity of $E[n(\mathbf{r})]$ and subsidiary condition Eq.16.7

$$\frac{\delta(E[n(\mathbf{r})] - \mu N[n(\mathbf{r})])}{\delta n(\mathbf{r})} = 0. \qquad (16.10)$$

This can be rewritten in the form:

$$\mu \equiv v(\mathbf{r}) + \frac{\delta F[n(\mathbf{r})]}{\delta n(\mathbf{r})} \qquad (16.11)$$

where μ is the Lagrange multiplier that can be identified with the zero-temperature limit of the chemical potential defined for the finite-temperature grand-canonical ensemble. Comprehensive discussion of the chemical potential in context of the density functional theory is contained in textbook by Parr and Yang [236].

Eq.16.11 is the basic equation of the density functional theory. It determines the electron density of the system. Note that the left–hand side of Eq.16.11 does not depend on $\mathbf{r}$.

The universal functional $F[n(\mathbf{r})]$, Eq.16.9, has an essential meaning in this theory but it is hard to come by in an explicit form. The large part of activity in DFT is concentrated on finding more or less approximate formula for $F[n(\mathbf{r})]$. A few efficient ways in this searching have been traced in the original Hohenberg-Kohn work [240] and the paper by Kohn and Sham [90].

16.2   The Kohn–Sham equations

The Kohn–Sham (KS) idea in its essence is to extract from the searched quantity $F[n(\mathbf{r})]$ the largest components which are known beforehand to be free to concentrate on further approximations to the remainder. The Hartree energy being the essential part of the interelectronic interaction energy $U[n]$ is a good example of such a component:

$$J[n] = \frac{1}{2} \int \frac{n(\mathbf{r})n(\mathbf{r}')}{|\mathbf{r}' - \mathbf{r}|} d\mathbf{r} d\mathbf{r}'. \tag{16.12}$$

In order to find other quantities of this type, Kohn and Sham have considered two exactly solvable N-electron systems:
(i) the reference system of non–interacting electrons,

$$\hat{H}_s = \hat{T} + \sum_i^N v_s(\mathbf{r}_i) \equiv \hat{T} + \hat{V}_s, \tag{16.13}$$

(ii) the homogeneous electron gas – a system of electrons moving under no external forces in cubical box of volume Ω with the density $n_0(\mathbf{r}) = N/\Omega$ remaining finite in the limit $N \rightarrow \infty$, $\Omega \rightarrow \infty$ (see [204])

$$\hat{H}_h = \hat{T} + \hat{U} - \sum_i^N \int \frac{\tilde{n}_0(\mathbf{r})}{|\mathbf{r}_i - \mathbf{r}|} d\mathbf{r}. \tag{16.14}$$

The last term in Eq.16.14 is the external potential due to the uniformly spread out positive charge density $\tilde{n}_0(\mathbf{r}) = N/\Omega$, where Ω that makes the system neutral everywhere, i.e. $\tilde{n}_0(\mathbf{r}) = n_0(\mathbf{r})$ at each point $\mathbf{r}$. This term is exactly cancelled by the greatest contribution to $\hat{U}$ – the Hartree potential. The homogeneous electron gas is the only model of a many-electron interacting system, the ground state energy of which is known precisely. Valence-electron gas in light metals is the closest realistic system, which can be approximated by the model fairly well.

First, we follow Kohn and Sham to incorporate the known properties of the non-interacting system (i). Their treatment is based on the HS theorems applied to this reference system. Let $n(\mathbf{r})$ be the true ground state electron density of the system described by the Hamiltonian Eq.16.1. Since, according to the first HK theorem, $v(\mathbf{r})$ is a unique functional of $n(\mathbf{r})$ for arbitrary many-particle system, there must exist also a system of non-interacting electrons with Hamiltonian $\hat{H}_s$, Eq.16.13, determined precisely by the same density $n(\mathbf{r})$:

$$v_s(\mathbf{r}) = v_{eff}(\mathbf{r}) \equiv v_{eff}(n(\mathbf{r}), \mathbf{r}). \tag{16.15}$$

From the elementary quantum mechanics we know that the exact ground state wave function for such a system has the form of single determinant,

$$\Psi_s = \frac{1}{\sqrt{N!}} \det[\psi_1 \psi_2 \ldots \psi_N], \tag{16.16}$$

where $\psi_i \equiv \psi_i(\mathbf{r}, \sigma)$ are the N lowest eigenstates of the one-electron Hamiltonian $\hat{h}_s$

$$\hat{h}_s \psi_i = \epsilon_i \psi_i, \tag{16.17}$$

$$\hat{h}_s = -\frac{\nabla^2}{2} + v_{eff}(\mathbf{r}). \tag{16.18}$$

The one-to-one correspondence between the electron density and the external potential in both the real system and the corresponding non-interacting system allows one to obtain the exact electronic structure data for the former by solving much simpler one-electron equations 16.17. Thus, the problem is reduced to a searching for an effective potential $v_{eff}(\mathbf{r})$. The electron density is the only common quantity connecting the initial and reference systems. It is expressible nicely in terms of Kohn-Sham orbitals $\psi_i(r, \sigma)$:

$$n(\mathbf{r}) = n_s(\mathbf{r}) = \sum_i^N \sum_\sigma |\psi_i(r, \sigma)|^2. \tag{16.19}$$

Formally, ψ_i are purely mathematical constructions. They do not describe the motion of the electrons as well as their total energy is not equal to the sum of the one-particle energies ϵ_i. Nevertheless, their physical sense can be seen in analogy with the single-particle wave-functions obtained from the Slater $X\alpha$ equations [247].

In order to find $v_{eff}(\mathbf{r})$, note first that there is another quantity that can be calculated very easily from the KS orbitals – the kinetic energy for the virtual non-interacting system:

$$T_s[n] = \left\langle \Psi_s[n] \left| \hat{T} \right| \Psi_s[n] \right\rangle = \sum_i^N \left\langle \psi_i \left| -\frac{\nabla^2}{2} \right| \psi_i \right\rangle. \tag{16.20}$$

$T_s[n]$ can be shown (see for example Ref.[241]) to be the dominating part of the true kinetic energy $T[n]$. Formally, the unknown functional $F[n]$, Eq.16.9, can be written now as

$$F[n] = T_s[n] + J[n] + E_{xc}[n] \tag{16.21}$$

where

$$E_{xc}[n] = T[n] - T_s[n] + U[n] - J[n] \tag{16.22}$$

is so called the *exchange-correlation energy*. $E_{xc}[n]$ contains the whole information beyond the mean field approximation about the exchange and correlation interaction of the electrons in the ground state. The variational principle of Hohenberg and Kohn Eq.16.11 with $F[n]$ given by Eq.16.21 can be now rewritten as:

$$\mu = \frac{\delta T_s[n]}{\delta n} + v(\mathbf{r}) + v_H(n; \mathbf{r}) + \frac{\delta E_{xc}[n]}{\delta n} \tag{16.23}$$

where all terms except $\frac{\delta E_{xc}[n]}{\delta n}$ are known including the well-known Hartree potential $v_H(n; \mathbf{r})$,

$$v_H(n; \mathbf{r}) = \int \frac{n(\mathbf{r'})}{|\mathbf{r'} - \mathbf{r}|} d\mathbf{r'}. \tag{16.24}$$

The corresponding equation for the non-interacting system with the same electron density distribution and the effective potential $v_{eff}(\mathbf{r})$ seems simpler:

$$\mu_s = \frac{\delta T_s[n]}{\delta n} + v_{eff}(\mathbf{r}). \tag{16.25}$$

Comparing Eq.16.25 and Eq.16.23 we see that, within the additive constant $\mu - \mu_s$,

$$v_{eff}(\mathbf{r}) = v(\mathbf{r}) + v_H(n, \mathbf{r}) + \frac{\delta E_{xc}[n]}{\delta n}. \tag{16.26}$$

The equations 16.17-16.19 and 16.26 assembled altogether represent the famous Kohn–Sham equations:

$$(-\frac{\nabla^2}{2} + v_{eff}(\mathbf{r}))\psi_i = \epsilon_i \psi_i, \tag{16.27}$$

$$n(\mathbf{r}) = \sum_i^N \sum_\sigma |\psi_i(r, \sigma)|^2, \tag{16.28}$$

$$v_{eff}(\mathbf{r}) = v(\mathbf{r}) + v_H(n; \mathbf{r}) + v_{xc}(n; \mathbf{r}), \tag{16.29}$$

where

$$v_{xc}(n; \mathbf{r}) = \frac{\delta E_{xc}[n]}{\delta n} \tag{16.30}$$

is the *exchange-correlation potential* defined as a functional derivative of $E_{xc}[n]$, Eq. 16.22. Like $E_{xc}[n]$, it contains extremely awkward terms being subjected to various approximations discussed further. The total energy of the system can be determined from the resultant density Eq.16.28 either directly from Eq.16.8 and Eq.16.21 or in terms of the eigenvalues ϵ_i for the non-interacting reference system according to the formula:

$$E = \sum_i^N \epsilon_i - J[n] + E_{xc} - \int n(\mathbf{r}) v_{xc}(n; \mathbf{r}) d\mathbf{r} \tag{16.31}$$

16.3   Local density approximation

For now no approximations have been made. The Kohn-Sham equations Eq.16.27-16.30 are non-linear in n and they have to be solved self-consistently starting with a guessed initial density. They resemble the Hartree equations but contrary to the latter they are exact - at least formally. In the density functional theory not only the exchange effects are taken into account as they are in the much more complicated Hartree-Fock theory but also the dynamic electron correlation effects. Of course, one needs to know $E_{xc}[n]$ precisely and this is a real problem. However, very existence of the exact theory provides impetus for searching new ideas and developing more and more accurate procedures.

The beginning in this work also belongs to Kohn and Sham [90]. Their widely celebrated *local density approximation* (LDA) is founded on the well-known properties of the reference electron system (ii), Eq.16.14, – the homogeneous electron gas. Practically, the exchange-correlation energy for this system is known within arbitrary precision for any uniform density n_0 [242, 243]. This is of major importance since in the local density approximation, the true $E_{xc}[n]$ is approximated *locally* with the *exchange-correlation energy-density* $g_{xc}^h[n_0]$ obtained for this reference system:

$$E_{xc}[n] \simeq E_{xc}^{LDA}[n] = \int g_{xc}^h[n_0 = n(\mathbf{r})]d\mathbf{r}. \qquad (16.32)$$

The explicit expression for the energy-density $g_{xc}^h[n_0]$ comprises the separated exchange and correlation contributions:

$$g_{xc}^h[n_0] = n_0(\varepsilon_x^h(n_0) + \varepsilon_c^h(n_0)), \qquad (16.33)$$

where $\varepsilon_x^h(n_0)$ is the exchange energy per electron given by the well-known formula [244]:

$$\varepsilon_x^h(n_0) = -\frac{3}{4}\left(\frac{3}{\pi}n_0\right)^{1/3} \qquad (16.34)$$

The corresponding expression for the correlation energy, $\varepsilon_c^h(n_0)$, as obtained by Ceperley and Alder [242] using quantum Monte Carlo method, has much more complicated form:

$$\begin{aligned}
\varepsilon_c^h(n_0) = {} & \frac{A}{2}\{\ln\frac{x^2}{X(x)} + \frac{2b}{Q}\tan^{-1}\frac{Q}{2x+b} \\
& - \frac{bx_0}{X(x_0)}\left[\ln\frac{(x-x_0)^2}{X(x)} + \frac{2(b+2x_0)}{Q}\tan^{-1}\frac{Q}{2x+b}\right]\}
\end{aligned} \qquad (16.35)$$

where

$$x = \left(\frac{3}{4\pi n_0}\right)^{1/6}, \quad X(x) = x^2 + bx + c, \quad Q = (4c - b^2)^{1/2}, \qquad (16.36)$$

and the numerical coefficients are following: $A = 0.0621814$, $x_0 = -0.409286$, $b = 13.0720$, $c = 42.7198$. This solution for the simplest many-electron system conveys also an idea of how complex the functional dependence must be for realistic systems.

The KS equations Eq.16.27-16.30 with approximation Eq.16.32 for the exchange-correlation energy and the corresponding $v_{xc}^{LDA}(n; \mathbf{r})$ potential,

$$v_{xc}^{LDA}(n; \mathbf{r}) = \varepsilon_{xc}^{h}(n(\mathbf{r})) + n(\mathbf{r})\frac{\delta\varepsilon_{xc}^{h}(n(\mathbf{r}))}{\delta n(\mathbf{r})} \qquad (16.37)$$

provide an efficient scheme for calculation of all properties of arbitrary many-electron system. Kohn and Sham have shown that, principally, LDA is justified in two limiting cases: slowly varying density and high-density limits. They have argued that the approximation is not expected to be very accurate in the highly inhomogeneous region in the vicinity of ionic sphere - the region of great importance for the crystal field effect. A considerable part of activity of theorists has been concentrated to extend the range of applicability of LDA, mainly by searching for more accurate representation of exchange correlation energy. The eventual justification for this kind of findings always stems from an appealing to a satisfactory agreement with the experimental data. Some achievements in this field are outlined in next sections followed by reformulation of the theory for more realistic systems.

16.4 Extensions

Among numerous examples of application of DFT, the most attractive ones in the context of CF effect, concern the intermetallic, semiconductive and ionic rare-earth and actinide compounds. It is clear that for these systems the original Hohenberg-Kohn-Sham formulation needs to be extended to account for degenerate ground state of arbitrary symmetry, relativistic effects and various types of particles in a system. In addition, a rigorous determination of the CF transitions requires incorporation of the electronic excitations. As we shall see in the following subsections, the density functional theory is very flexible in this matter.

16.4.1 Degenerate ground state and excited states

The localized states in solids behave like atomic states under an external potential. We have a sequence of degenerate CF levels each of which is formally associated with different electron density. Thus, extension of the density functional theory to the degenerate ground state has to be the first step towards the excited states. The corresponding reformulation of the theory is rather moderate. First, the requirement for the non-degenerate ground state imposed by Hohenberg and Kohn is not necessary to prove the basic theorem that maps uniquely n to v [246]. Their assumption ensures one-to-one correspondence between these two quantities and even between n and ground state Ψ. In order to restore the v to n mapping in the degenerate case it is enough to redefine the universal functional $F[n(\mathbf{r})]$ as a result of a *constrained search* for the minimum of the expectation value of $\hat{T} + \hat{U}$ over all N-electron wave functions that give the ground state density $n(\mathbf{r})$:

$$F[n(\mathbf{r})] \equiv \min_{\Psi \to n} \left\langle \Psi \left| \hat{T} + \hat{U} \right| \Psi \right\rangle. \tag{16.38}$$

Levy [245] and independently Lieb [246] have shown that the Hohenberg and Kohn theorems remain intact with this functional. Their constrained search comprises a wide class of wave functions but it selects only one set of degenerate wave functions corresponding to $n(\mathbf{r})$ – just one that minimizes the expectation value. Moreover, the Levy-Lieb functional Eq.16.38 is not restricted to the ground state. It can be considered for any subspace of the Hilbert space including these related to the excited states, thus making possible rigorous formulation of the excited state theory sketched below.

The first step towards excited states belongs to Slater [247] who introduced fractional occupation numbers for the single-particle states in his famous $X\alpha$ method. This idea has been adopted for DFT by Janak [248] and has found its justification in the subspace theory developed by Theophilou [249]. Earlier, Gunnarsson and Lundqvist [250] considered subspaces of the Hilbert space of specific angular momentum and spin symmetries. They extended the Hohenberg-Kohn-Sham (HKS) scheme so as to get the energy of the (degenerate) lowest state in each subspace, demonstrating the one-to-one correspondence between the density and the lowest-energy eigenstate in the subspace. Formally, such an approach makes E_{xc} and v_{xc} symmetry dependent. In addition, a problem arises with the single-determinant correspondence in the non-interacting system. In order to circumvent these difficulties von Barth [251] introduced states of mixed symmetry which correspond to Slater determinants of the non-interacting electrons with the same density. The mentioned general and mathematically rigorous formulation of DFT for excited states by Theophilou [249] justifies also the von Barth approach (see Ref.[252]).

The density-variable in the Theophilou subspace theory, called *ensemble density*, is defined as a sum of M lowest-energy eigenstate densities equally weighted. Herein we follow more general formulation developed by Gross, Oliveira and Kohn, [253], that admits non-equally weighted ensemble densities.

Let Ψ_{ik}, $(i = 1, 2, \ldots, \mathcal{I}, \quad k = 1, 2, \ldots, g_i)$ be the $\mathcal{M}_{\mathcal{I}} = \sum\limits_{i=1}^{\mathcal{I}} g_i$ orthonormal lowest-energy many-electron eigenstates of the Hamiltonian Eq.16.1:

$$\hat{H}\Psi_{ik} = E_i\Psi_{ik}, \quad E_1 < E_2 < \ldots < E_{\mathcal{I}}, \tag{16.39}$$

which correspond to $\mathcal{I}$ multiplets of degeneracies g_i. The ensemble of functions Ψ_{ik} spans a $\mathcal{M}_{\mathcal{I}}$-dimensional subspace of the Hilbert space that is invariant under $\hat{H}$. The original idea of Theophilou was to redefine for this subspace all the main quantities met in the HKS density functional theory that has been formulated for the one-dimensional subspace spanned by the non-degenerate ground state. According to Gross, Oliveira and Kohn [253], the ensemble density is defined as follows:

$$\tilde{n}_{\mathcal{I}}(\mathbf{r}, w) \equiv \sum_{i=1}^{\mathcal{I}} \sum_{k=1}^{g_i} w_{ik} n_{ik}(\mathbf{r}) \tag{16.40}$$

$$w_{ik} = w_{ik}(w) \equiv \left\{ \begin{array}{ll} \frac{1-wg_{\mathcal{I}}}{\mathcal{M}_{\mathcal{I}-1}} & \text{for} \quad i = 1, 2, \ldots, \mathcal{I} - 1 \quad \text{and} \quad k = 1, 2, \ldots, g_i \\ w & \text{for} \quad i = \mathcal{I} \quad \text{and} \quad k = 1, 2, \ldots, g_{\mathcal{I}} \end{array} \right\} \tag{16.41}$$

where $n_{ik}(\mathbf{r})$ is given by Eq.16.6 for Ψ replaced with Ψ_{ik}. w is a real parameter in the range $0 \leq w \leq 1/\mathcal{M}_{\mathcal{I}}$. The limit $w = 0$ corresponds to the equiensembles of $\mathcal{M}_{\mathcal{I}-1}$ states with $w_{i(\leq \mathcal{I}-1)k} = 1/\mathcal{M}_{\mathcal{I}-1}$, $w_{\mathcal{I}k} = 0$, whereas the other limit, $w = 1/\mathcal{M}_{\mathcal{I}}$, yields the equiensemble of $\mathcal{M}_{\mathcal{I}}$ states: $w_{ik} = 1/\mathcal{M}_{\mathcal{I}}$. Subsequently, the ensemble energy and the ensemble universal functional are defined as:

$$\mathcal{E}_{\mathcal{I}}[w, \tilde{n}_{\mathcal{I}}] \equiv \sum_{i=1}^{\mathcal{I}} \sum_{k=1}^{g_i} w_{ik} E_i = \int \tilde{n}_{\mathcal{I}}(\mathbf{r}, w) v(\mathbf{r}) d\mathbf{r} + \mathcal{F}_{\mathcal{I}}[w, \tilde{n}], \tag{16.42}$$

$$\mathcal{F}_{\mathcal{I}}[w, \tilde{n}_{\mathcal{I}}] \equiv \min_{\{\Psi_m\} \to \tilde{n}_{\mathcal{I}}} \left\{ \sum_{m=1}^{\mathcal{M}_{\mathcal{I}}} w_m \left\langle \Psi_m \left| \hat{T} + \hat{U} \right| \Psi_m \right\rangle \right\}, \tag{16.43}$$

where the weights are re-indexed according to the rule

$$w_m = w_{ik} \quad \text{for} \quad m = \sum_{i'=1}^{i-1} g_{i'} + k. \tag{16.44}$$

The minimum in Eq.16.43 is taken over all sets $\{\Psi_m\}$ of $\mathcal{M}_{\mathcal{I}}$ arbitrary orthonormal N-electron functions yielding the ensemble density $\tilde{n}_{\mathcal{I}}$.

It can be shown that the HK theorems hold for the above quantities: the potential $\hat{V}$ is uniquely (within a constant) determined by the ensemble density $\tilde{n}$, $\mathcal{F}_{\mathcal{I}}[w, \tilde{n}]$ is a unique functional of $\tilde{n}$, and the ensemble energy $\mathcal{E}_{\mathcal{I}}[w, \tilde{n}]$ attains its minimum for any density $\tilde{n}$ associated with the potential $\hat{V}$.

In order to derive the KS scheme for the ensemble density, the auxiliary noninteracting system with Hamiltonian Eq.16.13 has to be constructed so as to generate the $\mathcal{M}_{\mathcal{I}}$ determinantal eigenfunctions $\Psi_{s,m}$,

$$\hat{H}_s \Psi_{s,m} = E_{s,m} \Psi_{s,m} \quad E_{s,1} \leq E_{s,2} \leq \ldots \leq E_{s,\mathcal{M}_{\mathcal{I}}} \tag{16.45}$$

yielding the ensemble density equal to that of the interacting system:

$$\tilde{n}_{\mathcal{I}}(\mathbf{r}, w) = \tilde{n}_{s,\mathcal{M}_{\mathcal{I}}}(\mathbf{r}, w) \equiv \sum_{m=1}^{\mathcal{M}_{\mathcal{I}}} w_m n_{s,m}(\mathbf{r}) \tag{16.46}$$

where $n_{s,k}(\mathbf{r})$ is given by Eq.16.6 for Ψ replaced with $\Psi_{s,m}$. For this system the ensemble kinetic energy,

$$T_s[w, \tilde{n}_I] \equiv \min_{\{\Psi_{s,m}\} \to \tilde{n}_I} \left\{ \sum_{m=1}^{M_I} w_m \left\langle \Psi_{s,m} \left| \hat{T} \right| \Psi_{s,m} \right\rangle \right\}, \tag{16.47}$$

is a unique functional of the ensemble density as well.

The universal functional Eq.16.43 for the interacting system can be now written in the form analogous to Eq.16.21:

$$\mathcal{F}_I[w, \tilde{n}_I] = T_s[w, \tilde{n}_I] + J[\tilde{n}_I] + E_{xc}[w, \tilde{n}_I] \tag{16.48}$$

defining the exchange correlation energy $E_{xc}[w, \tilde{n}_I]$ for the ensemble. From the variational principle for the ensemble energies of the two systems one obtains

$$v_s(\mathbf{r}) = \tilde{v}_{eff}(\mathbf{r}) = v(\mathbf{r}) + v_H(\tilde{n}_I; \mathbf{r}) + v_{xc}(w, \tilde{n}_I; \mathbf{r}) \tag{16.49}$$

with

$$v_{xc}(w, \tilde{n}_I; \mathbf{r}) = \frac{\delta E_{xc}[w, \tilde{n}_I]}{\delta n_I}, \tag{16.50}$$

The eigen-problem Eq.16.45 being solved self-consistently with the density given in Eq.16.46 and the above potential $v_s(\mathbf{r})$ represents a generalized KS scheme for the ensemble of the excited levels. The ensemble energy Eq.16.42 can be evaluated for any w from the resulting $\tilde{n}_I$ and the eigen-values $E_{s,m}$ for the auxiliary non-interacting system using the following expression:

$$\mathcal{E}_I[w, \tilde{n}_I] = \sum_{m=1}^{M_I} w_m E_{s,m} - J[\tilde{n}_I] - \int \tilde{n}_I(\mathbf{r}, w) v_{xc}(w, \tilde{n}_I; \mathbf{r}) d\mathbf{r} + E_{xc}[w, \tilde{n}_I] \tag{16.51}$$

Finally, the energy E_I of the excited state can be obtained by evaluating $\mathcal{E}_I$ and $\mathcal{E}_{I-1}$ from the above expression and making use of the first equality in Eq.16.42 for the equiensemble limit, i.e. $w = 1/\mathcal{M}_I, 1/\mathcal{M}_{I-1}$, respectively. The resulting formula

$$E_I = \frac{\mathcal{M}_I}{g_I} \left[\mathcal{E}_I[1/\mathcal{M}_I, \tilde{n}_I] - \mathcal{E}_{I-1}[1/\mathcal{M}_{I-1}, \tilde{n}_{I-1}] \left(1 - \frac{g_I}{\mathcal{M}_I} \right) \right] \tag{16.52}$$

contains a difference of large quantities that must lead to large relative errors. Introduction of unequally weighted ensemble allows one to obtain an alternative expression that produces significantly more accurate excitation energies [253]:

$$E_I - E_1 = \frac{1}{g_I} \left\{ \frac{d\mathcal{E}_I(w)}{dw} \right\}_{w \le 1/\mathcal{M}_I} + \sum_{i=2}^{I-1} \frac{1}{\mathcal{M}_i} \left\{ \frac{d\mathcal{E}_i(w)}{dw} \right\}_{w \le 1/\mathcal{M}_i} \tag{16.53}$$

The procedure described above is exact but it requires repeated rather cumbersome calculations for each excited level. In many applications confined to few lowest-energy CF excitations, the DFT calculations are performed independently for each excited level simply by imposing appropriate symmetry constraints (see for example Ref. [254]).

16.4.2 Multicomponent system

In the intermetallic compounds of rare-earth (RE) and transition-metal, it is natural and common to distinguish two kinds of particles: 'core electrons' – the localized electrons including these in the 4f shell of the RE and the 'valence electrons' - 3d-electrons of the transition element of the itinerant character (cf. [254, 199, 183, 255, 256]. Besides, the electrons may be spin polarized due to, for example, external magnetic field. It is clear that in order to characterize such a system in details we need more information than just the total electron density. In general, our system may consist of several types of particles. We shall come back in subsection 16.7.2. to the specific situation of the mentioned above intermetallic compounds limiting the present discussion to principal only aspects of the KHS theory for the simplest case of two subsystems.

Suppose that there are two different kinds of particles, say A and B, characterized by number of particles, N_A, N_B, and corresponding external potentials, $v_A(\mathbf{r})$, $v_B(\mathbf{r})$. For simplicity we assume that there is no interaction that forces charge transfer transitions between the groups A and B. The Hamiltonian of the whole system can be written as

$$\hat{H} = \hat{H}_A + \hat{H}_B + \hat{U}_{AB}, \tag{16.54}$$

where each subsystem Hamiltonian $\hat{H}_{\alpha(=A,B)}$ has the form of Eq.16.1. The corresponding potentials $v_A(\mathbf{r})$ and $v_B(\mathbf{r})$ may be different. It can be shown by a simple extension of the Hohenberg-Kohn-Levy theory (cf. Ref.[257], [258], [259]) that subsystem densities $n_A(\mathbf{r})$, $n_B(\mathbf{r})$ determine the external potentials $v_A(\mathbf{r})$, $v_B(\mathbf{r})$ and all the ground state properties. There also exists a constrained search procedure (see previous subsection) that provides the ground-state wave functions $\Psi(\mathbf{x}_1, \mathbf{x}_2, \ldots, \mathbf{x}_{N_A}, \mathbf{x}_{N_A+1}, \ldots, \mathbf{x}_N)$ where $N = N_A + N_B$, densities $n_A(\mathbf{r})$, $n_B(\mathbf{r})$ and energy E. Namely,

$$E = \min_{n_A, n_B} \left\{ F[n_A, n_B] + \int n_A(\mathbf{r})v_A(\mathbf{r})d\mathbf{r} + \int n_B(\mathbf{r})v_B(\mathbf{r})d\mathbf{r} \right\}. \tag{16.55}$$

The minimization is over all pairs of $\{n_A, n_B\}$ and

$$F[n_A, n_B] = \min_{\Psi \to n_A, n_B} \left\langle \Psi \left| \hat{T} + \hat{U}_{AA} + \hat{U}_{BB} + \hat{U}_{AB} \right| \Psi \right\rangle. \tag{16.56}$$

Thus, one can proceed with the non-interacting counterpart to establish the KS scheme for the system under consideration. As a result, we obtain two sets of KS equations for $\alpha = A, B$:

$$(-\frac{\nabla^2}{2} + v_{eff,\alpha}(\mathbf{r}))\psi_{\alpha i} = \epsilon_{\alpha i}\psi_{\alpha i}, \qquad (16.57)$$

$$n_\alpha(\mathbf{r}) = \sum_i^{N_\alpha} \sum_\sigma |\psi_{\alpha i}(r, \sigma)|^2, \qquad (16.58)$$

$$v_{eff,\alpha}(\mathbf{r}) = v_\alpha(\mathbf{r}) + v_H(n; \mathbf{r}) + v_{xc,\alpha}(n_A, n_B; \mathbf{r}), \qquad (16.59)$$

where $n = n_A + n_B$ is the total density and

$$v_{xc,\alpha}(n_A, n_B; \mathbf{r}) = \frac{\delta E_{xc}[n_A, n_B]}{\delta n_\alpha}. \qquad (16.60)$$

The total energy is given by:

$$\begin{aligned}
E = {} & \sum_{\alpha=A,B} \sum_i^{N_\alpha} \epsilon_{\alpha,i} - \frac{1}{2} \int n(\mathbf{r}) v_H(n; \mathbf{r}) d\mathbf{r} + E_{xc}[n_A, n_B] \\
& - \sum_{\alpha=A,B} \int n_\alpha(\mathbf{r}) v_{xc,\alpha}(n_A, n_B; \mathbf{r}) d\mathbf{r}. \qquad (16.61)
\end{aligned}$$

The extension to three or more kinds of particles is elementary.

16.4.3 Local spin density approximation

As an example of multicomponent system we follow the von Barth and Hedin formulation of DFT for spin polarized system [260] first suggested by Kohn and Sham [90]. This generalization gives access to the spin density, which is a key quantity for all kinds of magnetic systems. It is capable of describing many-electron systems in the presence of a magnetic field acting on the spins of electrons or the systems with 'open core shell' electrons, which, as a rule, are not spin-compensated. Moreover, introduction of densities of different spin may enhance the quality of even rudimentary approximations for the exchange-correlation effects by allowing a greater flexibility and embedding more physics into the density functionals through theirs spin dependence. The theory gains additional degrees of freedom.

Suppose that a small external magnetic field $b(\mathbf{r})$ directed along the z-axis acts only on the spins of the electrons. Then the external potential in Hamiltonian Eq.16.1-16.4 becomes

$$v(\mathbf{r}_i) \rightarrow v(\mathbf{r}_i) + 2\beta b(\mathbf{r}_i)\hat{s}_{z,i}, \qquad (16.62)$$

where β is the Bohr magneton and $\hat{s}_{z,i}$ is the z-component of the electron i spin vector. In the absence of external magnetic field the system remains spin-compensated. Switching on the interaction in the right hand of Eq.16.62 causes it to be polarized with $N_\uparrow$ spin-up electrons and $N_\downarrow$ spin-down electrons, $N = N_\uparrow + N_\downarrow$. Subsequently, the external potential for the spin-up-electrons is $v_\uparrow(\mathbf{r}_i) = v(\mathbf{r}_i) + \beta b(\mathbf{r}_i)$ and for spin-down-electrons $v_\downarrow(\mathbf{r}_i) = v(\mathbf{r}_i) - \beta b(\mathbf{r}_i)$. The two-component formalism outlined above applies automatically in this case. It is enough to identify the spin orientations with subsystems A and B.

Introduction of a special kind of magnetic field that acts only on the electron spin seems to be artificial if the orbital magnetism is ignored. Usefulness of this approach, however, stems from the fact that, first, the initial spin-compensated model is more artificial and, second, there exists a precise exchange-correlation energy for the homogeneous-spin-polarized gas: $\varepsilon_{xc}^h(n_{0\downarrow}, n_{0\uparrow})$. The approximation

$$E_{xc}[n_\downarrow, n_\uparrow] \simeq E_{xc}^{LSDA}[n_\downarrow, n_\uparrow] = \int [n_\downarrow(\mathbf{r}) + n_\uparrow(\mathbf{r})]\, \varepsilon_{xc}^h[n_\downarrow(\mathbf{r}), n_\uparrow(\mathbf{r})]d\mathbf{r}$$

(16.63)

constitutes the *local spin density approximation* (LSDA) that has been proved to provide much more accurate results in comparison with the usual LDA calculations not only for the localized electrons in atoms, molecules and solids but also for the conduction electrons in metals (cf.[250]). The formulae for ε_x^h and ε_c^h in the case of polarized spins are similar to these given in Eq.16.34-16.36. They can be found in Ref. [242] or [236].

LSDA lowers the energy of the $4f$ electrons by a few electronovolts [238]. Nevertheless it still underestimates the exchange energy E_x by at least 10% while overestimating the E_c by a factor of 2 or more. The success of LSDA for calculating energy differences must be largely due to a cancellation of errors.

It is convenient sometimes to describe the spin-polarized system in terms of total electron density $n(\mathbf{r}) = n_\downarrow(\mathbf{r}) + n_\uparrow(\mathbf{r})$ and the spin magnetization or moment density $\Sigma(\mathbf{r}) = \mu_B[n_\uparrow(\mathbf{r}) - n_\downarrow(\mathbf{r})]$, where μ_B is the Bohr magneton, instead of the pair $n_\downarrow, n_\uparrow$. The density $\Sigma(\mathbf{r})$ appears in the relativistic formulation of the density functional theory presented in the next subsection as a natural consequence of introduction of the current vector density.

16.4.4 Relativistic effects

Necessity of relativistic extension for compounds containing heavy elements is well understood. In principle, since the kinetic energy of the f-electrons is rather low, these electrons can be described in the non-relativistic limit with relevant relativistic corrections accounted for the magnetic properties. However, one should also keep in mind a conspicuous radial expansion of the f-electron charge density due to the interaction with the inner-shells electrons being influenced by strong relativistic effects. In the wave-function quantum mechanics all these effects can be safely kept within the Dirac formalism. There also exists a relativistic version of the density functional theory.

The whole density functional apparatus in the relativistic approach is not much more complicated in its essence than the conventional one. Comprehensive presentation of this formalism would require an introduction of the advanced quantum electrodynamics formalism and terminology, what goes beyond the frames of this study. Therefore, we state here only the main points of the theory referring the reader interested in details to the original papers in which the subject is widely discussed - i.e. Refs [261], [262], [263], or in the context of magnetic anisotropy also to Refs [264], [265]. More fundamental introduction contains Eschrig textbook [237].

In the relativistic density functional theory the basic model parameters have to be replaced with the related time-independent four-quantities [266], the *four-vector current density*,

$$n(\mathbf{r}) \to \mathcal{J}(\mathbf{r}) = (n(\mathbf{r}), \mathbf{j}(\mathbf{r})) \tag{16.64}$$

and the *four-vector potential*,

$$v(\mathbf{r}) \to \mathcal{A}(\mathbf{r}) = (v(\mathbf{r}), \mathbf{A}(\mathbf{r})), \tag{16.65}$$

$\mathbf{j}(\mathbf{r})$ and $\mathbf{A}(\mathbf{r})$ refer to the space components of the current and vector potential operator. The ground state energy becomes a functional of the four-vector current $\mathcal{J}(\mathbf{r})$

$$E = E[\mathcal{J}(\mathbf{r})] = \int \mathcal{J}_\mu(\mathbf{r})\mathcal{A}^\mu(\mathbf{r})d\mathbf{r} + F[\mathcal{J}(\mathbf{r})] \tag{16.66}$$

and so the functional F, which expressed in terms of kinetic energy of the noninteracting system reads

$$F[\mathcal{J}(\mathbf{r})] = T_s[\mathcal{J}(\mathbf{r})] + J[n(\mathbf{r})] + E_{xc}[\mathcal{J}(\mathbf{r})]. \tag{16.67}$$

Rajagopal and Callaway [261] have shown that the energy Eq.16.66 reaches its minimum for the correct four-vector current $\mathcal{J}(\mathbf{r})$ and that the four-vector potential $\mathcal{A}(\mathbf{r})$ is a unique functional of $\mathcal{J}(\mathbf{r})$. By varying the energy functional with the subsidiary continuity condition,

$$\partial_\mu \mathcal{J}^\mu = 0 \tag{16.68}$$

one obtains the following *Kohn-Sham-Dirac* equations:

$$\left[-ic\hat{\alpha} \cdot \nabla + c^2(\hat{\beta} - 1) - v_{eff}(\mathbf{r}) + \hat{\alpha} \cdot \mathbf{A}_{eff}(\mathbf{r}) \right] \psi_i(\mathbf{r}) = \epsilon_i \psi_i(\mathbf{r}), \tag{16.69}$$

where c is the light velocity, $\hat{\alpha}$ and $\hat{\beta}$ are the standard 4×4 Dirac matrices and $\psi_i(\mathbf{r})$ are the four-spinors [266]. The effective scalar and vector potentials are given by

$$v_{eff}(\mathbf{r}) = v(\mathbf{r}) + v_H(n;\mathbf{r}) + \frac{\delta E_{xc}(\mathcal{J})}{\delta n}, \qquad (16.70)$$

$$\mathbf{A}_{eff}(\mathbf{r}) = \mathbf{A}(\mathbf{r}) + \frac{1}{2}\int \frac{\mathbf{j}(\mathbf{r})}{|\mathbf{r}-\mathbf{r}'|}d\mathbf{r}' + \frac{\delta E_{xc}(\mathcal{J})}{\delta \mathbf{j}}, \qquad (16.71)$$

where the corresponding electron and current densities are

$$n(\mathbf{r}) = \sum_i^N tr\{\psi_i(\mathbf{r})\psi_i(\mathbf{r})\}, \qquad (16.72)$$

$$\mathbf{j}(\mathbf{r}) = \sum_i^N tr\{\psi_i(\mathbf{r})\hat{\alpha}\psi_i(\mathbf{r})\}. \qquad (16.73)$$

The current density $\mathbf{j}(\mathbf{r})$ can be decomposed [264] into the curl of the total spin momentum density $\boldsymbol{\Sigma}(\mathbf{r})$, and the total orbital momentum density $\mathbf{L}(\mathbf{r})$

$$\mathbf{j}(\mathbf{r}) = c\mu_B \nabla \times [\boldsymbol{\Sigma}(\mathbf{r}) + \mathbf{L}(\mathbf{r})] \qquad (16.74)$$

These spin and orbital momentum densities have rather auxiliary character. The theory provides only the total current density $\mathbf{j}(\mathbf{r})$ similarly as the spin-independent HKS scheme gives the total charge density – not the spin densities. As previously, usefulness of the theory hinges on finding a good approximation to the exchange-correlation energy. However, the current density $\mathbf{j}(\mathbf{r})$ vanishes inside the homogeneous system. The method usually adopted is to neglect the orbital contribution and replace E_{xc} with its local spin density approximation known from the nonrelativistic case

$$E_{xc}(\mathcal{J}) = E_{xc}[n, \boldsymbol{\Sigma}, \mathbf{L}] \simeq E_{xc}^{LSDA}[n, \boldsymbol{\Sigma}]. \qquad (16.75)$$

This is justified to some extent since in atom, the energies related to the Hund's first rule - maximum spin polarization, are larger than those related to the Hund's second rule - maximum orbital polarization of the spin polarized atom. Moreover, certain non-zero orbital current density can be forced by the spin-orbit coupling if the time reversal symmetry is to be preserved. Nevertheless, a systematic discrepancy of the orbital moments in relation to these observed for f-electron systems [265], [267] points to the necessity to include an exchange-correlation enhancement of the orbital polarization someway. One of the methods [268], [269] relies on a scaling of the atomic orbital polarization energy anzatz that results in an additional term in E_{xc} being dependent on the orbital moment $\mu^t = \mu^t[\mathbf{L}]$ on each atom t,

$$E_{xc}[n, \boldsymbol{\Sigma}, \mathbf{L}] \simeq E_{xc}^{LSDA}[n, \boldsymbol{\Sigma}] + \frac{1}{2}\sum_t B^t \left(\mu^t[\mathbf{L}]\right)^2 \qquad (16.76)$$

in such a way that it leads to the following term in the effective KS potential

$$\frac{\delta E_{xc}[n, \mathbf{\Sigma}, \mathbf{L}]}{\delta \mathbf{L}} = \sum_t B^t \mu^t[\mathbf{L}] \hat{P}^t, \qquad (16.77)$$

where B^t is certain radial Slater integral for atom t and $\hat{P}^t$ is a projection operator that selects those local d or f states which contribute to μ^t.

The relativistic corrections to the exchange-correlation energy for the homogeneous gas are given in Refs. [270], [243]. They are certainly important for the electron orbitals 1s, 2s, 2p, ... especially of the heavy elements and indirectly, through the interelectron repulsion, for all the remaining electrons.

16.5 Exchange-correlation energy

The quality of the DFT calculations hinges on accuracy of determination of the exchange correlation energy E_{xc}. In general, there are two complementary methods in searching possibly most reliable representation of E_{xc}, both initiated in the original Hohenberg-Kohn-Sham works [240], [90]. The first method relies on introduction of successive improvements to the initial local density approximation. Incorporation of the spin and orbital momentum densities discussed in the previous section may serve here as a good examples of that practice. Regardless of more or less heuristic character of the corrections postulated there, satisfactory results obtained with them speak for themselves. A next important step of this kind, connected with elimination of the self-interaction problem, is presented in subsection below.

In the second method, the search for the exchange and correlation energy is extended beyond the framework of the local density approximation. For instance, an introduction of the non-local Hartree-Fock exchange energy as discussed already by Kohn and Sham [90] solves the problem of the exchange energy definitely. However, the price for this particular extension is high: the inherent simplicity of the DFT formalism, so important in the case of large systems, is lost. As we will see in second subsection, another idea inspired by basic Hohenberg-Kohn-Sham works - the density gradient expansion of the energy functional, has been successfully developed to a very efficient form.

16.5.1 Self-interaction correction

The Hartree energy $J[n]$ contains the self-interaction of the charge densities belonging to individual KS states, which, in the exact theory, is precisely cancelled by the respective terms in the exchange-correlation energy E_{xc}. In the Hartree-Fock approximation, this self-interaction in $J[n]$ is also entirely cancelled by self-exchange. However, there is no complete compensation between the self contributions to $J[n]$ and E_{xc} in L(S)DA [272]. This

202

is easily seen if we consider a single-particle system – the hydrogen atom as an example. The interelectronic repulsion for this system and dynamic correlations should obviously be equal to zero, i.e.

$$J[n] + E_{xc}^{exact}[n] = 0, \qquad (16.78)$$

for *any* single-electron density n or alternatively $v_H(\mathbf{r}) + v_{xc}^{exact}(\mathbf{r}) \equiv 0$. Let the ground state wave function be $\psi \sim e^{-r+i\varphi}$, where r and φ are radial and azimuthal coordinates of $\mathbf{r}$ and $i = \sqrt{-1}$. Subsequently, the electron density, $n \sim e^{-2r}$, and $v_H(\mathbf{r}) \sim 1/r - e^{-2r}/r + \int_r^\infty \frac{e^{-2r}}{r} dr$, where the two first terms tend to cancel each other in the small r limit and the two last terms tend to cancel each other with increasing r. The exact exchange correlation potential cancels this undesirable consequence of the construction of the theory but the LSDA potential, $v_{xc}^{LSDA}(\mathbf{r}) \sim -n^{1/3} \sim -e^{-2r/3}$, does not. As a result, the effective potential in KS equations $v_{eff}(\mathbf{r}) = v(\mathbf{r}) + v_H(n;\mathbf{r}) + v_{xc}(n;\mathbf{r})$ is deformed in the whole range of r in comparison with the exact potential: $v_{eff}^{exact}(\mathbf{r}) = -1/r$. It becomes less attractive both in the limit of small r: $v_{eff}(\mathbf{r}) \sim -1/r + \int_r^\infty \frac{e^{-2r}}{r} dr$ as well as for larger r: $v_{eff}(\mathbf{r}) \sim -e^{-2r/3}$. This situation is shown in Figure 16.1.

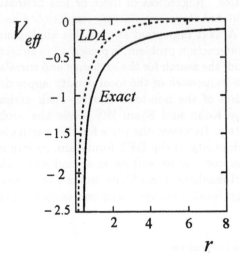

Fig. 16.1 Comparison of Exact and LDA effective potentials for hydrogen atom

Self-interaction problem becomes less important as a number of electrons increases. By no means, hydrogen is the worst case for LDA since it is a single particle system. Nevertheless, this is suspected to be a serious defect of the local density approximation for any localized electron system.

To overcome this, Perdev and Zunger [272] have suggested a *self-interaction corrected* (SIC) L(S)DA that restores the exact relation Eq.16.78 for localized single-electron states simply by subtracting self-Coulomb and self -xc energies for each occupied state from the total energy functional

$$E^{SIC}[n, \{\psi_i\}] = \sum_i \left\langle \psi_i \left| -\frac{\Delta}{2} \right| \psi_i \right\rangle + \int v(\mathbf{r}) n(\mathbf{r}) d\mathbf{r} + J[n] + E_{xc}^{LDA}$$
$$- \sum_i \left\{ J[|\psi_i|^2] + E_{xc}^{LDA}[|\psi_i|^2] \right\} \tag{16.79}$$

The energy E^{SIC} appears to be a functional of both the density and the set of occupied orbitals. For Bloch-like states it correctly coincides with E^{LDA}. The direct minimization of Eq.16.79 leads to the SIC equations [273]:

$$(-\frac{\nabla^2}{2} + v_{eff}^{LDA}(\mathbf{r}) + v_i^{SIC}(\mathbf{r})) \psi_i = \epsilon_i \psi_i + \sum_{i' \neq i} \lambda_{ii'} \psi_{i'}, \tag{16.80}$$

$$v_i^{SIC}(\mathbf{r}) = -v_H(|\psi_i|^2; \mathbf{r}) - v_{xc}^{LDA}(|\psi_i|^2; \mathbf{r}), \tag{16.81}$$

where $\lambda_{ii'}$ are the Lagrange multipliers given by

$$\lambda_{ii'} = \left\langle \psi_{i'} \left| -\frac{\Delta}{2} + v_{eff}^{LDA}(\mathbf{r}) + v_i^{SIC}(\mathbf{r}) \right| \psi_i \right\rangle. \tag{16.82}$$

Note, that in a metallic system any set of Bloch states that solves the KS equations is simultaneously a solution of the SIC equations. The KS equations provide thus a local minimum for the ground state energy. The SIC improves the LSDA results considerably. The gain in the total energy may reach even 600 eV in the case of $4f$-electron element [238]. So large difference comes mainly from inner core shells. The $4f$ contribution amounts to 6 eV 'only', i.e. it is almost the same as a difference between the LDA and LSDA solutions.

16.5.2 *Generalized gradient approximation*

Local density approximation and KS one-particle equations as well as various extensions and improvements, the examples of which have been presented in previous subsections, form a basic calculation scheme for numerous applications reported in the literature. Astonishing effectiveness of so crude, in principle, local density approximation has inclined one to extend its applicability to weaker and weaker interactions. Simultaneously, successive trails of going beyond LDA have been undertaken all over the time from the very beginning. LDA has been introduced at the level of the density of the exchange-correlation energy. A temptation at hand is to start with the exact DFT formulation and to proceed with (more) accurate approach (than LDA) a few steps farther. The *generalized gradient approximation* (GGA) [274], [275] seems to be one of the most promising ideas of this type [276]. In this logical step in developing the theory, apart from the local spin densities, their gradients are taken onto account explicitly. GGA has been derived using additional information about the electron gas of slowly varying density and enforcing the correct asymptotic behaviour for the exchange energy. A simple form of this correction is an additional advantage

$$E_x^{GGA}[n] = -\frac{3}{4} \left(\frac{3}{\pi}\right)^{1/3} \int n(\mathbf{r})^{4/3} F(s) d\mathbf{r}, \qquad (16.83)$$

where

$$s = \frac{|\nabla n(r)|}{2k_F n(r)}, \qquad k_F = \left(3\pi^2 n\right)^{1/3} \qquad (16.84)$$

and

$$F(s) = (1 + 0.0864 s^2/m + bs^4 + cs^6)^m, \qquad (16.85)$$

$m = \frac{1}{15}$, $b = 14$, $c = 0.2$. We see that $F(0)$ corresponds to LDA limit. GGA improves accuracy of determination of the exchange energy reducing error to 1%. Unlike SIC, this correction is fully consistent with the HKS scheme.

16.6 Mapping DFT on effective Hamiltonian

Phenomenological Hamiltonian being commonly employed in interpretation of the experimental data is simply a parametric version of the effective Hamiltonian (see chapter 17). DFT and effective Hamiltonian describe the same physical reality. To enable a comparison with experiment it is advisable to translate the DFT results to the language of parameters. This can be done in several ways. The DFT energies obtained for the ground and excited states according to the method presented in subsection 16.3.1 can simply be restored in terms of the parameters in the same way as experimental data are. This rather awkward method is conceptually clear: we need not consider further links between the two approaches apart from equating the energies, which obviously should be the same. It is much easier to get the CF parameters from one-electron KS energies as it is practiced in Slater $X\alpha$ and other one-electron methods. One has to assume, however, that KS energies do correspond to the CF one-electron levels. The construction of the theory does not ensure this property (see section 16.2). Similar doubts appear if the CF potential is identified with a non-spherical part of effective

KS potential $v_{eff}(\mathbf{r})$ so that the CF parameters are determined directly from the formula:

$$B_{kq} = \int R_{nf}^2(r)v_{eff}(\mathbf{r})C_q^{(k)*}(\mathbf{r}/r)d\mathbf{r}, \qquad (16.86)$$

where $R_{nf}(r)$ is a radial part of the KS orbital representing the nf-electron. Nevertheless, on the grounds of an independent reasoning, this formula has been argued to be correct. We will come back to this question below.

A next, more natural and transparent method relies on use of the real physical quantity determined in DFT calculations very precisely: the electron density. In the case of localized open-shell subsystem with a fixed number of electrons, suffice it to accept the nf-electron density n_{nf} given by the corresponding KS orbitals and ignore differences in their radial distributions. These do not seem to be very restrictive or unusual assumptions. Decomposition of the total electron density n into n_{nf} and the remainder n_{rest},

$$n = n_{nf} + n_{rest} \qquad (16.87)$$

allows one to follow the classical definition of the CF potential as an external potential acting on open shell electrons:

$$v^{CF}(\mathbf{r}) = v(\mathbf{r}) + \int \frac{n_{rest}(\mathbf{r}')}{|\mathbf{r}-\mathbf{r}'|}\,d\mathbf{r}', \qquad (16.88)$$

where $v(\mathbf{r})$ is the potential of the nuclei. The CF parameters are then given by (cf. chapters 7 and 10)

$$B_{kq} = \int\int R_{nf}^2(r)[n_{rest}(\mathbf{r}') + n_{nucl}(\mathbf{r}')]\frac{r_<^k}{r_>^{k+1}}\,C_q^{(k)*}(\mathbf{r}'/r')d\mathbf{r}'d\mathbf{r}, \quad (16.89)$$

where instead of $v(\mathbf{r})$, a charge density of the nuclei n_{nucl} is introduced.

The above equation would be correct if the n_{rest} density would not be polarized by the non-spherical part of n_{nf}. Fähnle [277] and Steinbeck et al [267] have shown that this polarization leads to an additional term in the CF potential – just the exchange-correlation potential $v_{xc}(\tilde{n})$ that is included in $v_{eff}(\mathbf{r})$. Here $\tilde{n}(\mathbf{r})$ is a total electron density of the system $n(\mathbf{r})$, the f-electron part of which is spherically averaged. Thus, instead of Eq.16.88 we have

$$v^{CF}(\mathbf{r}) = v(\mathbf{r}) + \int \frac{n_{rest}(\mathbf{r}')}{|\mathbf{r}-\mathbf{r}'|}\,d\mathbf{r}' + v_{xc}(\tilde{n}). \qquad (16.90)$$

Note that any spherically symmetric density added to $n_{rest}(\mathbf{r}')$ in Eq.16.89 does not influence the integral. Therefore, $n_{rest}(\mathbf{r}')$ in Eq.16.90 can be freely replaced with the density $\tilde{n}(\mathbf{r}')$. In this way we arrive at equation Eq.16.86 with n replaced with $\tilde{n}$.

16.7 Applications

Number of papers reporting DFT calculations of the CF parameters or energy levels systematically increases. Despite considerable development of the theory, a large part of the results still have a qualitative character only, playing a role of successive tests for various calculation schemes. Agreement with the experimental data happens so poor that correct signs and order of magnitude of the CF parameters are the measure of success. Fortunately, there exist very encouraging results as well. On the other hand, also the experimental data are not free of error. Essentially, this is the problem of false minima appearing in fitting procedures. It is encountered very often, among the others, in the interpretation of the inelastic neutron scattering measurements. The case of metallic systems is especially difficult since there are no other than INS experimental techniques that provide a direct access to the CF transitions. Moreover, unlike for ionic systems, there are no simple rules for parameters, which allow to exclude some of false solutions (see next section). In what follows, various effectiveness of DFT prediction of the CF effect is illustrated making use of some recent results reported for both ionic and intermetallic compounds.

16.7.1 Ionic compounds

Crystal field effect in ionic compounds can satisfactorily be described within the conventional perturbation theories. No wonder that DFT simulations are not very common yet for this group of compounds. Nevertheless, a number of calculations have already been published. One of the latest attempts has been undertaken for the superconducting rare-earth cuprates $Nd:R_2CuO_4$ and $Nd,Er:RBa_2Cu_3O_7$ (R = Y, La) [278]. The calculations have been performed using advanced codes for the LDA calculations from the commercial package WIEN95. Two various models have been employed, SIC-LDA and GGA-LDA, in two limiting valences, the neutral atom and trivalent cation, expecting the actual valence being somewhere in-between. The simulations have been confined to the second order parameters B_{20} and B_{22}. The obtained values of these parameters depend on the model used rather weakly: they are varied within only a few per cent. In comparison with the experimental data they are underestimated in the absolute values by a factor of about two, i.e. far too much to discuss a specific choice of the model in this context. Irrespectively of reliability of experimental data, the method of calculations may be suspected not to be very adequate. Presumably, inclusion of spin polarization would improve the results. Fractional coordinates of atomic positions in the crystal taken from the data for the isostructural compound might cause an additional error. Nevertheless the order of magnitude of the parameters and their signs are restored correctly.

The following contributions to the CF parameters have been distinguished:

- *Coulomb on-site* contribution arising mainly from the asphericity of the outer shells on the same atom. It has turned out to be dominating contribution. Probably, this contribution includes the shielding and contact shielding effects known from the perturbation model as well as part of the overlap contribution. Thus, its relatively large value in the case of not only the second order parameter could be expected indeed. If this analogy with the perturbation approach is true the magnitude of this contribution in the case of second order parameter is determined mainly by two opposing factors, so, its sign in relation to other contributions may vary from compound to compound.

- *lattice* contributions generated by all charges beyond a given atom. An order of magnitude of its absolute value has been the same as that of the corresponding Coulomb on-site contribution. This contribution corresponds closely to the classic CF potential.

- *correlation-exchange* contribution being of the order of 10% of the corresponding total (absolute) value.

16.7.2 Intermetallic compounds

Strength of the density functional theory is especially visible in the case of intermetallic f-electron compounds. The two kinds of particles, conduction electrons and localized electrons can be described simultaneously within a uniform formalism. DFT calculations of the CF effect in (inter)metallic systems has been intensively developed during the last decade. Main turning points on this way have been reflected in the applications reported in the mentioned earlier papers [254], [199], [183], [255], [256], [267]. Some of the results are precise enough to restore satisfactorily subtle temperature dependencies of the paramagnetic susceptibility and specific heat. This concerns among others one of very recent works in this series [279].

The methods applied in most of these papers have been developed from the relativistic LSDA formulation presented by Richter et al [199]. The latest version comprises the relativistic SIC-LSDA with the effective potential modified to account for Hund's first rule, i.e. the spin polarization of f-electrons is fixed so as to ensure actual multiplicity of the ground state also in paramagnetic phase. The agreement in some cases is almost perfect as far as one concerns *ab initio* results. For instance, the calculated and experimental (in parentheses) parameters obtained for $NdPd_2Ga_3$ [279] are following (Stevens parameters, all values in meV): $B_2^0 = 0.278(0.214)$, $B_4^0 = -0.145 \times 10^{-2}(-0.105 \times 10^{-2})$, $B_6^0 = -0.114 \times 10^{-4}(-0.224 \times 10^{-4})$, $B_6^6 = 0.116 \times 10^{-2}(0.0438 \times 10^{-2})$. Generally, the CF effect predicted in DFT simulations is overestimated by 30-50% with respect to the measured one. The results for parameters of second and fourth order are more accurate than for the sixth order parameters, which as a rule are overestimated by a factor of two.

The internal structure of these calculations is modified in relation to that in ionic compounds. Judging from the data reported, for instance, for RGa_2 (R = Ce or Er) [255], the role of the on-site contributions increases considerably. Also the exchange-correlation contribution becomes more important. This contribution is even dominating for the parameters of fourth and especially sixth order. A small lattice contribution in the metallic systems seems to be justified by the screening effect of the conduction electrons.

16.7.3 Final remarks

Accuracy of early DFT simulations as measured by error in spin band splitting or electron binding energy is contained in the range 0.1-0.5 eV [250], i.e. the error is larger than CF splitting energies in RE compounds and of the order of magnitude of CF energies in actinide compounds. Main improvements to the local density approximation, such as spin polarization, self-interaction correction or generalized gradient approximation lower the ground state energy falling to a magnetic nf-electron by several electronovolts. In other words the 'corrections' are at least one order of magnitude higher than the CF energies themselves. Among them, only generalized gradient approximation has a sound basis in the fundamental theory whereas the remaining have rather a heuristic character. Further corrections due to the relativistic effects are more subtle but they overlay only the former. These remarks lead to the conclusion that whenever satisfactory results in prediction of the CF effect are obtained they may ensue from a mutual cancellation of presumably large errors. It may also happen that introduction of a new correction to the exchange-correlation energy gives less satisfactory results if it throws these errors out of balance. The main question at the present stage of the density functional theory in its local density approximation is that we do not know exactly *why* the results are so good (if they are). A continued search for more accurate exchange-correlation functionals raises hopes of better understanding that issue in the future.

Another problem concerns the relation of the density functional theory to the effective Hamiltonian picture. The exact method of the mapping outlined in section 16.6 requires cumbersome calculations for excited states entailed by formalism sketched in subsection 16.4.1. To our best knowledge no such calculations for a more complex system than the He atom have been published yet. In the alternative, more common method, the electron density is the quantity that links the two approaches. However, to get the CF parameters we have to educe from it something that corresponds to the square of the radial part of the nf wave functions of the effective Hamiltonian. The actual question connected with this method concerns adequacy of the spherically averaged partial density obtained from the corresponding Kohn-Sham orbital.

The localized states in metallic systems have been considered herein as a subsystem with a fixed number of particles. An extension of this static formalism to account for hybridization effects encountered in f-electron systems is currently one of the most intensively developed sphere of the theory.

Finally, we have omitted in the present study a large block of questions connected with the translational invariance and construction of the Bloch states. Certainly, various numerical procedures known from the one-electron theories being adopted for the DFT calculations have their own margin of error. An optimal choice of the method especially in the case of two or more kinds of particles is not a trivial task.

CHAPTER 17

Analysis of the experimental data. Interpretation of crystal field parameters with additive models

Interpretation of the magnetic, spectroscopic and other electronic properties of the rare-earth and actinide compounds and their consistent explanation are far from trivial and require much experience, intuition and caution. The literature provides many examples of essential changes in interpretation after introducing some experimental or theoretical data, which could not be accounted for in the initial models. This is because the available experimental data remain usually too scanty to assure an unequivocal interpretation without additional more or less physically justified assumptions. The main parameters of the model Hamiltonian commonly used for localised electrons in a crystal – the Slater integrals F^k and the spin-orbit coupling constant ζ - can be tentatively predicted on the grounds of their general regularities [3, 280, 2, 281, 7, 8]. This, however, does not apply to the crystal field parameters, which may vary substantially from compound to compound depending rigorously on details of the metal ion co-ordination geometry. In this chapter we will attend to the question of an effective method of the interpretation of the experimental data in f-electron ionic systems paying a special attention to simplified CF models based on the idea of a partitioning of the CF potential. These simplified models have been of invaluable help not only in the burdensome ambiguity problem connected intimately with the standard parameteric analysis. Each of several further areas of application indicated in section 3 of this chapter is not less promising. First, the theoretical grounds of the models, their shortcomings and advantages are discussed. To start with, general frames of the interpretation routine are sketched.

17.1 Phenomenological Hamiltonian

Standard phenomenological Hamiltonian also termed parametric Hamiltonian, which is to be presented in the following, allows one to describe in a relatively simple way low energy electronic excitations in a localised system. This is a parametric form of the effective Hamiltonian discussed in chapter 4. It has been obtained by a projection of the Hilbert space of the true Hamiltonian onto certain definite function basis. A limitation of the excitations range due to the single-configuration approximation is a crucial simplifying assumption but not very restrictive in fact. For example, the energy of the first excited configuration for the Pr^{3+} ion - $5d4f$ is about 60000 cm^{-1} above the ground configuration $4f^2$ [281], i.e. high enough to ensure a reliable description of various observed properties of the compounds containing this type of element. Inclusion of certain well-known perturbation corrections due to the configuration interaction (CI) extends the range of the applicability of the model considerably. In particular, a thorough analysis of an essential part of optical absorption spectra, say up to 20000 cm^{-1}, of majority of the f-electron ionic compounds becomes realistic.

The excitations within the ground configuration being amenable to the parametric analysis are responsible for the magnetic, thermodynamic, spectroscopic and other properties of ionic systems as well as these intermetallic compounds which, apart from the conduction band(s), have partially occupied atomic-like shells. Though the problem of electronic structure of metallic systems is not considered here, it is worth of noticing that there are no obstacles to apply the single-configuration phenomenological Hamiltonian also for these states.

Restriction of the Hilbert space to the many-electron wave-functions that are built up from the spin-orbitals, the spatial part of which transforms according to the irreducible representation $D^{(l)}$ ($l = 3$ for the f electrons) of the rotation group O ensured compactness of the description. Of course, there exists some margin in determination of the function basis. A reasonable choice is to take the wave functions being as close to the eigenfunctions of the system as possible. In the case of the lanthanide ions, this requirement is fulfilled to some extent by the vectors $|\tau SLMJ\rangle$ defined in terms of the many-electron global spin (S), orbital (L) and total (J) momentum quantum numbers, the projection of the latter to the quantization axis (M_J) and additional quantum numbers τ necessary to identify different vectors of the same $SLJM_J$ for the electron number N greater than two. In view of apparently stronger CF and spin-orbit interactions, the $|\tau SLMJ\rangle$ vectors are less satisfactory for the actinide ions. On the other hand, there is no distinct function basis that could be justified more. Advanced simultaneous diagonalization techniques commonly applied and nearly unlimited computation possibilities make a particular choice of the function basis less important nowadays.

Due to the definite function basis any dependence on the radial co-ordinate in efective Hamiltonian can be eliminated by an integration over this co-ordinate weighted with the square of the radial part of the orbital wavefunction. Thereby, the effective parametric Hamiltonian $\mathcal{H}$ acts only on angular co-ordinates. In $\mathcal{H}$ one can distinguish the $\mathcal{H}_0$ part of spherical symmetry and the non-spherical part - the crystal field operator $\mathcal{H}_{CF}$, transforming according to the point group symmetry of the metal ion site in the crystal:

$$\mathcal{H} = \mathcal{H}_0 + \mathcal{H}_{CF} \qquad (17.1)$$

$\mathcal{H}_0$ comprises interactions, which can also be met in a free ion: the inter-electronic Coulomb repulsion, spin-orbit interaction, the second order configuration interaction terms $\mathcal{H}^{CI}$, corrections due to the relativistic effects of higher order $\mathcal{H}^{rel}$ and three- or more- electron interactions $\mathcal{H}^{3el}$:

$$\mathcal{H}_0 = \sum_k F^k f_k + \sum_i \zeta_i l_i \cdot s_i + \mathcal{H}^{CI} + \mathcal{H}^{rel} + \mathcal{H}^{3el} \qquad (17.2)$$

Following the conventional notation, f_k and $l_i \cdot s_i$ represent angular parts of the interelectronic repulsion and spin-orbit interaction, respectively. F^k ($k = 2, 4, 6$) and ζ_i are the corresponding radial integrals. A more comprehensive discussion of particular terms in $\mathcal{H}_0$ can be found elsewhere - see the Wybourne monograph [3] or the review paper by Goldschmidt [280]. Note that the above parametric Hamiltonian may represent a much wider class of the interactions than those from which it has been derived. For example, the first term in Eq.17.2 - the parametric expansion of the Coulomb interaction of electrons $1/r_{12}$ within the function basis as above represents also the screened Coulomb interaction of the type of $\exp(-r_{12}/\lambda)/r_{12}$, with an arbitrary screening length parameter λ. Strictly speaking, since $\mathcal{H}$ is now the parametric version of the effective Hamiltonian, each its term may include some trace of the higher-order perturbation corrections.

The effective CF potential $\mathcal{H}_{CF}$ has the well-known form of the expansion into the normalised spherical harmonics series:

$$\mathcal{H}_{CF} = \sum_i \sum_{k,q} B_{kq} C_q^{(k)}(\mathbf{r}_i/r_i) \qquad (17.3)$$

This expansion is restricted here to the one-electron terms but two-electron CF terms representing so called *correlation CF effect* (cf. chapter 19) may also be included if quality of the experimental data makes a more precise fitting possible.

We remember that the number of the independent CF parameters is strictly determined by two different factors: the transformation symmetry of the function basis in which the Hamiltonian Eq.17.1 has been defined and the point group symmetry of the metal ion site in the lattice. For the

f-electrons ($l = 3$) this number ranges to 27 in the case of no symmetry elements and decreases with the symmetry increasing. The rules concerning the restrictions imposed on the CF parameters by various symmetry elements and further information relevant to a choice of a proper set of the parameters are given in chapter 2.

In the standard fitting procedure, all the parameters in $\mathcal{H}_0$ and $\mathcal{H}_{CF}$ are varied until the resulting eigenvalues and eigenvectors of Hamiltonian Eq.17.1 give the closest fit to the available experimental data: electronic transition energies, their intensities, magnetic moments and others. The dependence of these quantities on the parameters has a non-linear character and an iterative self-consistent method of calculation has to be applied. The resulting parameters are called sometimes the *experimental parameters* in contradiction to the *theoretical parameters* derived in various *ab initio* calculations or estimated by means of quasi-phenomenological approaches with effective charges, bond lengths and so on. The quality of the energy levels fitting is characterised by the root mean square error (r.m.s.e.) of the calculated quantities in relation to the corresponding experimental values and the deviations of the parameters obtained by evaluation of their variational derivatives δ. Infinite δ for a given parameter means that the value of the parameter is immaterial, i.e. it does not influence r.m.s.e. The interactions included in the parametric Hamiltonian have various ranges. Usually the Coulomb interelectronic repulsion is the largest one. Also weight of parameters representing a given interaction is not the same. This establishes a specific order of the parameters, which are to be varied in consecutive steps of the fitting procedure. Specifically, this order may be the following: $F^k \rightarrow \zeta \rightarrow B_{kq} \rightarrow$ parameters in $\mathcal{H}^{CI} \rightarrow$ remaining parameters. If the iteration procedure is convergent in each step and the diagonalized matrices are algebraically well-defined, all the parameters can be varied simultaneously in final steps.

Sometimes it is advisable to compare various sets of CF parameters obtained in different fitting procedures or in different methods. The quality of the match between, say, experimental and theoretical values of the CFP can be estimated in terms of scale and reliability factors s_k and r_k [105]. They are derived in the following way: the B_{kq} are considered as the components of a vector $\vec{B}_k$ so that the lengths and directions of $(\vec{B}_k)_{exp}$ and $(\vec{B}_k)_{theor}$ may be compared. The scale factor is defined as

$$s_k = \frac{\left| (\vec{B}_k)_{exp} \right|}{\left| (\vec{B}_k)_{theor} \right|} \tag{17.4}$$

with

$$\left| \vec{B}_k \right| = \left(\sum_q B_{kq} B_{kq}^* \right)^{\frac{1}{2}} = \left(\vec{B}_k \cdot \vec{B}_k \right)^{\frac{1}{2}} \tag{17.5}$$

and consequently the reliability factor, which is the angle between the vectors $(\vec{B}_k)_{exp}$ and $(\vec{B}_k)_{calc}$ as

$$r_k = \arccos \frac{\left|(\vec{B}_k)_{exp} \cdot (\vec{B}_k)_{theor}\right|}{\left|(\vec{B}_k)_{exp}\right|\left|(\vec{B}_k)_{theor}\right|} \tag{17.6}$$

The s_k (three values for $k = 2, 4$ and 6, respectively) and r_k (three values also) are ideally equal to 1 and $0°$, respectively if a perfect match between experimental and calculated CFP's is obtained.

Another quantity commonly used to compare various sets of CF parameters is so called *CF strength parameter* [282, 283]:

$$N_\nu = \left[4\pi \sum_{k \neq 0} \frac{1}{2k + 1} \left| \vec{B}_k \right|^2 \right]^{1/2} \tag{17.7}$$

N_ν is invariant under rotation of the co-ordinate system but it is not a precise measure of the CF strength. One can show it to be dependent on various ligand-metal-ligand angles in a co-ordination polyhedron [284]. Nevertheless, N_ν is close to the root square of the sum of the squares of barycentred one-electron energies, $\bar{E}$, that can be expressed in the form [285]

$$\bar{E} = \left[7 \sum_{k \neq 0} \frac{1}{2k + 1} \begin{pmatrix} 3 & k & 3 \\ 0 & 0 & 0 \end{pmatrix}^2 \left| \vec{B}_k \right|^2 \right]^{1/2} \tag{17.8}$$

Abundance of the data concerning N_ν in the literature seems to be an additional recommendation for its continue use.

Large number of the CF parameters, especially for crystals of lower symmetries, makes a rational description of the experimental data difficult. In the case of non-transparent crystals, only very few electronic levels can be determined from the experiment. The problem with the excess parameters appears then just for the crystal symmetries lower than the cubic one. This, among the others, gives rise to an increasing interest in approximate models leading to the reduced number of independent parameters.

17.2 Simplified crystal field models

In principle, two different kinds of approximations to the CF parameterisation are commonly in use. One of them is based on the *symmetry descent technique* (SDT). The other relies on introduction of some further constraints imposed on the conventional phenomenological model with the Hamiltonian given by Eq.17.3.

SDT is used for systems, the relatively high symmetry of which is, as a matter of fact, only approximate. The higher approximate symmetry, the fewer parameters are to be determined in the initial steps of the fitting. This makes sense if only a low symmetry perturbation, as measured by a mean displacement of the atomic positions $\overline{\Delta R}$, is small enough. There exist, however, certain difficulty with a precise determination of the limit for $\overline{\Delta R}$. The CF parameters and so the energies of the electronic levels for a given co-ordination may appear to be very sensitive even to tiny changes in the atomic positions. In the extreme, an application of SDT, when suggested merely by the co-ordination polyhedron geometry details, may lead to completely erroneous interpretation. An instructive example of low symmetry system, UF_4, discussed in Ref. [286], evidences that the correlation between the mean displacement of the atomic positions $\overline{\Delta R}$ and the corresponding mean deviation of the energy levels $\overline{\Delta E}$ may be very weak indeed. Thus, only a complementary estimation of $\overline{\Delta E}$ seems to be a stringent test for validity of the SDT approximation in a given case. In other words $\overline{\Delta E}$ that can be treated as a complementary to $\overline{\Delta R}$ measure of a low symmetry distortion. Derivation of the quantities like $\overline{\Delta E}$ is just a one of the right fields of applicability of the second kind of approximate methods. The mentioned example of UF_4, [286], illustrates also how useful in estimation of $\overline{\Delta E}$, a simplified phenomenological model can be. Now we proceed with more detailed discussion of the problem at hand.

17.2.1 Decomposition of the CF potential. Virtual ligands

Simplified phenomenological models had been in use long before they were formulated in a more systematic way. It is still a common practice to preserve the ratios of the CF parameters of the same order obtained in the PCM calculations. This specific method has turned out to be astonishingly efficient despite of evident inadequacy of the oversimplified basal model. The fact of the matter is that PCM preserves certain general properties characteristic of the majority of mechanisms forming the crystal field potential. These properties have been articulated explicitly by Newman in his definition of the *superposition model* (SM) [27]. He followed a natural and simple idea, according to which the total ligand field potential may be represented by a superposition of local ligand contributions. Alternative formulation by Gerloch et al [26] is based on a partitioning of the neighbourhood of the metal ion into certain specified non-overlapping cells while the central ion site symmetry is preserved. Consequently, the ligand potential v has a form of a superposition of the cell (or ligand) potentials v_t centred at $\mathbf{R}_t$:

$$v(\mathbf{r}) = \sum_t v_t(\mathbf{r} - \mathbf{R}_t) \tag{17.9}$$

By using the rotation transformations, the matrix elements of v can be expressed in terms of the matrix elements of the single cell (ligand) potentials v_t, evaluated in the most convenient local co-ordinate system. Similarly, the global CF parameters have a form of linear combination of local ones, attributed to the particular cells (ligands).

Gerloch at al [26] considered the ligand field potential in a molecular system. Perhaps this is also a step towards phenomenological simplification of the theories based on the electron density (cf. previous chapter). Their cellular partitioning is arbitrary except for the symmetry requirements for v and an prerequisite minimal number of different v_t's. There are no obstacles to apply this kind of decomposition for the whole crystal field potential, the more so as an explicit form of v_t is meaningless from the point of view of developing of parameterisation schemes. Neither dimensions and shape of the cells nor identification of v_t's with individual ligand potentials is really crucial at this stage. These two approaches, Newman's and Gerloch's et al correspond to different physico-chemical pictures which need not be specified beforehand.

These arguments lead straightforwardly to a generalised formulation of the decomposition without specifying the origin or other details pertained to v_t's. We can follow the precise formulation by Gerloch et al extended for the total CF potential in all points except identifying v_t with any concrete physical object. Instead, it is convenient to introduce the term *virtual ligands*, i.e. certain abstractive objects, the potentials of which are summed up to the true CF potential. The virtual ligands are defined merely by the decomposition Eq.17.9 and certain transformational properties of v_t which are specified below. This definition v_t comprises both the case of empty cells in two dimensional complexes considered by Gerloch et al as well as the further neighbours contributions characteristic for solids omitted in ligand field theory. Simultaneously, we avoid a confusing problem of determination of borders of the cells or a space between cells that must appear if we intend to specify an approximate local pseudosymmetry of the cells. Finally, this formulation is strictly consistent with the phenomenology, in the sense that the local parameters connected with the components v_t derived from the fitting of the experimental data represent *de facto* our abstractive objects - virtual ligands, not actual ligands or the Gerloch's et al cells.

Such a generalised decomposition has a formal character and there is no loss in generality up to here. The main approximation that is commonly applied refers to the local pseudosymmetry of v_t. Namely, it is assumed to be axial - $C_{\infty v}$. This simplifies the model considerably but at the expense of its accuracy. It would be advisable to stop for a while at that point. Some arguments warning against and in favour of the approximation are set fort in subsection 17.2.3.

The generalised formulation causes some difficulties with the quantitative estimation of the model parameters from first principles. To facilitate discussion and further approximations we stay with the definition closer to the traditional one. Specifically, we assume that the virtual ligands are placed at the positions of actual ligands and that potentials v_t are represented by the corresponding true ligand potentials. We can always invoke the generalised formulation whenever needed.

17.2.2 Superposition model and angular overlap model

Several various parameterisation schemes based on the CF potential decomposition Eq.17.9 can be found in the literature. In majority, they resolve themselves into two the most common ones: the *angular overlap model* (AOM) [78, 289, 26, 287, 132] and the mentioned Newman superposition model (SM) [27]. Interesting example of alternative approaches, so called the *simple overlap model* has been proposed by Malta [288] for ionic systems and the *modified point charge model* by Żołnierek [110, 111] for both ionic and intermetallic systems. In these models the potentials v_t are generated by metal-ligand overlap charges distributed over small regions centred around either the middle distance between the metal and the ligand (Malta) or the distance determined in terms of the ligand and metal electronegativities (Żołnierek). Both the models have been developed on the grounds of some heuristic arguments and they have not been verified in a more rigorous approach up to now. Nevertheless, quasi-phenomenological calculations based on these models have allowed one to interpret or predict very easily the CF parameters for a large number of compounds. Since the models are basically similar to AOM and SM we will not discuss them in details here.

We have accepted temporarily the decomposition into ligand potentials v_t. The SM parameters $b_k^t \equiv b_k^t(R_t)$ are defined as coefficients in the expansion of v_t in terms of the spherical harmonics, averaged with the radial part of the nf-electron density:

$$\langle v_t \rangle_{nf} \equiv v_t(\mathbf{r}/r) = \sum_k b_k^t(R_t) C_0^{(k)}(\mathbf{r}/r) \qquad (17.10)$$

This expansion is similar to that given in Eq.17.3 but this time the co-ordinate system is specific for a given ligand t. Namely, its z-axis is directed towards the ligand. Hereafter we refer to this system as *local*. Due to the assumed axial symmetry, only terms with $q = 0$ are present in Eq.17.10. The parameters b_k^t correspond to the usual parameters $B_{kq}(q = 0)$ for so called *linear ligator*, i.e. the system of metal ion and single ligand with the quantization axis aligned along the ligator symmetry axis. The co-ordination cluster can be regarded as a structure composed of several ligators with a common metal ion. Simple rotation transformations of the co-ordinate systems corresponding to particular ligators to the central co-ordinate system lead to the following expression for the global parameters (see chapter 2 and appendix A):

$$B_{kq} = \sum_t b_k^t(R_t) D_{0q}^{(k)*}(0, \Theta_t, \Phi_t) \qquad (17.11)$$

R_t, Θ_t, Φ_t are the polar co-ordinates of ligand t in the central co-ordinate system and $D_{0q}^{(k)*}(0, \Theta_t, \Phi_t)$ is the matrix element of the k-order irreducible representation $D^{(k)}$ of the rotation group. Following the simple tensor algebra similar to that employed in chapter 2, one finds that b_k^t's are connected to the matrix elements $\langle a_{nf\pm\mu} | v_t | a_{nf\pm\mu} \rangle_t$ of the potential v_t determined in the local co-ordinate system:

$$b_k^t = \frac{2k+1}{7} \left[\begin{pmatrix} 3 & k & 3 \\ 0 & 0 & 0 \end{pmatrix} \right]^{-1} \times$$

$$\times \sum_{\mu=0}^{3} (-1)^\mu (2 - \delta_{\mu 0}) \begin{pmatrix} 3 & k & 3 \\ -\mu & 0 & \mu \end{pmatrix} \langle a_{nf\pm\mu} | v_t | a_{nf\pm\mu} \rangle_t \quad (17.12)$$

The matrix is diagonal and $\langle a_{nf,\mu} | v_t | a_{nf,\mu} \rangle_t = \langle a_{nf,-\mu} | v_t | a_{nf,-\mu} \rangle_t$ due to the assumed $C_{\infty v}$ symmetry of v_t.

The parameters in the angular overlap model, in turn, represent just the above matrix elements:

$$e_\mu^t \equiv e_\mu^t(R_t) = \langle a_{nf\pm\mu} | v_t | a_{nf\pm\mu} \rangle_t \qquad (17.13)$$

$\mu = 0(\sigma)$, $1(\pi)$, $2(\delta)$, $3(\phi)$, denotes the magnetic quantum number of nf-electron in the local co-ordinate system and R_t is a distance between the metal ion and the ligand t. Conventionally, the e_μ^t parameter with $\mu = 3$, which is the smallest one [26], is set to zero what is nothing more than a specific choice of the beginning of the energy scale. Making use of the transformational properties of the f orbitals, the matrix elements of the total CF potential v, Eq.17.9, can be expressed in terms of e_μ^t,

$$\langle a_{nfm} | v | a_{nfm'} \rangle = \sum_t \sum_\mu D_{\mu m}^{(3)*}(0, \Theta_t, \Phi_t) D_{\mu m'}^{(3)}(0, \Theta_t, \Phi_t) e_\mu^t(R_t) \quad (17.14)$$

where $D_{\mu m}^{(3)}(0, \Theta_t, \Phi_t)$ is the matrix element of the irreducible representation $D^{(3)}$ of the rotation group.

The angular overlap model is one of those simplified phenomenological models which provide a highly compact description of the crystal field effect irrespectively of particular crystalline structure. From among different propositions of this type AOM distinguishes itself by its simplicity, clear physico-chemical foundations, flexibility for further simplifications and systematization. It has been introduced by Jørgensen, Pappalardo and Schmidtke [78] and Schäffer and Jørgensen [289] for the d-electron systems. The name originates from the molecular orbital theory according to which the matrix elements of the ligand potential appear to be dependent on the squares of the respective overlap integrals between the metal and ligand orbitals. In the global co-ordinate system, the dependence on the overlap integrals implies the angular dependence of type Eq.17.14.

Within the perturbation approach developed in chapters 7 and 8, the role of the molecular potential connected with ligand t is played by the sum $v_t^{c.cov} + v_t^{cov}$. It is shown there that the matrix elements of this sum can also be expressed in terms of the squares of the overlap integrals:

$$
\begin{aligned}
e_\sigma^t|_{c.cov\&cov} &\approx \langle a_{nf0}|v_t^{c.cov} + v_t^{cov}|a_{nf0}\rangle_t \\
&\approx \left|\langle a_{nf0}|\chi_{tn's0}\rangle_t\right|^2 c_s + \left|\langle a_{nf0}|\chi_{tn'p0}\rangle_t\right|^2 c_p \quad (17.15)
\end{aligned}
$$

$$
\begin{aligned}
e_\pi^t|_{c.cov\&cov} &\approx \langle a_{nf\pm1}|v_t^{c.cov} + v_t^{cov}|a_{nf\pm1}\rangle_t \\
&\approx \left|\langle a_{nf\pm1}|\chi_{tn'p\pm1}\rangle_t\right|^2 c_p \quad (17.16)
\end{aligned}
$$

As previously, $n's$ and $n'p$ are assumed to be the only external electron shells of ligand, what corresponds to the case of simple ligands like F^-, Cl^-, O^{2-}. The coefficients c_s and c_p depend, among the others, on the binding energy of electrons $n's$ and $n'p$ and the Madelung energy at the metal and ligand ions. Since, by assumption, the ligands have no d-electrons, there is also no corresponding contribution to e_δ^t:

$$
e_\delta^t|_{c.cov\&cov} \approx \langle a_{nf\pm2}|v_t^{c.cov} + v_t^{cov}|a_{nf\pm2}\rangle_t \approx 0 \quad (17.17)
$$

It is interesting to notice that, in the considered case of simple ligands, the value of this particular parameter e_δ^t, when derived from the experimental data, may be treated as a measure of the other contributions than $v_t^{c.cov} + v_t^{cov}$. Due to the dominating role of $v_t^{c.cov} + v_t^{cov}$ all the above three relations 17.15-17.17 may serve as an easy preliminary test of validity of the phenomenological results.

Both the sets of the parameters, $\{b_k^t\}$ and $\{e_\mu^t\}$, describe a separated metal-ligand (ML) subsystem t. To distinguish them from the *global parameters* B_{kq}, they are called *intrinsic parameters*. Whole information concerning the crystal structure including details of the co-ordination polyhedron geometry is contained in structural factors expressed in terms of matrix elements of the respective rotation operators in Eqs 17.11 and 17.14. If a dependence of b_k^t and e_μ^t parameters on ML distance is known beforehand (see subsection 17.2.4), this factorisation into the structural and intrinsic components open wide possibilities to investigate various properties of solids: their behaviour under pressure, structural phase transitions, variation of the electronic structure in incommensurate phases, electron-phonon couplings, magnetoelastic phenomena, consequences of idealisation of the metal site symmetries including estimation of the corresponding energy deviation $\overline{\Delta E}$

mentioned at the beginning of this section and others. This is much more than only a clever reduction of the number of effective parameters.

Applicability of the considered models can be extended even more, merely by removing the assumption relating to the axial symmetry of v_t. For ligands of a complex internal structure or strongly polarised such a generalization may be of some interest. It is enough to introduce the non-axial parameters $b_{kq\neq0}^t$ or non-diagonal $e_{\mu\mu'\neq\mu}^t$ with a straightforward modification of Eqs 17.10, 17.11, 17.13, 17.14. This generalization, however, has naturally also a disadvantage. Namely, the number of the intrinsic parameters is undesirably increased, what for many cases makes the extended models practically useless. Notwithstanding, one can imagine such structures for which only one or two additional non-axial intrinsic parameters are effective. An instructive examples of complex ligands, i.e. water and PF_3 molecules, have been discussed by Schäffer [131]. Certainly, our choice concerning the simplified model in each particular case hinges on the total number of all the intrinsic parameters in relation to the number of the effective global parameters. Presumably, in the extended virtual ligands approach, the contribution of the whole lattice can be accounted for if non-axial intrinsic parameters are included. However, this speculation and its possible practical consequences require confirmation in more rigorous treatment on the group-theoretical basis.

17.2.3 Limitations

As stated in chapter 7, the contact covalency effect determines to a large extent the total CF parameters (cf. also [287], [290]). This mechanism, as well as the most of the others, which has been discussed earlier, obey the assumptions of the simplified models under consideration. Now we continue with adverse contributions provided by the second order perturbation theory, which break these assumptions. Two of them are the most important: the electrostatic interaction of further neighbours (*FN*) and non-axial components of multipoles induced on ligands (*NALP*, i.e. *non-axial ligand polarisation*) breaking the local pseudosymmetry. The numerical data available for a number of different lanthanide and actinide ionic compounds (see for instance[105], [132]) lead to a conclusion that majority of the absolute values of the contributions FN and $NALP$ are of the order of several or at most a dozen or so per cent of the total absolute values of B_{kq} depending on the compound and the parameter. Usually, lower symmetries and higher polarizabilities of anions promote higher absolute values of these contributions. Some characteristic examples of the calculations are presented in Table 17.1.

Table 17.1 Main contributions to the crystal field parameters (in cm^{-1}), which break the assumptions of the conventional superposition and angular overlap models[a]: further neighbours (FN) and non-axial dipole and quadrupole ligand polarisation ($NALP$). They are related to the total values (T) obtained within the perturbation model described in previous chapters. The data come from Ref. [132] and [291].

		B_{20}	B_{40}	B_{44}	B_{60}	B_{64}
UO_2	FN		610	370	120	-220
	$NALP$		0	0	0	0
	T		-7140	-4270	2890	-5410
UCl_4	FN	88	75	216	-21	10
	$NALP$	-154	-76	-307	277	-175
	T	-1370	2210	-6330	-710	310
UOS^b	$FN+LP$	-576	-1562	-785	45	-74
	T	-1455	-4392	-963	3419	842
$UOTe^b$	$FN+LP$	82	-418	-1137	53	361
	T	-1281	-4876	-279	3333	573

		B_{20}	B_{40}	B_{43}	B_{60}	B_{63}	B_{66}
$CsUF_6$	FN	-153	-125	-163	0	0	0
	$NALP$	63	-19	-22	-116	92	-6
	T	640	-17410	19290	3670	3810	2640
Cs_2UCl_6	FN	81	19	-56	-10	10	-10
	$NALP$	-74	83	-2	44	66	10
	T	-680	-4960	-7350	2070	-40	1210

[a] Shielded values (cf. chapter. 9). See also section 17.2.1 for a comment on generalised models.

[b] LP- total ligand polarisation including the axial components of the induced multipoles.

Among encouraging in general results, which fairly well fit the above limits, some exceptions, apparently, take place. As regards the data gathered in Table 17.1, this concerns the relatively large $FN+LP$ contribution to the B_{44} and B_{64} parameters for the uranium oxychalcogenides UOY (Y = S, Te), or the $NALP$ contribution to B_{63} for UCl_4 and the B_{63} parameter for Cs_2UCl_6. The mentioned cases may serve as typical examples of, respectively, the effect of quenching of the referred total values of the CF parameters due to cancellation of the essential contributions of two different ligand groups - in this particular case $4O^{2-}$ and $5Y^{2-}$ or two non-equivalent groups of chlorine ions or, eventually, specific geometry of the equidistant chlorine co-ordination polyhedron. The larger cancellation the more apparent the FN and $NALP(LP)$ contributions are. Nevertheless, in practice, these distinguished contributions are not more important than for the other cases - the only difference in relation to the ordinary results is that the corresponding absolute values of the reference parameters are small. From the point of view of the total CF effect they still are of little importance.

In the limiting cases of B_{44} for UOS and both B_{44} and B_{64} for UOTe the $FN + LP$ contributions are approximately equal to the corresponding total values or even greater in the absolute values. However, the axial components of the multipoles induced on ligands has not been excluded from the data presented for these compounds. More detailed results for Cs_2UCl_6 shown in Table 10.3 demonstrate that they may appear to have a dominating character even. Thus, the $FN + LP$ results for the oxychalcogenides should be considered rather as a crude upper limit of the actual values of the $FN + NALP$ data that, at most, warns against routine application of the simplified models.

As it has been argued in subsection 17.2.1, the influence of the further neighbours may be represented partly in the intrinsic parameters if they are treated as phenomenological quantities determined from the experiment. For the special case of UO_2 the FN contribution has to be represented even strictly. Indeed, as seen in Table 17.1, the ratio of the FN parameters to the corresponding total values is constant. Such a property is forced here by the high point group symmetry of the uranium ion site. In general, representability of the FN contributions in the intrinsic parameters is accounted for in a more consistent with the phenomenology formulation of the models referring to the virtual ligand idea introduced above.

The contact polarisation described in chapter 7 is the next lowest order mechanism foreseen in our perturbation approach that produce a non-axial contribution. This effect breaks also the superposition principle. Its precise evaluation, however, requires somewhat awkward calculation of three-centre integrals. Roughly estimated in terms of two-centre integrals this mechanism leads only to a negligible correction inasmuch as almost perfect cancellation of the direct and exchange contributions takes place.

17.2.4 Non-equivalent ligands

As seen from Eq.17.12 or specifically,

$$
\begin{aligned}
14b_2 &= 10e_\sigma + 15e_\pi \\
7b_4 &= 9e_\sigma + 3e_\pi - 21e_\delta \\
70b_6 &= 130e_\sigma - 195e_\pi + 78e_\delta
\end{aligned}
\tag{17.18}
$$

the models SM and AOM are algebraically equivalent for a single linear ligator. This automatically applies to a group of ligands provided they are identical and occupy the same crystallographically positions. We call them *equivalent ligands*. In practice two ligands are equivalent if they can be interchanged by a rotation only. Consequently, the whole CF effect generated by that group can be described by three intrinsic parameters, either b_k or e_μ. Now we consider a co-ordination cluster containing two non-equivalent ligands. Formally the number of the model intrinsic parameters increases twice. This would be so if ligands in the cluster were just different ions. However, if ligands differ merely in their distance from the central ion, then the corresponding sets of parameters cannot be treated as they were completely free and independent. This is because the intrinsic parameters vary with the metal-ligand (ML) distance in a certain specific for a given ML pair way. In the literature, there is a widely employed power function approximation:

$$b_k(R) \sim R^{-\alpha_k} \tag{17.19}$$

$$e_\mu(R) \sim R^{-\alpha_\mu} \tag{17.20}$$

To be more precise, various mechanisms discussed earlier give their own more or less complicated contribution and both the functions $b_k(R)$ and $e_\mu(R)$ are more complex. The above equations represent the simplest form to which any true dependence can be reduced in a sufficiently narrow range of R .

The exponents α_k and α_μ can be determined either from ab initio calculations or from the experimental data if they are treated as additional free parameters varied in certain limited range. Under some restrictions specified below they are considered the quantities characteristic of a given linear ligator. Accordingly, their values can be transferred from one compound to another containing the same metal-ligand pair. To some extent, the ordinary AOM and SM parameters, b_k or e_μ, manifest a similar universal character. We need only to scale them according to Eqs 17.19, 17.20 to account for the actual ML distance.

17.3   Towards applications

Note that the approximations Eqs 17.19,17.20 are not consistent with each other. If, for example, $e_\mu(R)$ decreases with increasing R according to Eq.17.20 then the relations Eq.17.18 force $b_k(R)$ to have a form of a sum of three power functions with three different exponents and *vice versa*. Moreover, within the AOM approximation, the characteristic for SM fixed ratios $B_{kq'}/B_{kq}$ become dependent on the e_μ parameters. In this sense the AOM and SM cannot be considered equivalent.

In principle, different is the physico-chemical context of these models. The superposition model can be seen as a phenomenological generalization of PCM or more precisely the electrostatic model. The two main AOM parameters e_σ and e_π , in turn, can be interpreted as energies of the anti-bonding states σ and π in analogy to the molecular orbital approach. The dominating character of $v_t^{c.cov} + v_t^{cov}$ observed in various numerical simulations seems to confirm the latter interpretation.

Unfortunately, most of the AOM results are scattered in the literature. Their systematization would certainly give a deeper insight into nature of the CF effect. Nevertheless even a cursory look at fragmentary data reveals encouraging properties. The data collected in Table 17.2 illustrate them pretty well.

Spectrochemical ordering of ligands, strong dependence on ionisation degree, monotonic variation along nf-electron series - all of these properties are clearly reflected in the AOM results. Moreover, the AOM parameters determined for the average ML distance R_0^t and the corresponding power exponents seems to be specific quantities for a given ML pair:

$$e_\mu^t|_{R_0^t} \simeq const \quad \alpha_\mu|_{R_0^t} \simeq const \qquad (17.21)$$

Their universal character allows one to apply a set of the parameters derived for one compound to another, containing the same pair of ions with appropriate scaling according to Eq.17.20 to account for actual ML distances. The ratios of the e_μ^t parameters corresponding to two different ligands t and t' or the ratios of the parameters of different μ for a specified ligator can also be treated as characteristic quantities:

$$\frac{e_\mu^t}{e_\mu^{t'}} \simeq const \quad \frac{e_\mu}{e_{\mu'}} \simeq const \qquad (17.22)$$

At least for simple ligands the following order of the parameters holds:

$$e_\sigma > e_\pi > |e_\delta| \qquad (17.23)$$

These distinguishing features of the e_μ^t parameters open wide possibilities of further reduction of the number of independent parameters or verification of results obtained within the standard model. Even if all the three parameters are allowed to vary in a particular fitting procedure the area of the acceptable solutions has to be limited to account for all the above well-established, not only in the microscopic theory, rules.

The mentioned general properties of the AOM parameters are not precisely observed for e_δ except perhaps the one: the absolute value of e_δ always is smaller than the values of the remaining two parameters. The ratio of the e_δ parameter to e_σ or e_π as well as the ratio $e_\delta^t/e_\delta^{t'}$ happen to be approximately fixed for isostructural series at most. We have already noticed that the main CF contribution represented by $v_t^{c.cov} + v_t^{cov}$ does not influence e_δ in the case of simple ligands.

226

Table 17.2 AOM parameters derived from experimental data mainly for various compounds and the power exponents characterising their dependence on ML distance calculated from first principles (given in parentheses).

M	L	Compound	R	e_σ (α_σ)	e_π (α_π)	e_δ (α_δ)	Ref.
Pr^{3+}	F^-	$Pr^{3+}:LiYF_4$	2.246	552	103		[292][a]
	Cl^-	$Cs_2NaPrCl_6$	2.73	455	185		[293][a]
Tb^{3+}	F^-	$LiTbF_4$	2.246	435	125		[292][a]
	Cl^-	$Cs_2NaTbCl_6$	2.69	324	135		[293][a]
U^{5+}	F^-	$CsUF_6$	2.057	3380 (6.8)	1090 (8.0)	-400	[132]
		β-UF_5	2.190	2660	840	430	[132]
U^{4+}	Cl^-	UCl_4	2.65	1892 (6.6)	710 (7.9)	-150 (8.6)	[132][b]
			2.88	981	355	20	
	Br^-	$U^{4+}:ThBr_4$	2.986	1210 (7.2)	420 (8.9)	280	[132]
	O^{2-}	UOS	2.357	2743 (4.9)	1698 (6.1)	507 (7.0)	[291][b,c]
		UO_2	2.369	1970	660	90	[132][d]
	Se^{2-}	UOSe	3.026	1702 (5.0)	763 (6.3)	260 (7.1)	[291][b,c]
	Te^{2-}	UOTe	3.099	1647 (4.4)	811 (6.3)	280 (6.9)	[291][b,c]
U^{3+}	Cl^-	UCl_3	2.932	596 (6.2)	226 (8.1)	32	[132]
	Br^-	UBr_3	3.090	459 (6.3)	177 (8.2)	24 (9.3)	[132]
	I^-	UI_3	3.277	461 (6.1)	166 (8.3)	63	[132]
Np^{4+}	Cl^-	$Np^{4+}:ThCl_4$	2.718	1132 (7.1)	632 (8.8)	251	[294][b]
	O^{2-}	NpO_2	2.355	1755 (6.8)	590 (7.7)	100	[132][d]
Pu^{4+}	O^{2-}	PuO_2	2.336	1530 (7.0)	510 (7.9)	100	[132][d]

[a] Obtained within conventional two-parameter AOM (e_σ, e_π).
[b] The AOM parameters have been determined from first principles calculation.
[c] Variation of the Madelung energy with the ML distance has been accounted for also in generation of the one-electron free-ionic wave functions.
[d] Roughly estimated on the ground of experimental data for UO_2 and ab initio calculations for the whole series AnO_2 with the constraint $e_\pi \approx e_\sigma/3$.

Hence, other mechanisms must determine its value, among which the important role is played by the ligand polarisation and the potential generated by further neighbours, i.e. the mechanisms, which strongly depend on details of the crystalline structure. This lattice-sensitivity of the e_δ parameter is clearly manifested in the model first principles calculations. As seen in Table 17.2, both its sign and value are irregular. Probably the e_δ parameter for the systems with d-electron ligands would behave in a more predictable way.

Some characteristic examples of application of angular overlap model, which exploit the regularity of the e_μ^t parameters are presented in Refs. [295, 296, 132, 297, 291]. They reveal an encouraging range of problems amenable to the analysis in terms of the intrinsic parameters:

- initial fitting of any experimental data for systems of arbitrary symmetry,

- simultaneous fitting of the energy spectra observed for several different compounds containing the same metal-ligand pair,

- simulation of the variation of the CF splitting induced by a continuous distortion of the co-ordination polyhedron geometry either in the incommensurate phase or by applying the "chemical" or hydrostatic pressure,

- calculation of Jahn-Teller energies, estimation of the strength of electon-phonon coupling (cf. chapter 18)

- verification of some artificial and unphysical solutions provided by the standard parametric analysis,

- interpretation of the inelastic neutron scattering spectra - prediction of the energies and eigenvectors corresponding to both visible and invisible in the spectra electronic transitions, which are crucial in explanation of the complex magnetic properties,

- estimation of the distance dependence of the CF parameters on the grounds of phenomenological data merely.

This list of possible applications of the simplified models is certainly not complete but it conveys an idea of their usefulness. Efficiency of the twin superposition model may be similar in some cases but in light of the ab initio calculations the AOM scheme provides the parameters, which seem to be more natural and more appropriate for both comparison and classification as well as for anticipation of their values.

It would be much-desired to adopt AOM or SM to more complicated systems with complex ligands, d-electron ligands or, in particular, to the intermetallic compounds. Known problems with a reliable determination of the CF effect in these systems would make so simple and efficient models especially attractive. We have already noticed the possibility of breaking down the axial metal-ligand symmetry. Very existence of the closed d-electron shell needs no a special extension of the models yet. Real problems are connected with presence of several kinds of particles in metallic systems. Interactions of the conduction electrons, a specific role played by the virtual bound state, other higher order terms in the perturbation expansion, which are expected to be more important for the intermetallic compounds than for the ionic systems, indicate that the simplified models under consideration cannot be retraced automatically. The modified point charge model by Żołnierek [111] can be considered the first step in this direction. Independently of further attempts of this kind, the basic idea of the decomposition of the CF potential in metallic systems needs to be established on sounder foundations. The perturbation scheme presented in previous chapters provides the right frames for this task. An alternative approach based on the density functional theory seems also to be promising in this respect provided its relation to the effective Hamiltonian for the localised states in metals is correctly set. Still quite fundamental question concerning validity of the efective Hamiltonian method for systems with nearly adjacent localised and band states has to be justified.

CHAPTER 18

Lattice Dynamics Contribution

So far we have taken into account the electronic Hamiltonian only and ignored the nuclear one as if they were separate entities. However, some elementary questions inevitably arise. How the static crystal field picture is influenced by lattice (or nuclear framework) dynamics? And, in what degree does the static approach remain adequate?

The basic assumption of our analysis is that it is possible to write down the total potential energy of the system of electrons and nuclei as an explicit differentiable function of the coordinates, $V(\mathbf{r}, \mathbf{Q})$, where $\mathbf{r}$ and $\mathbf{Q}$ denote the whole set of coordinates of the electrons, $\mathbf{r}_i$, $i = 1, 2, \ldots, n$, and nuclei, $\mathbf{Q}_\alpha$, $\alpha = 1, 2, \ldots, N$, respectively. Thus, the total Hamiltonian of the Schrödinger equation may be divided into three terms:

$$\mathcal{H} = \mathcal{H}_r + \mathcal{H}_Q + V(\mathbf{r}, \mathbf{Q}) \qquad (18.1)$$

where $\mathcal{H}_r$ is the pure electronic component, $\mathcal{H}_Q$ is the kinetic energy of the nuclei and $V(\mathbf{r}, \mathbf{Q})$ is the energy due to interaction of the electrons with the nuclei and internuclear repulsion. The instantenous potential seen by an electron in some nuclear framework (crystalline or molecular) can be divided into two parts. There is a static part $V(\mathbf{r}, 0)$ which is a function only of the mean nuclear position and is the basis of discussion in the remaining sections. There is also a dynamic part which depends on the displacements $\mathbf{Q}$ of the nuclei from their mean positions ($\mathbf{Q}_0 = 0$). The potential $V(\mathbf{r}, \mathbf{Q})$ can be expanded as a series of small displacements of the nuclei about the origin ($\mathbf{Q}_0$, i.e. $\mathbf{Q}_{\alpha 0} = \mathbf{Q}_{\beta 0} = \ldots = 0$)

$$
\begin{aligned}
V(\mathbf{r}, \mathbf{Q}) \;=\; & V(\mathbf{r}, 0) + \sum_\alpha \left(\frac{\partial V}{\partial \mathbf{Q}_\alpha} \right)_0 \mathbf{Q}_\alpha + \\
& + \frac{1}{2} \sum_{\alpha, \beta} \left(\frac{\partial^2 V}{\partial \mathbf{Q}_\alpha \partial \mathbf{Q}_\beta} \right)_0 \mathbf{Q}_\alpha \mathbf{Q}_\beta + \ldots
\end{aligned} \qquad (18.2)
$$

where $\mathbf{Q}_\alpha$ denotes here the displacement of α nucleus.

Solving the electronic part of the Schrödinger equation

$$[\mathcal{H}_r + V(\mathbf{r}, 0) - \varepsilon'_k]\varphi_k(r) = 0 \tag{18.3}$$

a set of energies ε'_k and wavefunctions $\varphi_k(r)$ for a given nuclear configuration corresponding to the point $\mathbf{Q}_0$ is obtained.

In order to see how these solutions vary under nuclear displacements, the full Schrödinger equation

$$(\mathcal{H} - E)\Psi(\mathbf{r}, \mathbf{Q}) = 0 \tag{18.4}$$

must be solved. The total wave function $\Psi(\mathbf{r}, \mathbf{Q})$ can always be expanded in terms of the complete set of electronic functions $\varphi_k(\mathbf{r})$, Eq.18.3.

$$\Psi(\mathbf{r}, \mathbf{Q}) = \sum_k \chi_k(\mathbf{Q})\varphi_k(\mathbf{r}) \tag{18.5}$$

where the expansion coefficients $\chi_k(\mathbf{Q})$ are functions of the nuclear coordinates. Substituting Eq.18.5 into Eq.18.4 one obtains the following system of coupled equations for the functions $\chi_k(\mathbf{Q})$:

$$\{\mathcal{H}_Q + [\varepsilon'_k + W_{kk}(\mathbf{Q})] - E\}\chi_k(\mathbf{Q}) + \sum_{m \neq k}' W_{km}(\mathbf{Q})\chi_m(\mathbf{Q}) = 0 \tag{18.6}$$

where $W_{km}(\mathbf{Q})$ denotes the electronic matrix element of so called vibronic interactions, i.e. that part of the electron-nuclear interaction $V(\mathbf{r}, \mathbf{Q})$ which depends on $\mathbf{Q}$, i.e.

$$\begin{aligned} W(\mathbf{r}, \mathbf{Q}) &= V(\mathbf{r}, \mathbf{Q}) - V(\mathbf{r}, 0) = \sum_\alpha \left(\frac{\partial V}{\partial \mathbf{Q}_\alpha}\right)_0 \mathbf{Q}_\alpha + \\ &+ \frac{1}{2}\sum_{\alpha, \beta} \left(\frac{\partial^2 V}{\partial \mathbf{Q}_\alpha \mathbf{Q}_\beta}\right)_0 \mathbf{Q}_\alpha \mathbf{Q}_\beta + \dots \end{aligned} \tag{18.7}$$

Let us note that

$$\varepsilon_k(\mathbf{Q}) = \varepsilon'_k + W_{kk}(\mathbf{Q}) \tag{18.8}$$

is the potential energy of the nuclei in the mean field of the electrons in state $\varphi_k(\mathbf{r})$.

18.1 Adiabatic and harmonic approximations

If vibronic mixing of different electronic states can be ignored the system of equations (18.6) decomposes into a set of simple equations

$$[\mathcal{H}_Q + \varepsilon_k(\mathbf{Q}) - E_k]\chi_k(\mathbf{Q}) = 0 \tag{18.9}$$

In other words, the motions of the nuclei and electrons are separated and the total wave function has the form $\Psi(\mathbf{r}, \mathbf{Q}) = \varphi_k(\mathbf{Q})\chi_k(\mathbf{Q})$ and its energy is a sum of ε_k' and the corresponding vibrational energy of the nuclear framework. This is the simple adiabatic approximation, or the Born-Oppenheimer approximation [298, 299, 300], and $\varepsilon_k(\mathbf{Q})$ plays a role of the adiabatic potential. The adiabatic approximation is the crucial point of the approach. According to it the large difference between the masses of the electron and nucleus causes a significant difference in their motion and a stationary electronic state is attained for every instantaneous nuclear configuration. In other words, the electrons adiabatically and noninertially follow the motions of the nuclei whereas the latter are moving in the averaged field produced by the electrons. The error introduced by the simple adiabatic approximation is of order of $(m/M)^{1/2} \simeq 0.02$ [299] where m and M are the electron and nuclear masses, respectively, which is sufficiently small. The most essential feature of the Born-Oppenheimer adiabatic approximation is the form of eigenfunctions which are the individual products $\varphi_k(\mathbf{r})\chi_k(\mathbf{Q})$, where $\varphi_k(\mathbf{r})$ is the electronic eigenfunction for the equilibrium nuclear configuration $(\mathbf{Q}_0)$ and $\chi_k(\mathbf{Q})$ describes the distribution of the nuclei for the electronic distribution $\varphi_k(\mathbf{r})$. Since the vibronic interactions are rather weak in comparison with the electronic energy differences $\varepsilon_m' - \varepsilon_k'$ the perturbation approach can be applied. As is well known, the perturbation theory for singlet states is different from that for degenerate states. In the former case it leads to the adiabatic approximation whereas for degenerate states it results in the vibronic Hamiltonian which allows for the nonadiabatic mixing of electronic states having the same or close energies. Then, the simple Born-Oppenheimer products are not, in general, the proper eigenfunctions but their superpositions (Eq.18.5) are as accurate as the Born-Oppenheimer functions in the nondegenerate case [306].

The next essential step is the harmonic approximation, i.e. expansion of the $V(\mathbf{r}, \mathbf{Q})$ up to and including quadratic terms. This implies that the nuclear displacements from the stable configurations are small as compared with the interatomic distances. Nevertheless, the error of the harmonic approximation is much greater than that of the adiabatic approximation and introduces an error of the order of $(m/M)^{1/4} \simeq 0.15$ [299]. In other words, in order to improve the solutions obtained in the adiabatic and harmonic approximations, the anharmonicity has to be taken into account first.

In the adiabatic approximation for nondegenerate electronic states normal vibrations (oscillator solutions) can be determined by Eq.18.9 in which the adiabatic potential (Eq.18.8) is taken in the harmonic approximation. The quadratic form of $W_{kk}(\mathbf{Q})$ expressed as

$$W_{kk}(\mathbf{Q}) = \sum_{\alpha} \left(\frac{\partial V_{kk}}{\partial \mathbf{Q}_\alpha}\right)_0 \mathbf{Q}_\alpha + \frac{1}{2}\sum_{\alpha,\beta} \left(\frac{\partial^2 V_{kk}}{\partial \mathbf{Q}_\alpha \mathbf{Q}_\beta}\right)_0 \mathbf{Q}_\alpha \mathbf{Q}_\beta \qquad (18.10)$$

can be reduced by means of normal coordinates to canonical (diagonal) form and the kinetic energy operator maintains its additive form $\mathcal{H}_Q = -(\hbar^2/2)\sum_\alpha M_\alpha^{-1}\left(\frac{\partial}{\partial Q_\alpha}\right)^2$. Here V_{kk} stands for the diagonal matrix element of V and $\varphi_k(\mathbf{r})$ state. With normal coordinates, i.e. the appropriate combinations of the $Q_{\alpha i}$ denoted as $Q_{\bar\Gamma\bar\gamma}$ (see below), Eq.18.9 for nuclear motion separates into equations of individual harmonic oscillators

$$-\frac{\hbar^2}{2M_{\bar\Gamma}}\frac{\partial^2\chi_{k,\bar\Gamma\bar\gamma,n}}{\partial Q_{\bar\Gamma\bar\gamma}^2} + \frac{1}{2}\omega_{k,\bar\Gamma}^2 Q_{\bar\Gamma\bar\gamma}^2 \chi_{k,\bar\Gamma\bar\gamma,n} = \epsilon_{k,\bar\Gamma\bar\gamma,n}\chi_{k,\bar\Gamma\bar\gamma,n} \tag{18.11}$$

where $M_{\bar\Gamma}$ is the reduced mass of the $\bar\Gamma$ normal vibration, $\omega_{k,\bar\Gamma}^2$ its frequency (both independent on $\bar\gamma$), the subscript k specifies the coupled electronic state, $\bar\Gamma\bar\gamma$ denotes the vibrational mode (can be degenerate) and n is the oscillator quantum number. One should remember that each $\omega_{k,\bar\Gamma}$ has its own quantum numbers n, so strictly n ought to be written as n_F. The solutions of Eq.18.11 are well known [301]

$$\epsilon_{k,\bar\Gamma\bar\gamma,n} = \hbar\omega_{k,\bar\Gamma}\left(n + \tfrac{1}{2}\right) \qquad n = 1, 2 \ldots$$
$$\chi_{k,\bar\Gamma\bar\gamma,n}(Q_{\bar\Gamma\bar\gamma}) = (n!\,2^n\,\sqrt{\pi})^{-1/2} H_n(Q_{\bar\Gamma\bar\gamma})\exp\left(-\frac{M_{\bar\Gamma}\omega_{k,\bar\Gamma}Q_{\bar\Gamma\bar\gamma}^2}{2\hbar}\right) \tag{18.12}$$

where $H_n(Q_{\bar\Gamma\bar\gamma})$ are Hermite polynomials.
The total energy of k-th Born-Oppenheimer product state $\varphi_k(\mathbf{r})\chi_k(\mathbf{Q})$ amounts to

$$E_k = \varepsilon_k' + \sum_{\bar\Gamma,n}\epsilon_{k,\bar\Gamma,n} \tag{18.13}$$

where the summation runs over all possible normal vibrations $\bar\Gamma$ (including degeneracy) of corresponding quantum numbers n. The simple adiabatic approximation is valid if and only if the terms of the vibronic mixing of different electronic states in Eq.18.6, W_{km}, can be ignored. A simple criterion of applicability of the adiabatic approximation is [299]

$$\hbar\omega \ll |\varepsilon_m' - \varepsilon_k'| \tag{18.14}$$

where $\hbar\omega$ is the energy quantum of vibrations in the electronic state under consideration (k or m) and ε_m' and ε_k' are the energy of the electronic levels.

Combined adiabatic and harmonic approximations enable us to solve the vibronic problem for nondegenerate states with a satisfactory accuracy. The Born-Oppenheimer products of static electronic functions $\varphi_k(\mathbf{r})$ (for stable nuclear configurations) and oscillator nuclear functions $\chi_k(\mathbf{Q})$ are the solution of the problem. So, these approximate eigenstates may be called as electron-vibrational ones. For degenerate electronic states the adiabatic approximation is invalid and in consequence the solution of the problem is more complicated. Before we pass to the problem let us consider the transformational properties of all quantities entering the vibronic Hamiltonian starting from the coordinates.

18.2 Collective (normal) coordinates and the "quasi-molecular" model

A short comment should be given now on methods of presentation of distortions of nuclear framework resulting from its motion. Since the Cartesian nuclear coordinates $Q_{\alpha i}$ (α label the nuclei and i the Cartesian directions of displacement) do not transform in a simple way under the operations of the symmetry group of the system their using in description of V is not very convenient in practice. It is always possible to choose linear combinations of $Q_{\alpha i}$ which do transform according to the irreducible representations of the point symmetry group [302]. These are the collective or symmetrized coordinates. Those of them with the type of symmetry which occurs only once in the group-theoretical classification for the system under consideration are automatically the normal coordinates. For repeated types of symmetrized displacements the normal coordinates are linear combinations of the symmetrized ones. Thus, the nuclear coordinates (displacements) are classified by the irreducible representation Γ of the appropriate symmetry group G and its row γ (if degenerate) and are denoted by $Q_{\Gamma\bar{\gamma}}$ (overlined symbols are reserved for operators). Normal coordinates can be related to normal vibrations which considerably simplifies the problem. Both the total Hamiltonian and the potential V are invariants (scalars) of the point symmetry group of the states under consideration. This is not trivial. Obviously, the Hamiltonian has to be invariant of its own point symmetry group but the states on which it acts can transform according to representations of a more general group. For example, the crystal field potential is the invariant of its own definite point group but is not it for the $O(3)$ group (free ion point group) and hence a splitting may occur.

If the matrix element of an operator transforming as $\bar{\Gamma}$ between states transforming as Γ and Γ', respectively, is not to vanish identically, the direct product $\Gamma \times \bar{\Gamma} \times \Gamma'$ must contain the identity representation A_1, that is, $\bar{\Gamma}$ must be contained in the decomposition of the simple product $\Gamma \times \Gamma'$, i.e. $\bar{\Gamma} \in \Gamma \times \Gamma'$. In order to complete the group-theoretical classification of all quantities forming the Hamiltonian and its matrix elements we have to know the transformational properties of first and second derivatives of the potential V. Since V is invariant, $\partial V/\partial Q_{\Gamma\bar{\gamma}}$ transforms in the same way as $Q_{\Gamma\bar{\gamma}}$, i.e. as $\bar{\Gamma}\bar{\gamma}$. Let us introduce the notation

$$V_{\bar{\Gamma}\bar{\gamma}}(\mathbf{r}) = \left(\frac{\partial V(\mathbf{r}, \mathbf{Q})}{\partial Q_{\bar{\Gamma}\bar{\gamma}}} \right)_0 \qquad (18.15)$$

while the quantities $\partial^2 V/\partial Q_{\bar{\Gamma}_1\bar{\gamma}_1} \partial Q_{\bar{\Gamma}_2\bar{\gamma}_2}$ have the same properties of the "second rank" tensor $\Gamma_1\Gamma_2$. There are as many symmetrized combinations of

$$V_{\bar{\Gamma}_1\bar{\gamma}_1\bar{\Gamma}_2\bar{\gamma}_2}(\mathbf{r}) = \left(\frac{\partial^2 V(\mathbf{r}, \mathbf{Q})}{\partial Q_{\bar{\Gamma}_1\bar{\gamma}_1} \partial Q_{\bar{\Gamma}_2\bar{\gamma}_2}} \right)_0 \qquad (18.16)$$

as components in decomposition of $\bar{\Gamma}_1 \times \bar{\Gamma}_2$. Now, we can present Eq.18.7 as a sum of scalar convolutions of irreducible tensors

$$
\begin{aligned}
W(\mathbf{r},\mathbf{Q}) &= \sum_{\bar{\Gamma}\bar{\gamma}} V_{\bar{\Gamma}\bar{\gamma}}(\mathbf{r}) Q_{\bar{\Gamma}\bar{\gamma}} + \\
&+ \frac{1}{2} \sum_{\bar{\Gamma}_1,\bar{\Gamma}_2} \sum_{\substack{\bar{\Gamma}\bar{\gamma} \\ \Gamma \in \bar{\Gamma}_1 \times \bar{\Gamma}_2}} \{V(\bar{\Gamma}_1 \times \bar{\Gamma}_2)(\mathbf{r})\}_{\bar{\Gamma}\bar{\gamma}} \times \\
&\times \{\mathbf{Q}_{\bar{\Gamma}_1} \times \mathbf{Q}_{\bar{\Gamma}_2}\}_{\bar{\Gamma}\bar{\gamma}}
\end{aligned}
\tag{18.17}
$$

where

$$
\{V(\bar{\Gamma}_1 \times \bar{\Gamma}_2)(\mathbf{r})\}_{\bar{\Gamma}\bar{\gamma}} = \sum_{\bar{\gamma}_1,\bar{\gamma}_2} V_{\bar{\Gamma}_1\bar{\gamma}_1\bar{\Gamma}_2\bar{\gamma}_2}(\mathbf{r}) \langle \bar{\Gamma}_1\bar{\gamma}_1\bar{\Gamma}_2\bar{\gamma}_2 | \bar{\Gamma}\bar{\gamma} \rangle
$$

$$
\{\mathbf{Q}_{\bar{\Gamma}_1} \times \mathbf{Q}_{\bar{\Gamma}_2}\}_{\bar{\Gamma}\bar{\gamma}} = \sum_{\bar{\gamma}_1,\bar{\gamma}_2} Q_{\bar{\Gamma}_1\bar{\gamma}_1} Q_{\bar{\Gamma}_2\bar{\gamma}_2} \langle \bar{\Gamma}_1\bar{\gamma}_1\bar{\Gamma}_2\bar{\gamma}_2 | \bar{\Gamma}\bar{\gamma} \rangle
$$

(18.18)

and $\langle \bar{\Gamma}_1\bar{\gamma}_1\bar{\Gamma}_2\bar{\gamma}_2 | \bar{\Gamma}\bar{\gamma} \rangle$ are the Clebsch-Gordan coefficients of the point group G. As is seen, Eq.18.7 differs from Eq.18.18 only in a suitable rearrangement of terms of the second sum.

In crystal lattice the point symmetry of central ion allows only a limited number of possible normal vibrations (modes) to be coupled to the localized electronic states of specified symmetry (irreducible representation) as in isolated molecule.

As a first approximation in the crystal lattice only motions of the central atom and its immediate neighbors forming a quasi-molecule contribute appreciably to $V(\mathbf{Q})$. And although in a real crystal the $Q_{\bar{\Gamma}\bar{\gamma}}$ are not exactly normal coordinates we find the "quasi-molecular" approximation to be quite adequate at least for the electronic part of the Hamiltonian. In a perfect crystal the normal coordinates are plane waves and each one makes only an infinitesimal contribution to the displacement at any particular lattice site. In such a crystal we have [303]

$$
Q_{\bar{\Gamma}\bar{\gamma}} = (3N)^{-1} \sum_{\mathbf{k},r} \left[S_{\bar{\Gamma}\bar{\gamma}}(\mathbf{k},r) a_{\mathbf{k},r} + S_{\bar{\Gamma}\bar{\gamma}}^*(\mathbf{k},r) a_{\mathbf{k},r}^\dagger \right]
$$

(18.19)

$$
\text{and} \qquad \mathcal{H}_{\text{latt}} = \sum_{\mathbf{k},r} \hbar\omega_{\mathbf{k},r} \left[a_{\mathbf{k},r}^\dagger a_{\mathbf{k},r} + \tfrac{1}{2} \right]
$$

where $a_{\mathbf{k},r}^\dagger$, $a_{\mathbf{k},r}$ are the creation and destruction operators for the phonon from branch r with the wave vector $\mathbf{k}$ and frequency $\omega_{\mathbf{k},r}$, N is the number of ions in the crystal, $S_{\bar{\Gamma}\bar{\gamma}}(\mathbf{k},r)$ is the projection of this phonon onto $Q_{\bar{\Gamma}\bar{\gamma}}$, that is, a measure of the contribution of that particular phonon to the nuclear displacement described by $Q_{\bar{\Gamma}\bar{\gamma}}$. Thus, $Q_{\bar{\Gamma}\bar{\gamma}}$ has associated with it not a single frequency $\omega_{\bar{\Gamma}}$ but the whole range of the phonon spectrum weighted by $|S_{\bar{\Gamma}\bar{\gamma}}|^2$ and of course, by the phonon density of states. In the case of doped ion or impurity certain modes may have exceptionally large amplitudes. These are local modes if its ω is outside the allowed frequency band for the perfect crystal and quasi-local or resonance modes if within it.

18.3   The Jahn-Teller effect

In the case of degenerate or pseudo-degenerate states the situation differs radically. Degenerate electronic energy levels are reasonably sensitive to distortions of the nuclear framework. In other words, the vibronic interaction, Eq.18.17, i.e. the nuclear motion causes intense mixing the electron-vibrational states leading to so-called vibronic states, and the appropriate energy spectrum is called vibronic, as distinguished from the electron-vibrational states and corresponding electron-vibrational spectrum considered above.

While for a nondegenerate electronic state $\varphi_k(\mathbf{r})$ the adiabatic potential $\varepsilon_k(\mathbf{Q})$ in the form of an individual sheet in the $(\varepsilon(\mathbf{Q}), \mathbf{Q})$ space is given by Eq.18.8, for f-fold degenerate state the adiabatic potential has f sheets $\varepsilon_k(\mathbf{Q})$ for $k = 1, 2, \ldots, f$, which intersect at the point of degeneracy. The functions $\varepsilon_k(\mathbf{Q})$ can be obtained as solutions of the secular equation

$$
\begin{vmatrix}
W_{11} - \varepsilon & W_{12} & \cdots & W_{1f} \\
W_{21} & W_{22} - \varepsilon & \cdots & W_{2f} \\
\vdots & \vdots & & \vdots \\
W_{f1} & W_{f2} & \cdots & W_{ff} - \varepsilon
\end{vmatrix} = 0 \tag{18.20}
$$

where W_{ij} are the matrix elements of the vibronic interaction operator (Eq.18.17) calculated with the wave functions of the degenerate electronic term.

This is a general and immanent property of degenerate electronic levels (exceptions will be specified below) known commonly as the Jahn-Teller (JT) effect [299, 300, 303, 304]. The Jahn-Teller effect is the intrinsic instability of an electronically degenerate complex against distortions that remove the degeneracy in first order. Owing to fundamentality of the effect the last formulation has been raised to a standing of theorem and is known as the Jahn-Teller theorem. There is a straightforward formal proof of the theorem [299]. It is also a consequence of the Hellmann-Feynman theorem [303, 305] which states that any electronic state φ_k acts on the nuclei in the sense to increase the coordinate $Q_{\Gamma\bar{\gamma}}$ by the force

$$
F_{k,\Gamma\bar{\gamma}} = -\left\langle \varphi_k \left| \frac{\partial V}{\partial Q_{\Gamma\bar{\gamma}}} \right| \varphi_k \right\rangle \tag{18.21}
$$

If $F_{k,\Gamma\bar{\gamma}}$ is nonzero for a certain nuclear configuration the configuration is unstable and will spontaneously distort until $F_{k,\Gamma\bar{\gamma}}$ is zero. The JT theorem states, that if φ_k is a degenerate state when $Q_{\bar{\Gamma}\bar{\gamma}} = 0$, the matrix element, Eq.18.21, exists for some nontotally symmetric coordinate $Q_{\bar{\Gamma}\bar{\gamma}}$. The instability resulting from existing nonzero matrix elements of the linear vibronic operator, i.e. from acting the distorting or JT force is responsible for so-called static instability. This is the dominating part of the JT effect. The spontaneous distortion balanced against the "elastic force" (covalent and electrostatic) will occur until a new equilibrium is reached at some finite value of $Q_{\bar{\Gamma}\bar{\gamma}} = Q_{0,\bar{\Gamma}\bar{\gamma}}$. Another kind of instability – so named dynamic instability is also possible when the linear terms are equal to zero and the quadratic ones are respectively strong.

There are two exceptions of the JT effect. The first refers to the Kramers degeneracy which is due to symmetry with respect to inversion of time and which cannot be violated by any crystal structure distortion.

The second is a linear configuration of the atoms at the equilibrium point $\mathbf{Q}_0$, i.e. linear molecules whose degeneracy no distortion can remove in first order. The latter case is obviously unlikely to be of interest in solids. However, in this case the dynamic (quadratic) instability or so named Renner effect [300, 307] can be effective.

The JT effect is an inevitable consequence of non-existence of an extremum point of the adiabatic potential at its degeneracy point. Hence, an equivalent formulation of the JT theorem sounds: If the adiabatic potential of the system which is formal solution of the electronic part of the Schrödinger equation has several crossing sheets then at least one of these sheets has no extremum at the crossing point. This version of the theorem seems to be even more appropriate than the previous one because it does not directly hint at the necessity of a low symmetry distortion of the nuclear framework. We should realize that whereas the JT effect affects always any degenerate electronic state a static low symmetry distortion arising from it is observed rather rarely and that under some additional conditions.

Let us investigate the sheets $\varepsilon(\mathbf{Q})$ in a small region surrounding the point of degeneracy $\mathbf{Q}_0$. To do this it is enough to take into account only the linear vibronic terms. In the immediate vicinity of the point of degeneracy the roots of the secular equation (Eq.18.20) are also linearly dependent on $\mathbf{Q}$ and the $\varepsilon(\mathbf{Q})$ sheets are conically shaped surfaces with the apex at the point $\mathbf{Q}_0$ (Fig.18.1). It follows that at least one of the eigenvalues of Eq.18.20 decreases initially linearly with increasing the corresponding $Q_{\bar{\Gamma}\bar{\gamma}}$. By the terms of Wigner-Eckart theorem all matrix elements of any operator within a degenerate state can be presented as a product of two factors: the so-called reduced (double bar) matrix element which is the same for all the elements and is independent on the choice of reference system and the corresponding Clebsch-Gordan coefficient varying from element to element. The reduced matrix element is a measure of the strength of the electron-lattice coupling and is called the coupling constant. So, the interaction matrix is proportional to the Clebsch-Gordan coefficients matrix and the

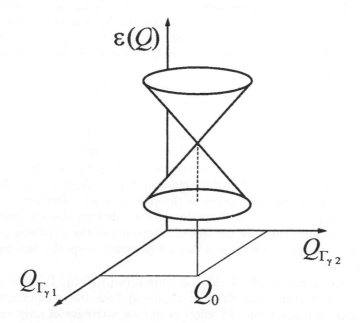

Fig. 18.1 Jahn-Teller splitting of the adiabatic potential in the vicinity of the degeneracy point. Schematic diagram for a two-fold degenerate distortion mode

reduced matrix element is the proportionality factor. The Clebsch-Gordan coefficients matrices distinguish themselves by an important characteristic – for any nontotally symmetric mode $Q_{\bar{\Gamma}\bar{\gamma}}$ their trace is equal to zero [308].

$$\mathbf{Tr}\left\langle \varphi_i \left| \frac{\partial V}{\partial Q_{\bar{\Gamma}\bar{\gamma}}} \right| \varphi_j \right\rangle = 0 \qquad (18.22)$$

where φ_i and φ_j are orthogonal substates of Γ state. Thus, the centroid of Γ is not shifted to first order in $Q_{\bar{\Gamma}\bar{\gamma}}$ [303], i.e. the nontotally symmetric modes are in no condition to shift position of the energy center of gravity of degenerate levels.

However, the question remains – what is about the splitting of degenerate levels? The JT instability forces the nuclear system to be spontaneously distorted so that the electronic levels undergo splitting and become nondegenerate. Therefore we should never observe the degenerate levels. On the other hand the degenerate levels are commonly observed if the symmetry conditions are obeyed and that the JT effect is mysteriously elusive in experiments and low symmetry distortions attributed to it are met in isolated cases only and are relatively weak, often considerably weaker than the expectation effects. No less than fifteen years had to elapse from theoretical prediction (1937) of the phenomenon and its first unambiguous evidence (1952) [309]. This situation is well characterized by a sentence from the book *"Paramagnetic Resonance in Solids"* by W. Low [310] – "it is a property of the Jahn-Teller effect that whenever one tries to find it, it eludes measurements".

The clue to an understanding this apparent paradox is realization of the fact that the vibronic Hamiltonian is a scalar of the point symmetry group of the states on which it acts and therefore the overall degeneracy of a level, i.e. that degeneracy associated with the multiplicity of a given irreducible representation cannot be reduced by the JT effect. However, the original electronic degeneracy in which an operation of the group changes the electronic and vibrational wave functions independently is replaced by vibronic degeneracy in which transformation of the electronic wave function is inextricably accompanied by transformation of the vibrational wave function. It is clear that knowledge on the lack of extremum of the adiabatic potential at the degeneracy point is not enough in order to specify the nuclear behaviour.

The properly conceived JT effect is widespread indeed. It refers to all degenerate levels (excluding Kramers doublets) both the ground and the excited ones. Moreover, the JT effect is not an attribute of only exactly degenerate levels originating from the same irreducible representation. A similar effect occurs in the case of close-in-energy (quasi- or pseudo- degenerate) electronic levels. This is so-called pseudo JT effect. The vibronic mixing the interacting levels is the driving force of the effect. In other words, the degree of departure from the adiabatic approximation is the decisive factor. Obviously, in the case of true degeneracy the most outlined effect takes place. Its magnitude varies within a very wide range. It can reach even 10^4 cm^{-1} for the transition metal ions [311, 312, 38, 313] and be small or very small, from 0.01 to 1 cm^{-1} for e.g. pseudo-degenerate systems including the Kramers ones or for S-state ions, e.g. Mn^{2+}, where only higher order spin-orbit perturbations are effective [303]. Similarly, the energy levels of an ion in an f^n configuration, i.e. all trivalent rare earths, most divalent ones in or near the ground state and most highly ionized actinides are rather insensitive to the nuclear configuration and the JT energy is usually small. The JT effect can also be observed in the case of pseudo-degenerate Kramers doublets as it occurs in orthovanadates, e.g. in $DyVO_4$ [314].

Confusing the JT energy which is the reduction in the potential due to the JT effect with the splitting of actual energy levels is a captions trap which leads to the error frequently made.

The aforementioned lack of the splitting the degenerate electronic-lattice states by the vibronic potential may be compared with the effect of acting the spin-orbit coupling operator $(\lambda \mathbf{L} \cdot \mathbf{S})$ within the basis of $|LSM_L M_S\rangle$ functions, when similarly the total "degeneracy" is preserved because

$(2L + 1)(2S + 1) = \sum_J (2J + 1)$. However, there is an evident difference between the two cases that consists in the fact that in the latter case there are several scalar values of $(\lambda \mathbf{L} \cdot \mathbf{S})$ according to different resultant J values (the angles between $\mathbf{L}$ and $\mathbf{S}$). The non-existence of the adiabatic potential extremum at the degeneracy point of electronic states is automatically equivalent to existing the distorting force which in balance with the elastic restoring force yields a new position of equilibrium point $Q_{0,\Gamma\bar{\gamma}}$ where the distorting force vanishes and the potential reaches its minimum value, i.e. the JT or stabilization energy, E_{JT} (measured in relation to the energy at $\mathbf{Q}_0$ point).

$Q_{0,\Gamma\bar{\gamma}}$ varies linearly with the coupling constant $\langle \Gamma || \frac{\partial V}{\partial \mathbf{Q}_\Gamma} || \Gamma \rangle$ and inversely proportionally with the oscillator effective force constant $M_{\bar{\Gamma}} \omega_{\bar{\Gamma}}^2$. The stabilization energy, E_{JT}, depends in the same way on the force constant but is proportional to the square of the coupling constant. After such a distortion ($\mathbf{Q}_0 \rightarrow \mathbf{Q}_{0,\Gamma\bar{\gamma}}$) the corresponding adiabatic potential attains several equivalent minima (with respecting the point group symmetry) at different points in the distorting mode space. In some simplified models their number can be infinite and they form continuous curves. It should be emphasized that, in general, the JT distorting modes are degenerate too. A good illustration of this situation is shape of the adiabatic potential for a two-fold degenerate E term interacting linearly with a doubly degenerate e-type vibrations (linear E-e problem) [315, 316] – so-called "Mexican hat" (Fig.18.2).

It is convenient to use polar coordinates (ϱ, θ) in the (Q_{e_1}, Q_{e_2}) plane. Then $Q_{e_1} = \varrho \cos \theta$ and $Q_{e_2} = \varrho \sin \theta$. The "Mexican hat" is simple the surface generated by rotating the parabola $\varepsilon(\varrho) = \pm \langle E || \frac{\partial V}{\partial \mathbf{Q}_e} || E \rangle \varrho + \frac{1}{2} M_e \omega_e^2 \varrho^2$ about the $\varrho = 0$ axis. The adiabatic potential reaches its minimum at all points of the circle $\varrho = |\mathbf{Q}_{0,e}|$ in the plane of (Q_{e_1}, Q_{e_2}) displacements. Since all directions of the distortion are equivalent in this simplified linear model approach the JT effect causes no symmetry change and may be overlooked at all. This is a typical example of so-called dynamic JT effect.

In fact the curve consisting of minima of the adiabatic potential surface is more complicated. The smooth paraboloidic $\varepsilon(\varrho)$ sheet (which, in general, is not necessarily solid of revolution) warps and the curve runs over three wells alternating regularly with three humps (the potential barriers). In addition its axial symmetry $C_{\infty v}$ is reduced to C_{3v}. It is roughly shown with the dashed line in Fig.18.2. This profile is shaped by mechanisms which are of secondary meaning with respect to the linear JT effect, viz. by anharmonicity and the quadratic JT effect [317]. Thus, the system can be frozen into one of the minima leading to a permanent distortion. This is called the static JT effect. At high temperatures thermal activation over the potential barrier that separates the minima can occur. It is also possible for the complex to go over from one potential minimum to another, even at zero

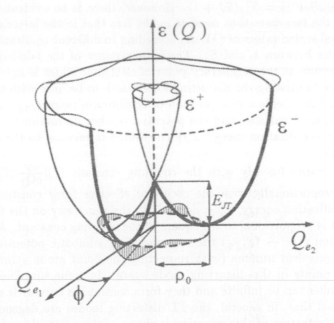

Fig. 18.2 The adiabatic potential sheets (*Mexican hat*) for a linear *E-e* problem. Q_{e_1}, Q_{e_2} coordinates describe the two-fold *e*-type vibration. The dotted lines show the deformations when anharmonic (or quadratic) terms are included and the axial symmetry is replaced by C_{3v} one. The lower dotted line links the saddle points and the minima on the $\varepsilon(Q)$ surface

temperature, by quantum-mechanical tunneling. This is because the wave functions describing the system as localized in one or another minimum are not orthogonal and are therefore not true eigenfunctions. Such averaging by whatever means achieved we call just the dynamic JT effect.

There are three general ways in which the JT effect might be expected to have observable consequences [303]:

- direct lowering of site symmetry,

- difference between the vibronic energy levels and the vibrational levels the system would have if there were no JT effect,

- effect on electronic operators of replacing electronic degeneracy by vibronic degeneracy, this can drastically affect the values of the matrix elements.

To be detectable, the effect, e.g. the mean distortion, must be finite when averaged over a characteristic time of the experiment and then such a "permanent" lowering of the site symmetry is the static JT effect. If the distortion averages out to zero we have a dynamic JT effect. The same system can, in principle, show a static effect in an experiment with a short characteristic time, e.g. ESR or a dynamic effect in an experiment with a relatively long characteristic time such as NMR.

If the JT active ions are far enough apart not to interact the dynamic distortions around different centers are not mutually correlated and the initial symmetry is preserved. When the interaction between the centers is taken into account the mean "molecular field" produced by the environment and influencing each center is not isotropic and distortions equivalent in free centers become nonequivalent (soft modes). New properties of the crystal arising from the ordering (correlation) of the JT center distortions, including the formation of new crystal structures and structural phase transitions are called cooperative JT effect. The structure of such a crystal is considerably different from that expected without vibronic interaction. Strictly, such a crystal state is not stationary, but due to the infinitely large barrier heights it may remain there for an infinitely long time. The temperature dependent breakdown of the JT center distortion correlations is nothing else than a structural phase transition which formally is isomorphic with the magnetic ordering transition (the second order phase transition with characteristic plots of splitting the vibronic level versus temperature, heat capacity vs. temperature etc.). The cooperative splitting of the vibronic level plays here a role of magnetization in the magnetic molecular field approach and the soft modes a role of easy directions of magnetization.

The JT structural phase transitions may yield important physical consequences. The ordering of the JT distortions in some noncentrosymmetrical crystal may lead to spontaneous polarization of the lattice and to ferroelectricity [318]. This is the right moment to finish the short review on the JT effect and come back to our main question – how the energy level spectrum of an paramagnetic ion in crystal, and in consequence the crystal field parameters are modified by the lattice dynamics?

18.4 Lattice dynamics and the crystal field effect

On the ground of the short analysis of the JT effect given above one can state that nontotally symmetric modes (vibrations) are effective for degenerate states only leaving nondegenerate ones intact in the frame of the adiabatic model. As long as we deal with the dynamic JT effect which is omnipresent for degenerate levels (beyond perhaps the low temperature range) the actual energy level system is practically unperturbed with reference to the corresponding solely electronic one, at least if the nontotally symmetric modes are taken into account. This somewhat surprising conclusion results from preservation of degeneracy of the levels and maintaining their energy centers of gravity in fixed positions. In turn, when the static JT occurs (below a certain temperature) which in crystal lattice manifests itself as the cooperative effect, a qualitatively new situation with a new point group of lower symmetry takes place.

What remains to be considered is the impact of the nuclear framework dynamics on the curvature of the adiabatic potential of individual electronic states. This curvature represents their vibrational energy.

Let us remember that the operator of vibronic interactions in normal coordinates $Q_{\Gamma\tilde{\gamma}}$ may be written in the form of Eq.18.17. The coefficients of this expansion are derivatives of the electron-nuclear interaction and are named vibronic constants. The vibronic constants characterize the measure of influence of the nuclear displacements on the electron distribution and, conversely, the effect of the changes in the vibronic structure upon nuclear dynamics. Since the vibronic interactions within degenerate electronic systems do not obey the adiabatic assumption the degenerate and nondegenerate states have to be clearly distinguished.

Suppose that the electronic states transform as $\Gamma, \Gamma' \ldots$ irreducible representations or more strictly as some their components $\gamma, \gamma' \ldots$, respectively, and similarly, the normal coordinates of nuclear displacements as $\bar{\Gamma}\bar{\gamma}, \bar{\Gamma}'\bar{\gamma}' \ldots$ components of the irreducible representations of the same point symmetry group G of the system. The matrix element

$$F_{\bar{\Gamma}}^{\Gamma\Gamma'} = \left\langle \Gamma \left\| \left(\frac{\partial V}{\partial \mathbf{Q}_{\bar{\Gamma}}} \right) \right\| \Gamma' \right\rangle \qquad (18.23)$$

is the linear vibronic constant. $F_{\bar{\Gamma}}^{\Gamma\Gamma'}$ is nonzero if and only if $\bar{\Gamma} \in \Gamma \times \Gamma'$. Let us consider first the diagonal element ($\Gamma = \Gamma'$) and the case of nondegenerate level. Then, we always have $\Gamma \times \Gamma = A_1$ and hence $\bar{\Gamma} = A_1$, i.e. the totally symmetric representation. So, the electrons can distort the nuclear configuration only in the direction of totally symmetric displacements without change of the symmetry. This is so-called radial or "breathing" mode. It corresponds to an uniform shift of energy of all the states by a certain constant value provided their wave functions have identical radial distribution. If Γ is degenerate the symmetrical product $[\Gamma^2]$ [303] contains representations nontotally symmetric along with the symmetric one. As is well known, these nontotally symmetric modes are just the JT active modes considered above. Let us consider the consequences of the symmetric mode A_1. The appropriate matrix composed of the Clebsch-Gordan coefficients is here the unity matrix. This means that any degenerate state as a whole is uniformly shifted in energy. The diagonal linear constant $F_{\bar{\Gamma}}^{\Gamma}$ has the sense of the force with which the electrons in state Γ affect the nuclei in the direction of normal displacement $\mathbf{Q}_{\bar{\Gamma}}$.

Nonzero off-diagonal matrix elements $F_{\bar{\Gamma}}^{\Gamma\Gamma'}$ can be readily found by means of the group-theoretical selection rules. They are ignored in the adiabatic approximation but they influence, in second order, the curvature of the adiabatic potentials $K_{\bar{\Gamma}}^{\Gamma}$ or the corresponding force constants (at equilibrium points).

Among the quadratic (or second order) vibronic constants the most important are their totally symmetric combinations of the form $\left\{ \left(\frac{\partial^2 V}{\partial Q_{\bar{\Gamma}}^2} \right)_0 \right\}_{A_1} =$ $\sum_{\bar{\gamma}} \left(\frac{\partial^2 V}{\partial Q_{\bar{\Gamma}\bar{\gamma}}^2} \right)_0$, where the symbol $\{...\}_{A_1}$ denotes the proper convolution, since the product of only equal irreducible representations contains A_1.

Their diagonal matrix element contributes to the curvature of the adiabatic potential according to

$$K_{0\bar{\Gamma}}^{\Gamma} = \left\langle \Gamma \left| \left\{ \left(\frac{\partial^2 V}{\partial Q_{\bar{\Gamma}}^2} \right)_0 \right\}_{A_1} \right| \Gamma \right\rangle \qquad (18.24)$$

and correspondingly, the scalar convolution of the coordinates reads $\{Q_{\bar{\Gamma}}^2\}_{A_1} = \sum_{\bar{\gamma}} Q_{\bar{\Gamma}\bar{\gamma}}^2$. The totally symmetric part of the diagonal matrix element $K_{0\bar{\Gamma}}^{\Gamma}$ has a clear physical meaning, together with the nondiagonal linear vibronic constants $F_{\bar{\Gamma}}^{\Gamma\Gamma'}$ it forms the curvature $K_{\bar{\Gamma}}^{\Gamma}$ of the adiabatic potential of electronic state Γ in the direction $Q_{\bar{\Gamma}}$ [300]

$$K_{\bar{\Gamma}}^{\Gamma} = \left\langle \Gamma \left| \left\{ \left(\frac{\partial^2 V}{\partial Q_{\bar{\Gamma}}^2} \right)_0 \right\}_{A_1} \right| \Gamma \right\rangle + 2\sum_{\Gamma'}{}' \frac{\left| \left\langle \Gamma \left| \left(\frac{\partial V}{\partial Q_{\bar{\Gamma}}} \right)_0 \right| \Gamma' \right\rangle \right|^2}{\varepsilon_{\Gamma} - \varepsilon_{\Gamma'}} \qquad (18.25)$$

where $\varepsilon_{\Gamma} - \varepsilon_{\Gamma'}$ is the appropriate energy difference between the states Γ and Γ' (beware of the order in the denominator). This value of $K_{\bar{\Gamma}}^{\Gamma}$ has been obtained as a coefficient of $\{Q_{\bar{\Gamma}}^2\}_{A_1}$ in the expression for the energy as a function of nuclear coordinates calculated by means of perturbation theory (by considering the vibronic interaction terms as small perturbations). At the point of stable equilibrium configuration where $F_{\bar{\Gamma}}^{\Gamma} = 0$, $K_{\bar{\Gamma}}^{\Gamma}$ equals the force constant.

The first term in $K_{\bar{\Gamma}}^{\Gamma}$, the so-called primary force constant, represents the contribution to $K_{\bar{\Gamma}}^{\Gamma}$ from the nuclear displacement in a rigid electronic distribution that does not follow the nuclei. The second term in Eq.18.25 accounts for the contribution of changes in electronic states due to the nuclear displacements (the "relaxation" term) [300].

The remaining diagonal matrix elements of nontotally symmetric operators

$$\left\{ \left(\frac{\partial^2 V}{\partial Q_{\bar{\Gamma}_1} \partial Q_{\bar{\Gamma}_2}} \right)_0 \right\}_{\bar{\Gamma}},$$

where $\bar{\Gamma} \in \bar{\Gamma}_1 \times \bar{\Gamma}_2$ and $\bar{\Gamma} \neq A_1$, yield the quadratic corrections to the JT effect which, in general, are of secondary meaning.

At last, the nondiagonal quadratic matrix elements define so-called quadratic vibronic constants. More precisely, only one reduced constant due to the Wigner-Eckart theorem for every representation $\bar{\Gamma} \in \Gamma \times \Gamma'$, where Γ and Γ' refer to the electronic states, can be introduced. These vibronic constants must be distinguished from the force constants.

As mentioned above, the curvature of the adiabatic potential corresponding to an electronic state Γ is equal, at the equilibrium point, to the force constant of a definite oscillator mode of the nuclear framework or crystal lattice.

From the knowledge of the force constant, i.e. the vibration frequency, the nuclear (or lattice) part of the energy of every state can be found. This is just the quantity which if differentiated can lead to modifications of the static crystal field splitting pattern. Assume, for simplicity, the radial distributions of all states originating from a given electronic term to be identical. Anyway, the same assumption is introduced to the classical static approach. Nevertheless, it is to be expected that the excited state distributions may be a little more spatially extended, the more excited the more extended. Under this assumption the primary force constants are the same for all these states.

From the point of view of changes generated by the nuclear framework dynamics in the system of energy levels of a paramagnetic ion in the crystal field, the second component in expression for curvature, Eq.18.25 – the "relaxation" term is effective only. Contributions to it coming from the states of lower energy ($\varepsilon_{\Gamma'} < \varepsilon_{\Gamma}$) are negative whereas from those of higher energy – positive. Moreover, taking into account the selfevident differences in $|\varepsilon_{\Gamma'} - \varepsilon_{\Gamma}|$ values it is seen that the term is in a position to modify in different degree the curvatures of particular adiabatic potentials and in consequence the force constants of the corresponding eigenstates.

A general tendency is obvious. The ground state and possibly some lower excited levels should be characterized by less curvatures (the second component in Eq.18.25 is negative) and smaller force constants. On the other hand higher excited levels are characterized by greater K, respectively. This is why the actual splitting scheme should be somewhat "swollen" in relation to the static pure electronic scheme. This effect ought to be taken into account in exact fitting procedures of experimental and calculated patterns.

The set of Γ' states being effective in Eq.18.25 goes beyond the ground term states. Therefore to estimate the contribution of the second term in K_{Γ}^{Γ} more exactly the interaction with all higher states obeying the selection rules should be included. However, their weight rapidly drops due to the energy denominator. Under the previous assumption regarding the invariability of the radial distribution and ignoring the dependence of the energy denominators $\varepsilon_{\Gamma} - \varepsilon_{\Gamma'}$ on Q the curvature of the adiabatic potential would be corrected by the relaxation term but it would remain independent on Q and the harmonic description would be valid. However, an inevitable dependence of the $\varepsilon_\Gamma - \varepsilon_{\Gamma'}$ on Q implies K to be dependent on Q, i.e. the anharmonicity. This is so-called vibronic anharmonicity – the main component of the anharmonicity more important than so-called proper anharmonicity resulting from the presence of the third and higher degree terms in the power expansion of the potential.

There is a standing out analogy between the curvature of the adiabatic potential K and the paramagnetic susceptibility χ of a system with the free energy $\mathcal{F}$. This analogy is helpful in understanding the variations of the force constants of different energy levels induced by the nuclear framework vibrations. The formal analogy is obvious, these two problems are "isomorphic". The curvature K is the coefficient at Q^2 in expansion of energy as a function of nuclear coordinates whereas the susceptibility χ is the coefficient at H^2 in expansion of the free energy F with respect to the magnetic field intensity H. Thus, in the dynamic approach to the crystal field effect each vibration mode $Q_{\Gamma_{\bar{\gamma}}}$ plays a role of H in the theory of paramagnetism and the first derivative $\left(\frac{\partial V}{\partial Q_{\Gamma_{\bar{\gamma}}}}\right)_0$ plays a role of the magnetic moment.

A certain difference with respect to the Van Vleck formula for paramagnetic susceptibility [319] results from the fact that in the case of diagonal elements $F_{\bar{\Gamma}}^{\Gamma} = 0$ (equilibrium states) i.e. there are no so-called low-frequency terms. There is also difference in sign of the "relaxation" term because $\mu = -\partial F/\partial H$.

If the energy difference $|\varepsilon_\Gamma - \varepsilon_{\Gamma'}|$ were too small the system ceases to fulfill the conditions of the adiabatic approximation. The total energy effect arising from the lattice dynamics is a sum over all allowed independent vibration modes. The vibronic structure of a central ion environment (or molecule) can be estimated within the frame of the MO approach by means of quantum chemistry methods [300, 320]. For each individual molecular orbital we are able to find then so-called orbital vibronic constants [300, 321, 322] and hence the distorting force and force constant in which we are particularly interested.

To sum up, according to expectation the lattice dynamics has no first order influence on the electronic energy levels system in the static crystal field. A certain constant shift of all levels is inessential from the crystal field splitting point of view. However, on the other hand the lattice dynamics modifies the force constant of the nuclear oscillator part each of the eigenstates individually diminishing the energy of the lower states and increasing the energy of excited ones enlarging somewhat the total splitting.

CHAPTER 19

Extension of the crystal field potential beyond the one–electron model

19.1 Two–electron correlation effect in the crystal field model

The only one formal limitation of the conventional parameterization of the crystal field potential is the assumption that this is one–electron potential exclusively conditioned by the angular momentum of the individual unpaired electrons. In the frame of this one–electron approach many convincing interpretations of several physico–chemical characteristics of various transition, lanthanide and actinide ions have been given, particularly those concerning the electronic spectra of the ions in crystalline solids by means of relatively small number of parameters and with the non–accidently small mean standard deviation [323, 3, 1]. However, some characteristic divergences in the fitting patterns suggesting a dependence of the crystal field parameters on the electron terms of the paramagnetic ions have been observed.

There is no doubt that any generalization of the model has to rely upon deleting the assumption on the one–electron character of the crystal field potential. Taking into account the two–electron crystal field potential U seems to be a natural first step extension of the model. Certainly, there are also upper n-electron correlations but since they are connected with the higher order terms in the perturbational series, they are considerably less and can be neglected.

The two–electron correlation crystal field U describes an orbital correlation between pairs of the open–shell electrons of a bound ion, i.e. the correlation induced by its crystalline surroundings [346]. In other words, the potential U expresses an extension of the spherically symmetric Coulomb interaction between the open–shell electrons for anisotropic contributions.

In a free ion, considered the lack of a favoured direction, only the isotropic (scalar) contributions of the interelectron Coulomb repulsion are

247

included. And so, for the l^N configuration they are presented in the form:

$$\sum_{\substack{i=j \\ i>j}} \sum_{\substack{k\leq 2l \\ \text{even}}} \sum_{g=-k}^{k} F^k(C_q^{(k)}(i) \cdot C_q^{(k)}(j)) +$$

$$+ \sum_{\substack{i=j \\ i>j}}^{N} \sum_{\substack{k\leq 2l \\ \text{odd}}} \sum_{g=-k}^{k} T^k(C_q^{(k)}(i) \cdot C_q^{(k)}(j)), \qquad (19.1)$$

where F^k are the Slater, and T^k – the Trees parameters, respectively. In the case of f-electrons there are four Slater parameters (for $k = 0, 2, 4$ and 6) and three Trees parameters (for $k = 1, 3$ and 5).

In the crystal field, i.e. for a bound ion all remaining non–scalar contributions which are the results of all admissible non–scalar couplings of the $C^{(k_1)}(i)$ and $C^{(k_2)}(j)$ tensors ought to be included. The formal parameterization of U [177, 324, 349] is, in principle, analogous to that in the case of the one–electron potential. However, the number of independent parameters (and operators) arising in the parameterization is so large that its using is practically impossible. Thus, reducing this number to a reasonable level is the basic problem. The clue as to the problem can be either a recognition of the predominating physical mechanisms of the two–electron correlations or experimental observations based on sufficiently comprehensive material.

The basic mechanisms leading to the electron correlations are:

- the configuration interactions [325, 326],

- the covalency and contact covalency [177, 327, 328],

- the polarization of ligands [329, 324],

- the effect of mutual orientation of electron spins within the open shell [328, 330, 331].

Two of the mechanisms have been introduced to the parameterization practice by now. These are: the method of term dependent crystal field parameters and the spin correlated crystal field. The last method is exceptionally simple and convenient for experimentalists.

19.2 Parameterization of the two–electron potential

The two–electron correlation potential can be described, in the most convenient way, in the form of expansion into the series of double products of tensor operators

$$U = \sum_{i>j} U_{ij} = \sum_{i>j} \sum_{k_1} \sum_{k_2} \sum_{q_1} \sum_{q_2} A_{q_1 q_2}^{k_1 k_2} t_{q_1}^{(k_1)}(i) \cdot t_{q_2}^{(k_2)}(j) =$$

$$= \frac{1}{2} \sum_{k_1} \sum_{k_2} \sum_{q_1} \sum_{q_2} A_{q_1 q_2}^{k_1 k_2} \, T_{q_1}^{(k_1)} \cdot T_{q_2}^{(k_2)} \tag{19.2}$$

where i and j denote different electrons of the open shell, $\mathbf{t}^{(k)}(i)$ is a tensor operator, e.g. $\mathbf{C}^{(k)}(i)$ acting on the i-th electron coordinates, $T_q^{(k)} = \sum t_q^{(k)}(i)$, and $A_{q_1 q_2}^{k_1 k_2}$ are the parameters of the expansion.

The hermiticity of the potential U and its invariance with respect to the time–reversal require the sum $k_1 + k_2$ to be even whereas the symmetry in relation to the interchange of the electrons is ensured if $A_{q_1 q_2}^{k_1 k_2} = A_{q_2 q_1}^{k_2 k_1}$.

Since the potential U enters the matrix elements accompanied, in general, with electron wave functions $|\gamma L S J M\rangle$ of the l^N configuration it is recommended to express U by means of double tensor operators $X_Q^{(K)}(k_1 k_2)$ defined by

$$T_{q_1}^{(k_1)} \cdot T_{q_2}^{(k_2)} = (-1)^Q (2K+1)^{1/2} \sum_{K,Q} \begin{pmatrix} k_1 & k_2 & K \\ q_1 & q_2 & -Q \end{pmatrix} X_Q^{(K)}(k_1 k_2) \tag{19.3}$$

Consequently, Eq.19.2 may be rewritten in the form

$$U = \sum_{K,Q} \sum_{k_1} \sum_{k_2} B_{KQ}(k_1 k_2) X_Q^{(K)}(k_1 k_2) \tag{19.4}$$

where

$$B_{KQ}(k_1 k_2) = (-1)^Q (2K+1)^{1/2} \sum_{q_1} \sum_{q_2} \begin{pmatrix} k_1 & k_2 & K \\ q_1 & q_2 & -Q \end{pmatrix} A_{q_1 q_2}^{k_1 k_2} \tag{19.5}$$

The restrictions referring to the indices k_1 and k_2 of the parameters $A_{q_1 q_2}^{k_1 k_2}$ affect the $B_{KQ}(k_1 k_2)$ parameters as follows

$$B_{KQ}^*(k_1 k_2) = (-1)^{K+Q} B_{K-Q}(k_1 k_2) \tag{19.6}$$

$$B_{KQ}(k_1 k_2) = (-1)^K B_{KQ}(k_2 k_1) \tag{19.7}$$

Since k_1 and $k_2 \leq 2l$ and $K \leq k_1 + k_2$, K can take values from the range $[0, 4l]$. In addition, if $k_1 = k_2$, K has to be even as it results from Eq.19.7. The correlation potential defined in this way contains two types of contributions which already have been included either in the conventional parameterization scheme of the free ion, i.e. 7 parameters (4 Slater and 3 Trees ones) for $K = Q = 0$, or in the one–electron crystal field parameterization, i.e. 27 parameters in the most general case (no symmetry elements) for k_1 or k_2 being equal to zero.

The complete matrix of the two–electron potential U for f electrons is of order of 49 and the total number of its elements amounts to $49 \times 49 = 2401$. The requirements of hermiticity and time–reversal symmetry reduces the number of independent parameters to 1225, and that of symmetry in relation to interchange of the electrons to 637 including 490 parameters of even K, respectively. Thus, this is the number of the independent parameters in the extended parameterization which should be compared with 27 parameters in the case of the one–electron crystal field. For higher point symmetries of the central ion the number of independent parameters is subject to a noticeable reduction but is still too large from the viewpoint of a practical parameterization.

As is well known employing the superposition model [27] (if valid) leads to an effective decrease of the number. Within a local coordinate system of $C_{\infty v}$ symmetry the potential produced by the individual ligand can be expressed merely by means of the tensor operators (and parameters) with K even and $Q = 0$ [177]. Then, there are 30 parameters with both K_1 and K_2 even and 20 with k_1 and k_2 odd. Subtracting from their total number, i.e. from 50, the 7 free–ion parameters and 3 intrinsic parameters of the one–electron potential, the number of 40 parameters remains, which however still is too large for a phenomenological parameterization of any experimental data.

Although formally, the two–electron correlation field can be parameterized based on the additivity of the potential U [177] the problem of legitimacy of the superposition approach to the correlation field has not been solved so far. It seems that the extending the superposition rule over the correlation crystal field may be a source of divergences greater then those occurring in the case of the one–electron potential.

A more effective reducing the number of independent parameters may be attained owing to restrictions imposed by a mechanism or mechanisms recognized as the dominating ones and leading to the correlation.

19.3   The term dependent crystal field

The two–electron correlation field can be relatively simply expressed in the form of an effective one–electron potential being dependent on the open–shell electron term [177, 27]. The simplest method resolves itself into finding the term dependent parameters for the initial l^2 configuration. Next, reconstructing step by step the genealogy of the considered term of the l^N configuration with the help of the fractional parentage coefficient method [3], the contributions of the l^2 configuration terms are calculated.

According to the Wigner–Eckart theorem a matrix element of the double tensor operator $X_Q^{(K)}(k_1 k_2)$ within the basis of $|LM_L\rangle$ eigenfunctions of the l^2 configuration has the form

$$\langle LM_L | X_Q^{(K)}(k_1 k_2) | L'M_L' \rangle =$$

$$= (-1)^{L-M_L} \begin{pmatrix} L & K & L' \\ -M_L & Q & M'_L \end{pmatrix} \langle L||X^{(K)}(k_1 k_2)||L' \rangle \quad (19.8)$$

where

$$\langle L||X^{(K)}(k_1 k_2)||L' \rangle = [(2l+1)(2K+1)(2L'+1)]^{1/2} \times$$

$$\times \begin{Bmatrix} l & l & k_1 \\ l & l & k_2 \\ L & L' & K \end{Bmatrix} \langle l||t^{(k_1)}||l \rangle \langle l||t^{(k_2)}||l \rangle \quad (19.9)$$

and $\langle \ldots || \ldots \rangle$ denotes the reduced (double bar) matrix element, and

$$\begin{Bmatrix} \cdot & \cdot & \cdot \\ \cdot & \cdot & \cdot \\ \cdot & \cdot & \cdot \end{Bmatrix}$$

the 9-j symbol (see Appendix B).

Therefore, it is convenient to define a new set of parameters explicitly connected with the terms of the l^2 configuration

$$B_{KQ}(LL') = \sum_{k_1} \sum_{k_2} B_{KQ}(k_1 k_2) \langle L||X^{(K)}(k_1 k_2)||L' \rangle \quad (19.10)$$

In turn, the above expression can be transformed to its equivalent form

$$B_{KQ}(LL') = A_{KQ}(LL') \sum_{k_1} \sum_{k_2} \langle L||X^{(K)}(k_1 k_2)||L' \rangle \quad (19.11)$$

The effect is that the two–electron correlation potential U within the l^2 configuration can be presented as an effective "one–electron" potential

$$V_{\mathrm{CF}}(LL') = \sum_K \sum_Q A_{KQ}(LL') T_Q^{(K)} \quad (19.12)$$

It means that for given L and L' the experimental values of the crystal field parameters should be expressible in the form

$$B_{kq} + A_{KQ}(LL') \quad (19.13)$$

where the second component stands for the correlation effects. However, one ought to keep in mind that now $K \leq 4l$, and not as previously, in the one–electron model, $k \leq 2l$.

In general, for any l^N configuration there is always, as mentioned, the chain calculation procedure allowing its electron terms to be expressed as superpositions of their parent l^2 configuration term contributions. Just the same refers to the genealogy of the parameters. The term dependent crystal field method has already been introduced to the spectroscopy practice [332] affording possibilities for a noticeable improvement in interpretation of the spectra.

In order to successively reduce the number of independent parameters several limitations of the general parameterization are assumed. For instance, one can include only K even and $\leq 2l$ or take only the term with $K = 2$ which supposedly should produce the dominating contribution to the correlation crystal field. In most cases, especially for well separate terms the parameters $A_{KQ}(LL')$ with $L \neq L'$ are ignored. Moreover, the off–diagonal matrix elements with $L \neq L'$ are effective only if both the terms of the l^N configuration contain contributions of different $l^2 LS$ terms but of the same spin multiplicity [27]. Invariability of the crystal field parameters for all pure Russel–Saunders terms with the same L and S quantum numbers, i.e. their independence on J quantum number is the characteristic feature of the discussed approach. This property was spectacularly corroborated in the case of $^4I_{9/2}$, $^4I_{11/2}$ and $^4I_{13/2}$ multiplet [333].

Probably, one of the reasons of adequacy of the conventional one–electron crystal field model is the fact that most of the available experimental data comes from low lying terms of the maximum spin multiplicity. The terms of other types making allowance for which could introduce some greater divergences are naturally discriminated, as e.g. in the case of $^1D_2$ term of f^2 configuration. Thereby, the experimental data, extort, in fact, the term dependent interpretation.

In spite of unquestionable merits and successes of the method treating the crystal field parameters as dependent on particular electron terms the question if the correlation crystal field so defined yields the adequate parameterization of the real physical effects responsible for the departure from the one–electron description is still open.

19.4 Spin correlated crystal field (SCCF)

This approach to the correlation problem is based on the hypothesis pointing towards the spin polarization of the open–shell electrons, i.e. their mutual orientation as the dominating correlation mechanism. The attractive exchange forces between electrons of parallel spins, $s_i \cdot s_j > 0$, lead to a certain contraction of their radial distributions compared to those of opposite spins. Thereby, the spin correlated system corresponds to a weaker interaction with the surroundings and hence to smaller crystal field parameters. This conclusion was an inspiration of the phenomenological SCCF model [328, 330, 331] in which the one–electron crystal field potential is replaced by

$$\sum_i \sum_k \sum_q B_{kq} C_q^{(k)}(i) \longrightarrow \sum_i \sum_k \sum_q B_{kq} \left[1 + \frac{b_{kq}}{B_{kq}} s_i \cdot S \right] C_q^{(k)}(i) \quad (19.14)$$

where b_{kq} is the measure of the correlation correction, and $\mathbf{S}$ is the total spin moment.

Within the superposition model and under assumption of identical distance dependence of both b_{kq} and B_{kq} parameters Eq.19.14 may be simplified to the form

$$b_{kq}/B_{kq} = c_k \tag{19.15}$$

One should notice that the number of independent parameters of the model exceeds that of the one–electron model only twice. The character of dependence of the modified potential (Eq.19.14) on spins of individual electrons, explicitly seen in expansion of the scalar product $\mathbf{s}_i \cdot \sum_j \mathbf{s}_j$, is of the same type as that in the Coulomb exchange interaction, i.e. this is not a real spin dependence in the direct sense.

Not all correlation effects can be included in the phenomenological SCCF formula (Eq.19.14). It has been shown that the SCCF is only a part of the full anisotropic correlation crystal field [334, 335] which means that there are some mechanisms elusive in the model [328].

Two basic mechanisms contributing to the SCCF have been identified [335, 336]

- The first one connected with the dependence of the radial distribution of the open–shell wave functions on their spin polarization. It produces negative contributions to all c_k parameters (Eq.19.15) which fulfill roughly the following inequalities: $c_2 < c_4 < c_6$.

- The second one resulting from the Coulomb exchange energy contribution into the energy denominator of the covalency term. It produces positive contributions to all c_k parameters with opposite than previously hierarchy of their magnitude: $c_6 > c_4 > c_2$ with relatively small values of c_2.

The experimentally estimated values of the c_k parameters for trivalent lanthanide and actinide ions in anhydrous chlorides and several other crystalline matrices [337, 338] seem to be a result of competition between these two mechanisms. The c_2 and c_4 parameters are characterized by a rather large scattering of their values whereas the c_6 parameter varies within pretty narrow limits, but may be very important [349]. Their tentative magnitudes are, respectively

- for the lanthanide ions

$$-0.20 < c_2 < +0.29$$
$$-0.10 < c_4 < +0.21$$
$$+0.09 < c_6 < +0.16$$

- for the actinide ions
$$c_2 \simeq -0.33$$
$$c_4 \simeq -0.15$$
$$c_6 \simeq +0.32$$

By analogy to the SCCF the orbitally correlated crystal field (LCCF) model has also been introduced [339]. This is the field which depends on relative orientation of the angular momentum of an individual electron of the open–shell with respect to the total angular momentum of the shell $\mathbf{L}$. This correction may be included exactly in the same way as in Eq.19.14 with replacing however $\mathbf{s}_i \cdot \mathbf{S}$ by the scalar product $\mathbf{l}_i \cdot \mathbf{L}$. Occurrence of such correlations in the case of the lanthanide ion compounds has been experimentally evidenced [339].

19.5 Many–electron approach to the crystal field effect

Being influenced by the central idea of the previous chapters one might be impressed that the crystal field effect is, from the very nature of things, the one–electron effect at most supplemented with the two–electron correlation correction and the total potential of the field is expressible in the form of expansion into the spherical tensor series and the main problem resolves itself, in principle, into finding the exact coefficients of the expansion. Therefore one should, first of all, recall to mind that such a form of the effective crystal field potential has been a priori postulated and many convincing successes of the model in interpretation of electronic spectra, magnetic properties etc. of open–shell (paramagnetic) bound ions evidence only that this is a good approximation indeed, nothing else. The clue of the problem is always a theoretical reconstruction of energy level spectrum of a bound ion or the crystal field effect but its parameterization in the one–electron model, as the simplest one, should be treated as a secondary problem.

A marked development of numerical methods in the last quarter of the century has made allowance for giving up the simple one–electron model and a more general approach to the subject [340, 341]. However, preservation some of the simplifying assumptions referring to the way of including the interelectron interactions or to applying the superposition model is still, owing to the complexity of the system, recommended or even indispensable. From these simplifying assumptions several corrections arise among which the interconfiguration interactions play a key part [347, 348].

Already for the free ion, in which the interelectron repulsion manifesting itself in a splitting of the configuration into LS terms is taken into account on the level of particular configurations, a certain part of the interactions is lost in the first order perturbation calculation and incorporated only in the second order as the interconfiguration interactions. For the l^N configuration the Coulomb interconfiguration interactions, effective only between configurations of the same parity, lead, apart from a linear shift of all terms of configuration and weak three–electron interactions, to $2l+1$ two–electron scalar interactions including $l+1$ renormalized Slater integrals and l extra parameters for each LS term, e.g. for the f-electrons these are α, β and γ parameters [342].

In the case of the bound paramagnetic ion its crystalline surroundings interaction introduces additional complications into the interconfiguration interactions. On the level of the second order approximation a potentiality of mixing various configurations through the crystal field potential occurs and that without the limitation regarding the parity of the configurations as well as through the electrostatic interaction including the crystal field potential (the electrostatic correlated crystal field) effective this time again between configurations of the same parity only. Consequently, for the l^N configuration the above extension of the approach leads to a simple renormalizing the crystal field parameters, i.e. to including the screening and antiscreening effect and to renormalizing the Slater integrals as well as to certain extra changes of the crystal field parameters which are different for various terms [343]. In the light of these facts one can easily realize how would the parameterization of the crystal field effect look like in the frame of the one–electron model making allowance for the configuration interactions.

In the approach neglecting the interconfiguration interactions and concerning one of the l^N type configurations only even terms (k even) in the crystal field potential expansion are effective. Hence, irrespective of the point group symmetry of the central ion the given expansions of the potentials are usually of this type. Nevertheless, for all these point groups which have no inversion, the potential contains both the even and odd terms. The latter play a vital role in the interconfiguration mixing [344].

Within the model extended by every possible interconfiguration interactions the entire crystal field effect can be reconstructed. However, one ought to employ the electron configurations of the whole ionic complex (and not only that of the central ion) and consider among the excited configurations those of charge–transfer type and those referring to the ligands themselves. The initial one–configurational model corresponding to absolutely fixed configurations, i.e. to infinitely large excitation energies, leads to the crystal field effect representation discussed in detail in chapter 7. All the remaining contributions are expressible by means of appropriate interconfigurational interactions [345]. In practice, the calculations are carried out under some simplifying assumptions, e.g. within a local coordinate system for two ions only using the superposition model, for selected most important configuration interactions, and with utilizing some free–ion data. From these calculations the spectrum of energy levels in the crystal field is available. This result can be, on the other hand, compared with the experimental data to verify adequacy of the method but on the other the spectrum can be put to the test of effective parameterization within the one–electron model.

Thereby, we obtain information regarding both the validity of the model and possible divergences from it, i.e. the correlation corrections.

CHAPTER 20

Appendices

APPENDIX A
Transformation from a local to the global coordinate system

The general or global coordinate system for a given ionic complex (cluster) is usually defined based on its symmetry elements. The maximum multiplicity axis is most often designated as the z-axis whereas the x- and y-axes are oriented according to other symmetry elements, i.e. reflection planes or rotation axes perpendicular to the main axis. For the coordinate system chosen properly (see chapter 2) the crystal field potential takes the form as simple as possible which shows the point group symmetry of the central ion.

Local systems are the axial binary systems with the z-axis joining the central ion nucleous with a chosen ligand. The local x- and y-axes are directed in the most natural but consistent way.

Conventional crystal field parameters B_{kq}, i.e. the global ones, connected with the whole surroundings can be expressed by means of the intrinsic parameters b_k^t connected with local systems t with the expression [27, 28]

$$B_{kq} = \sum_t D_{0q}^{(k)^*}(\omega_t) b_k^t = \sum_t C_q^{(k)^*}(v_t, \varphi_t) b_k^t \qquad (A-1)$$

where the summation runs over all ligands t, $D_{0q}^{(k)}(\omega_t)$ is the kth-order matrix element of rotation, ω_t stands for three Euler angles defining the rotation: α_t, β_t and γ_t measured with respect to the rigid main coordinate system (a first rotation γ about the z-axis, a second β about the y-axis and a third α about the z-axis again), v_t and φ_t are the angles which the local z-axis forms with the z- and x-axes of the main system, respectively.

In general, the matrix elements of a finite rotation can be calculated from the expression [28]

$$D_{m'm}^{(k)}(\alpha, \beta, \gamma) = e^{im'\gamma} d_{m'm}^{(k)}(\beta) e^{im\alpha} \qquad (A-2)$$

where

$$d^{(k)}_{m'm}(\beta) = [(k+m)!(k-m)!(k+m')!(k-m')!]^{1/2} \times$$

$$\times \sum_{\mu=\mu_{min}}^{\mu_{max}} (-1)^{k-m'-\mu} \frac{(\cos\frac{\beta}{2})^{2\mu+m+m'}(\sin\frac{\beta}{2})^{2k-2\mu-m-m'}}{(m+m'+\mu)!(k-m'-\mu)!(k-m-\mu)!\mu!} \qquad (A-3)$$

and $\mu_{min} = \max(0, -m - m')$, $\mu_{max} = \min(l - m', l - m)$.

The second equality in Eq.(A-1) results from the fact that the angles v_t, φ_t of the z-axis of the main system after the rotations with respect to the initial system are identical with the corresponding β and α Euler angles, respectively (the z-axis is invariant with respect to the γ rotation). This equivalence is illustrated in Fig.20.1

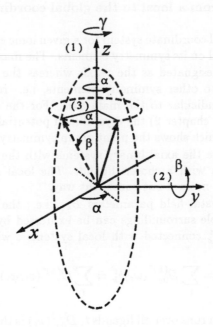

Fig. 20.1 Equivalence of v_t and φ_t angles which the z-axis after the rotation forms with the initial z- and x-axes and the β and α Euler angles, respectively.

It is directly seen from Eq.(A-1) that the parameters B_{kq} of the same k but various q remain in constant mutual proportions when geometry of the surroundings is preserved. This property valid for the axial local symmetry is one of features of the superposition model based on the additivity of the crystal field potential.

APPENDIX B
$3n$-j symbols

The aim of this short appendix is a straightforward presentation of the real meaning of 3-j, 6-j and 9-j symbols making possible a proper understanding formulas containing these symbols.

Their various symmetry properties and rich relations between them are presented with full particulars in many monographs, e.g. [28, 31, 44].

The eigenvectors $|\gamma j_i m_i\rangle$ of a set of commuting operators $\hat{\Gamma}$, $\hat{J}_i^2$, $\hat{J}_{iz}$, where $\hat{\mathbf{J}}_i$ is the angular momentum operator, $\hat{J}_{iz}$ the operator of projection of the angular momentum on the z-axis and $\hat{\Gamma}$ stands for the remaining operators being invariant with respect to the rotations, can be coupled in different manners leading to various final states which remain the eigenstates of the operators $\hat{\Gamma}$, $\hat{J}^2$, $\hat{J}_z$ but the resultant $\hat{\mathbf{J}}$ is now one of the allowed vector sums $\sum_i \hat{\mathbf{J}}_i$ and $m = \sum_i m_i$.

The states corresponding to the various coupling schemes are independent and their representations are mutually related with unitary transformations elements of which, when described in the form being characterized by the maximum symmetry with respect to permutation of the arguments, are called the $3n - j$ symbols if the number of the vector coupled is $n + 1$. In the case of coupling two eigenvectors $|\gamma j_1 m_1\rangle$ and $|\gamma j_2 m_2\rangle$ to the final state $|\gamma j_1 j_2 j m\rangle$ we have

$$|\gamma j_1 j_2 j m\rangle = \sum_{m_1} \sum_{m_2} |\gamma j_1 m_1 j_2 m_2\rangle \langle j_1 m_1 j_2 m_2 | j_1 j_2 j m\rangle \qquad (B-1)$$

where the second factor on the right side of Eq.(B-1) is the vector coupling coefficient (Clebsch–Gordan or Wigner coefficient). Simply, this is the common part or intersection of the states $|\gamma j_1 j_2 j m\rangle$ and $|\gamma j_1 m_1 j_2 m_2\rangle$.

The 3-j symbols being a symmetrized form of the vector coupling coefficient is given by

$$\begin{pmatrix} j_1 & j_2 & j_3 \\ m_1 & m_2 & m_3 \end{pmatrix} =$$

$$\qquad (B-2)$$

$$= (-1)^{j_1 - j_2 - m_3} (2j_3 + 1)^{-1/2} \langle j_1 m_1 j_2 m_2 | j_1 j_2 j_3 - m_3\rangle$$

In other words, 3-j symbol value shows what part of the resultant state $|\gamma j_1 j_2 j_3 - m_3\rangle$ is the component state $|\gamma j_1 m_1 j_2 m_2\rangle$.

In the case of coupling three eigenvectors: $|\gamma j_1 m_1\rangle$, $|\gamma j_2 m_2\rangle$ and $|\gamma j_3 m_3\rangle$ there are two different ways to reach the same final state $\gamma J M\rangle$. The unitary transformation linking together both these representations is given by

$$|\gamma j_1, (j_2 j_3) j_{23}, JM\rangle = \sum_{j_{12}} |\gamma_1 (j_1 j_2) j_{12}, j_3, JM\rangle \times$$

$$\qquad (B-3)$$

$$\times \langle (j_1 j_2) j_{12}, j_3, J | j_1, (j_2 j_3) j_{23}, J\rangle$$

The recoupling coefficient (the second factor on the right side of Eq.B-3) does not depend on M due to the Wigner–Eckart theorem. Its symmetrized form, i.e. the 6-j symbol is defined by

$$\left\{ \begin{array}{ccc} j_1 & j_2 & j_{12} \\ j_3 & J & j_{23} \end{array} \right\} =$$

$$= [(2j_{12}+1)(2j_{23}+1)]^{-1/2}(-1)^{j_1+j_2+j_3+J} \times \qquad (B-4)$$

$$\times \langle (j_1 j_2) j_{12}, \ j_3, \ J | j_1, \ (j_2 j_3) j_{23}, \ J \rangle$$

and, as seen, its value is determined by the common part of both alternative coupled states.

In the case of coupling four eigenvectors: $|\gamma j_1 m_1\rangle$, $|\gamma j_2 m_2\rangle$, $|\gamma j_3 m_3\rangle$ and $|\gamma j_4 m_4\rangle$ two types of the resultant states are important

$$|\gamma, \ (j_1 j_2) j_{12}, \ (j_3 j_4) j_{34}, \ JM\rangle$$

$$\text{and} \qquad (B-5)$$

$$|\gamma, \ (j_1 j_3) j_{13}, \ (j_2 j_4) j_{24}, \ JM\rangle$$

The 9-j symbol is defined by the common part of these two states according to

$$\left\{ \begin{array}{ccc} j_1 & j_2 & j_{12} \\ j_3 & j_4 & j_{34} \\ j_{13} & j_{24} & J \end{array} \right\} =$$

$$\qquad (B-6)$$

$$= [(2j_{12}+1)(2j_{34}+1)(2j_{13}+1)(2j_{24}+1)]^{-1/2} \times$$

$$\times \langle (j_1 j_2) j_{12}, \ (j_3 j_4) j_{34}, \ J | (j_1 j_3) j_{13}, \ (j_2 j_4) j_{24}, \ J \rangle$$

We always deal with the 3-j symbols during reducing matrix elements of tensor opeartors to the double–bar form with employing the Wigner–Eckart theorem.

The 6-j symbols appear in relations between the reduced (double–bar) matrix elements of a product of two tensor opeartors acting in the same set of states and the product of the reduced matrix elements of the individual operators.

The 9-j symbols appear, in turn, in relations being similar to those of 6-j symbols but when the operators consisting the product act within separate sets of the states.

APPENDIX C
Methods of orthogonalization of functions

1. The symmetric orthogonalization technics by Löwdin

If the initial functions, i.e. the free–ion, non–orthogonal ones are denoted by $|\Phi_p\rangle$, their new orthogonal combinations $|h_n\rangle$ can be expressed with [218]

$$|h_n\rangle = \sum_p (\tilde{1} + \tilde{S})^{-1/2}_{n_p} |\Phi_p\rangle \qquad (C-1)$$

where $\tilde{1}$ stands for the unity matrix order of which is equal to the dimension of the function set, and $\tilde{S}$ is the overlap matrix elements of which are defined by

$$S_{np} = \begin{cases} \langle \Phi_p | \Phi_n \rangle & \text{if} \quad n \neq p \\ 0 & \text{if} \quad n = p \end{cases} \qquad (C-2)$$

Thus, finally

$$|h_n\rangle = |\Phi_n\rangle - \frac{1}{2} \sum_p S_{sp} |\Phi_p\rangle + \frac{3}{8} \sum_\alpha \sum_p S_{n\alpha} S_{\alpha p} |\Phi_p\rangle + \ldots \qquad C-3$$

2. The Schmidt orthogonalization process

Starting from the set of non–orthogonal functions Ψ_1, Ψ_2 ..., Ψ_g for which $\int \Psi_s^* \Psi_k d\tau \equiv \langle \Psi_s | \Psi_k \rangle$ one can convert it to the set of orthogonal functions Ψ_1', Ψ_2',..., Ψ_g' by means of the following transformation

$$\Psi_k' = \begin{vmatrix} \langle \Psi_1 | \Psi_1 \rangle & \langle \Psi_1 | \Psi_2 \rangle & \ldots & \langle \Psi_1 | \Psi_{k-1} \rangle \Psi_1 \\ \langle \Psi_2 | \Psi_1 \rangle & \langle \Psi_2 | \Psi_2 \rangle & \ldots & \langle \Psi_2 | \Psi_{k-1} \rangle \Psi_2 \\ \ldots & \ldots & \ldots & \ldots \\ \langle \Psi_k | \Psi_1 \rangle & \langle \Psi_k | \Psi_2 \rangle & \ldots & \langle \Psi_k | \Psi_{k-1} \rangle \Psi_k \end{vmatrix} (G_k G_{k-1})^{-1/2} \quad (C-4)$$

where $k = 1, 2, \ldots, g$, and

$$G_k = \begin{vmatrix} \langle \Psi_1 | \Psi_1 \rangle & \langle \Psi_1 | \Psi_2 \rangle & \ldots & \langle \Psi_1 | \Psi_k \rangle \\ \langle \Psi_2 | \Psi_1 \rangle & \langle \Psi_2 | \Psi_2 \rangle & \ldots & \langle \Psi_2 | \Psi_k \rangle \\ \ldots & \ldots & \ldots & \ldots \\ \langle \Psi_k | \Psi_1 \rangle & \langle \Psi_k | \Psi_2 \rangle & \ldots & \langle \Psi_k | \Psi_k \rangle \end{vmatrix} \quad ; \quad G_0 = 1 \qquad (C-5)$$

References

1. Dieke, G. H., *Spectra and Energy Levels of Rare Earth Ions in Crystals.* ed. H. M. Crosswhite and H. Crosswhite. Wiley, New York, 1968.

2. Hüfner, S., *Optical Spectra of Transparent Rare Earth Compounds.* Academic Press, New York, San Francisco, London, 1978.

3. Wybourne, B. G., *Spectroscopic Properties of Rare Earths.* John Wiley, New York, 1965.

4. Morrison, C. A. and Leavit, R. P., Spectroscopic properties of triply ionized lanthanides in transparent host crystals. In *Handbook of the Physics and Chemistry of Rare Earths*, Vol. 5, ed. K. A. Gschneider Jr and L. Eyring. North-Holland, Amsterdam, 1982, pp. 461 692.

5. Carnall, W. T., Goodman, G. L., Rajnak, K. and Rana, R. S., A systematic analysis of the spectra of the lanthanides doped into single crystal LaF_3. *Journal of Chemical Physics*, 1988, **90**(7), 3443 3457.

6. Metcalf, D. H., Hopkins, T. A., Richardson, F.S., Electronic spectra, optical activity and crystal field energy level structure of Dy^{3+} in trigonal $Na_3[Dy(oda)_3]2NaClO_4 \cdot 6H_2O$ crystals. *Inorganic Chemistry*, 1995, **34**, 4868 4878.

7. Edelstein, N., Optical and magnetic properties of tetravalent actinide ions and compounds. *Journal of Less-Common Metals*, 1987, **133**, 39 51.

8. Krupa, J. C., Spectroscopic properties of tetravalent actinide ions in solids. *Inorganica Chimica Acta*, 1987, **139**, 223 241.

9. Hutchison, C. A. Jr and Candela, G. A., Magnetic susceptibilities of uranium (IV) ions in cubic crystalline fields. *Journal of Chemical Physics*, 1957, **27**(3), 707

10. Wang, Y. L, and Cooper, B. R., Crystal field effects and anomalous susceptibility of antiferromagnets: application to Ce – group V compounds. *Physical Review B*, 1970, **2**(7), 2607 2617.

11. Mulak, J., Crystal field effect and temperature dependence of paramagnetic susceptibility. *Journal of Less-Common Metals*, 1986, **121**, 141 150.

12. Abragam, A. and Bleaney, B., *Electron Paramagnetic Resonance of Transition Ions*. Oxford University Press, London, 1970.

13. Elliot, R. J. and Stevens, K. W. H., The theory of magnetic resonance experiments on salts of the rare earths. *Proceedings of Royal Society Ser A*, 1953, **218**, 553 566.

14. Birgeneau, R. J., Transition probabilities for f-electron J-multiplets in cubic crystal fields. *Journal of Physical Chemistry of Solids*, 1972, **33**, 59 68.

15. Furrer, A., Bührer, W., Heer, H., Purwins, H.-G., and Walker, E., Neutron crystal field spectroscopy in $Tm_{.25}Y_{.75}Al_2$. *International Journal of Magnetism*, 1973, **4**, 63 69.

16. Furrer, A., Crystal field and exchange interactions in ErAg. *Journal of Physics C: Solids State Physics*, 1975, **8**, 824 838.

17. Osborn, R., Lovesey, S. W., Taylor, A. D. and Balcar, E., Intermultiplet transitions using neutron spectroscopy. In *Handbook of the Physics and Chemistry of Rare Earths*, Vol. 14, ed. K. A. Gschneider Jr and L. Eyring. Elsevier Science Publishers B.V., 1991, chpt. 93, pp. 1 61.

18. Holland-Moritz, E. and Lander, G. H., Neutron inelastic scaterring from actinides and anomalous lanthanides. In *Handbook of the Physics and Chemistry of Rare Earths*, Vol. 19, ed. K. A. Gschneider Jr, L. Eyring, G. H. Lander and G. R. Choppin. Elsevier Science B. V., 1994, pp. 1 121.

19. Biegański, Z., Low temperature specific heats and related thermodynamical functions of neodynium dihydride $NdH_{2.60}$. Crystal field effects. *physica status solidi (b)*, 1971, **47**, 93 97.

20. Rao, V. U. S., and Wallace, W. E., Calculation of the influence of the crystalline electric field on the spin–disorder resistivity of rare-earth alloys and comparison with results on $CeAl_2$. *Physical Review B*, 1970, **2**(11), 4613 4616.

21. Andersen, H., Gregers-Hansen, P. E., Holm, E., Smith, H. and Vogt, O., Temperature–dependent spin–disorder resistivity in a Van Vleck paramagnet. *Physical Review Letters*, 1974, **32**, 1321 1324.

22. White, J. A. and Van Vleck, J. H., Sign of Knight shift in samarium intermetallic compounds. *Physical Review Letters*, 1961, **6**(8), 412 413.

23. Takagi, S., Oyamada, A. and Kasuya, T., ^{31}P NMR studies of the magnetically ordered heavy–electron compound YbP. *Journal of Physical Society of Japan*, 1988, **57**(4), 1456 1465.

24. Levy, P. M. and Zhang, S., Crystal field splitting in Kondo systems. *Physical Review Letters*, 1989, **62**(1), 78 81.

25. Mesot, J. and Furrer, A., The crystal field in rare earth based high-temperature superconductors. *Journal of Superconductivity*, 1997, **10**(6), 623 643.

26. Gerloch, M., Harding, J. H. and Wooley R. G., The context and application of ligand field theory. In *Structure and Bonding*, Vol. 46, ed. M. J. Clarke, J. B. Goodenough, P. Hemmerich, J. A. Ibers, C. K. Jørgensen, J. B. Neilands, D. Reinen, R. Weiss and R. J. P. Williams. Springer Verlag, Berlin, Heidelberg, New York, 1981, pp. 1 46.

27. Newman, D. J., Theory of lanthanide crystal field. *Advances in Physics*, 1971, **20**(84), 197 256.

28. Edmonds, A. R., *Angular Momentum in Quantum Mechanics*. Princeton University Press, Princeton, New York, 1957.

29. Racah, G., Theory of complex spectra.II. *Physical Review*, 1942, **62**, 438 462.

30. Kaplan, I. G., *Symmetry of Many-Electron Systems*. Academic Press, New York, 1975.

31. Rotenberg, M., Bivins, R., Metropolis, N. and Wooten, J. K. Jr, *The 3-j and 6-j Symbols*. MIT Press, Cambridge, Massachusetts, 1959.

32. Buckmaster, H. A., Tables of matrix elements for the opeartors $O_2^{\pm1}$, $O_4^{\pm1}$, $O_6^{\pm1}$, $O_6^{\pm5}$. *Canadian Journal of Physics*, 1962, **40**, 1670 1677.

33. Smith D. and Thornley, J. H. M., The use of 'operator equivalents'. *Proceedings of Royal Society*, 1966, **89**, 779 781.

34. Rudowicz, C., Concept of spin Hamiltonian, forms of zero field splitting and electronic Zeeman Hamiltonians and relations between parameters used in EPR. Critical review. *Magnetic Resonance Review*, 1987, **13**, 1 89.

35. Stevens, K. W. H., Matrix elements and operator equivalents connected with the magnetic properties of rare earth ions. *Proceedings of Royal Society*, 1952, **65**, 209 215

36. Bacquet, G., Dugas, J., Escribe, C., Gaite, J. M. and Michoulier, J., Comparative electron paramagnetic resonance study of Fe^{3+} and Gd^{3+} ions in monoclinic zirconia. *Journal of Physics C*, 1974, **7**, 1551 1563.

37. Rudowicz, C., Transformation relations for the conventional O_k^q and normalised O'^q_k Stevens operator equivalents with $k = 1$ to 6 and $-k \leq q \leq k$. *Journal of Physics C: Solid State Physics*, 1985, **18**, 1415 1430.

38. Ballhausen, C. J., *Introduction to Ligand Field Theory*. McGraw-Hill, New York, 1962.

39. Hutchings, M. T., Point charge calculations of energy levels of magnetic ions in crystalline electric field. *Solid State Physics*, 1964, **16**, 227 273.

40. Kibler, M., Ionic and paramagnetic energy levels algebra. *Journal of Molecular Spectroscopy*, 1968, **26**, 111 130.

41. Rudowicz, C., On the derivation of the superposition-model formulae using the transformation relations for the Stevens operators. *Journal of Physics C: Solid State Physics*, 1987, **20**, 6033 6037.

42. Altshuler, S. and Kozyrev, B. M., *Electron Paramagnetic Resonance in Compounds of Transition Elements*. John Wiley, New York, 1974.

43. Beem, K. M., Hornyak, E. J. and Rasmussen, P. G., Spin Hamiltonian of low-symmetry crystalline systems. *Physical Review B*, 1971, **4**(7), 2124 2132.

44. Judd, B. R., *Operator Techniques in Atomic Spectroscopy*. McGraw-Hill, New York, 1963.

45. Nielson, C. W. and Koster, G. F., *Spectroscopic Coefficients for p^n, d^n and f^n Configurations*. MIT Press, Cambridge, Massachusetts, 1963.

46. Lea, K. R., Leask, M. J. M. and Wolf, W. P., The raising of angular momentum degeneracy of f-electron terms by cubic crystal fields. *Journal of Physical Chemistry of Solids*, 1962, **23**, 1381 1405.

47. Buckmaster, H. A., Chatterjee, R. and Shing, Y. H., Ethylsulphate host lattice effects in the EPR spectra of Gd^{3+} ions. *Canadian Journal of Physics*, 1972, **50**, 991 1001.

48. Freeman, A. J. and Watson, R. E., Theoretical investigation of some magnetic and spectroscopic properties of rare-earth ions. *Physical Review*. 1962, **127**(6), 2058 2075.

49. Wakim, F. G., Synek, M., Grossgut, P. and Da Mommio, A., Expectation values of $\langle r_{4f}^n \rangle$ for rare-earth ions. *Physical Review A*, 1972, **5**(3), 1121 1124.

50. Newman, D. J. and Urban, W., Interpretation of S-state ion EPR spectra. *Advances in Physics*, 1975, **24**, 793 844.

51. Wooley, R. G., The angular overlap model in ligand field theory. *Molecular Physics*, 1981, **42**(3), 703 720.

52. Troup, G. J. and Hutton, D. R., Paramagnetic resonance of Fe^{3+} in kyanite. *British Journal of Applied Physics*, 1964, **15**, 1493 1499.

53. Wickman, H. H., Klein, M. P. and Shirley, P. A., Paramagnetic resonance of Fe^{3+} in polycrystalline ferrichrome A. *Journal of Chemical Physics*, 1965, **42**(6), 2113 2117.

54. Rudowicz, C. and Bramley, R., On standardization of the spin Hamiltonian and the ligand field Hamiltonian for orthorhombic symmetry. *Journal of Chemical Physics*, 1985, **83**(10), 5192 5197.

55. Rudowicz, C., On standardization and algebraic symmetry of the ligand field Hamiltonian for rare-earth ions at monoclinic symmetry sites. *Journal of Chemical Physics*, 1986, **84**(9), 5045 5058.

56. Jones, D. A., Baker, J. M. and Pope, D. F. D., Electron spin resonance of Gd^{3+} in lanthanum fluoride. *Proceedings of Physical Society*, 1959, **74**, 249 256.

57. Mistra, S.K., Evaluation of spin Hamiltonian parameters from ESR data of single crystals. *Magnetic Resonance Review*, 1986, **10**, 285 331.

58. Segal, E. and Wallace, W.E., Rare earth ions in a hexagonal field.I. *Journal of Solid State Chemistry*, 1970, **2** 347 365.

59. Mulak, J. and Czopnik, A., Splitting of the $^3H_4$ term of U^{4+} ion in a ligand field of Archimedean antiprism symmetry – magnetic susceptibility of $U(SO_4)_2 \cdot 4H_2O$ and $U(CH_3COCHCOCH_3)_4$. *Bull. Acad. Polon. Sci., Ser. Sci. Chim.*, 1972, **20**(3), 209 216.

60. Rudowicz, C., Algebraic symmetry and determination of the "imaginary" crystal field parameters from optical spectra of f^n-ions. Tetragonal symmetry. *Chemical Physics*, 1985, **97**, 43 50.

61. Low, W., Paramagnetic resonance spectrum of trivalent gadolinium in the cubic field of calcium fluoride. *Physical Review*, 1958, **109**, 265 271.

62. Lindgard, P. A., Tables of Product of Tensor Operators and Stevens Operators. *Journal of Physics C: Solid State Physics*, 1975, **8**, 3401 3407.

63. Silver, B.L., *Irreducible Tensor Methods*, Academic Press, New York, 1976.

64. Pappalardo, R., Absorption spectra of Er^{3+} in yttrium gallium garnet *Zeitshrift für Physik*, 1963, **173**, 374 391.

65. Newman D.J. and Ng, B., the superposition model of crystal field. *Reports on Progress in Physics*, 1989, **52**, 699 763.

66. Clark, M. G., Algebraic symmetry and quantum secular equations. *Journal of Chemical Physics*, 1971, **54**(2), 697 702.

67. Judd, B. R., A crystal field of icosahedral symmetry. *Proceedings of Royal Society Ser A*, 1957, **241**, 122.

68. Bethe, H., Termaufspaltung in Kristallen. (Splitting of terms in crystals). *Annalen der Physik*, 1929, **3**, 133 208.

69. Kramers, H. A., *Proceedigs of Academy of Science (Amsterdam)*, 1929, **32**, 1176 .

70. Christodoulos, F., Crystal fields for Er^{3+} in gold and zero-field splittings for Gd^{3+} in lanthanium ethylsulphate. Ph.D. thesis, University of Warwick, UK, 1987.

268

71. Anderson, P. W., Theory of magnetic exchange interactions: exchange in insulators and semiconductors. *Solid State Physics*, 1963, **14**, 99 214.

72. Penney, W. G. and Schlapp, R., *Physical Review*, 1932, **42**, 666 .

73. Kleiner, W. H., Crystalline field in chrome alum. *Journal of Chemical Physics*, 1952, **20**(11), 1784 1791.

74. Sugano, S. and Shulman, R. G., Covalency effects in $KNiF_3$. III.Theoretical studies. *Physical Review*, 1963, **130**(2), 517 530.

75. Anderson, P. W., Exchange in insulators: superexchange, direct exchange, and double exchange. In *Magnetism*, Vol. 1, ed. G. T. Rado and H. Suhl. Academic Press, New York, London, 1982, pp. 25 83.

76. Stevens, K. W. H., Exchange interactions in magnetic insulators. *Physics Reports, Physics Letters C*, 1976, **24**(1), 1 75.

77. Dixon, J. M. and Wardlaw, R. S., Crystalline electric fields in metals. *Physica A*, 1986, **135**, 105 138.

78. Jørgensen, C. K., Pappalardo, R. and Schmidtke, H. H., Do the "ligand field" parameters in lanthanides represent weak covalent bonding? *Journal of Chemical Physics*, 1963, **39**(6), 1422 1430.

79. Mc Weeny, R. and Sutcliffe, B. T., *Methods of Molecular Quantum Mechanics*, Academic Press, New York, 1969.

80. Gajek, Z., Mulak, J. and Faucher, M., Crystal field effect in the uranium compounds – model calcultions for $CsUF_6$, Cs_2UCl_6 and UCl_4. *Journal of Physical Chemistry of Solids*, 1987, **48**(10), 947 955.

81. Hubbard J., Rimmer, D. E. and Hopgood, F. R. A., Weak covalency in transition metal salts. *Proceedings of Physical Society (London)*, 1966, **88**, 13 36.

82. Mc Weeny, R. and Sutcliffe, B. T., *Methods of Molecular Quantum Mechanics*, Academic Press, New York, 1969, pp. 76 92.

83. Slater, J. C., A simplification of the Hartree–Fock method. *Physical Review*, 1951, **81**, 385 390.

84. Slater, J. C., Atomic shielding constants. *Physical Review*, 1930, **36**(1), 57 64.

85. Herman, F. and Skillman, S. *Atomic Structure Calculations*. Prentice Hall, Englewood Clifs, New York, 1963.

86. Shibuya, M., Modification of Slater orbitals. *Journal of Physical Society of Japan*, 1973, **34**, 567 .

87. Fisher, Ch. F., *The Hartree–Fock method for atoms*. John Wiley, New York, 1977.

88. Desclaux, J. P. and Freeman, A. J., Dirac–Fock studies of some electronic properties of actinide ions. *Journal of Magnetic Materials*, 1978, **8**, 119 129.

89. Liberman, D. A., Waber, J. T. and Cromer D. T., Self-consistent-field Dirac-Slater wave functions for atoms and ions. I. Comparison with previous calculations. *Physical Review A*, 1965, **137**(1), 27 34.

90. Kohn, W. and Sham, L. J., Self-consistent equations including exchange and correlation effects. *Physical Review A*, 1965, **140**(4), 1133 1138.

91. Van Vleck, J. H., Theory of the variations in paramagnetic anisotropy among different salts of the iron group. *Physical Review*, 1932, **41**, 208 215.

92. Van Vleck, J. H., The puzzle of rare earth spectra in solids. *Journal of Physical Chemistry*, 1937, **41**, 67 80.

93. Penney, W. G. and Schlapp, R., The influence of crystalline fields on the susceptibilities of salts of paramagnetic ions. I.The rare earths, especially Pr and Nd. *Physical Review*, 1932, **41**, 194 207.

94. Schlapp, R. and Penney, W.G., Influence of crystalline fields on the susceptibilities of salts of paramagnetic ions. II.The iron group,especially Ni, Cr and Co. *Physical Review*, 1932, **42**, 666.

95. Abragam, A. and Pryce, M. H. L., The theory of paramagnetic resonance in hydrated cobalt salts. *Proceedings of Royal Society Ser. A*, 1951, **206**, 173.

96. Hellwege, K. H., Elektronenterme und Strahlung von Atomen in Kristallen. *Annalen der Physik*, 1949, **4**, 95 127.

97. Bleaney, B. and Stevens, K. W. H., Paramagnetic resonance. *Reports on Progress in Physics*, 1953, **16**, 108 154.

98. Low, W., Paramagnetic resonance in solids. Theory of the crystalline field. In *Solid State Physics*, Suppl. 2, ed. F. Seitz and D. Turnbull. Academic Press, New York, London, 1960, pp. 8 75.

99. Margenau H. and Murphy G. H., *The Mathematics of Physics and Chemistry*. 2nd edn. Van Nostrand, Princeton, New Jersey, 1956, p. 100.

100. Griffith J. S., *The Theory of Transition-Metal Ions*. Cambrige University Press, London, New York, 1961, p. 75 .

101. Prather J. L., *Atomic Energy Levels in Crystals*. N.B.S. Monograph No. 19, 1961, p. 6.

102. Edmonds, A. R., *Angular Momentum in Quantum Mechanics*. Princeton University Press, Priceton, New York, 1957, pp. 10 29.

103. Lenander, C. J., Parameterized Slater modified Hartree-Fock method applied to actinide ions. *Physical Review*, 1963, **130**, 1033 1035.

104. Freeman, A. J., Desclaux, J. P. Lander, G. H. and Gaber, J. Jr, Neutron magnetic form factors of uranium ions. *Physical Review B*, 1976, **13**(3), 1168 1176.

105. Faucher, M. and Garcia, D., Electrostatic crystal field contributions in rare earth compounds with consistent multipolar effects. I.Contribution to k-even parameters. *Physical Review B*, 1982, **26**, 5451 5468.

106. Birgeneau, R. J., Bucher, E., Majta, J.P., Passel, L. and Turberfield, K.C., Crystal fields and the effective point charge model in the rare earth pnictides. *Physical Review B*, 1973, **8**(12), 5345 5347.

107. Ellis, M. M. and Newman, D. J., Crystal field in rare earth trichlorides. I.Overlap and exchange effects in $PrCl_3$. *Journal of Chemical Physics*, 1967, **47**(6), 1986 1993.

108. Margolis, J. S., Energy levels of $PrCl_3$. *Journal of Chemical Physics*, 1961, **35**(4), 1367 1373.

109. Johnston D. R., Satten, R. A., Schreiber, C. L. and Wong E. Y., Covalency effects in U^{4+} halide complexes. *Journal of Chemical Physics*, 1966, **44**(8), 3141L 3143L.

110. Żołnierek, Z., Crystal field parameters in a modified point charge model. *Journal of Chemical Physics*, 1984, **45**(5), 523 528.

111. Żołnierek, Z., π-Bonding approach in the modified charge model. *Solid State Communications*, 1985, **56**(7), 609 614.

112. Kasuya, T., S-d and s-f interaction and rare earth metals. In *Magnetism*, Vol. 2B, ed. G. T. Rado and H. Suhl. Academic Press, New York, London, 1966, pp. 215 294.

113. Gajek, Z., Quantum-chemical model of the crystal field in the uranium compounds. Ph.D. thesis, W. Trzebiatowski Institute of Low Temperature and Structure Research, Polish Academy of Sciences, Wrocław, 1984.

114. Freeman, A. J. and Desclaux, J. P., Dirac-Fock studies of some electronic properties of rare earth ions. *Journal of Magnetism and Magnetic Materials*, 1979, **12**(1), 11 21.

115. Devine, R. A. B. and Berthier, Y., The variation of $\langle r^4 \rangle$ and $\langle r^6 \rangle$ for $4f$ electrons across the rare-earth series. *Solid State Communications*, 1978, **26**, 315 317.

116. Sharma, R. R., Expansion of a function about a displaced center for multicenter integrals: A general and closed expression for the coefficients in the expansion of a Slater orbital and for over integrals. *Physical Review A*, 1976, **13**(2), 517 527.

117. Shen, Y. and Bray, K. L., Ab initio calculation on crystal fields of Sm^{2+} in solids with Cl and F ligands. *Physical Review B*, 1998, **58**(9), 5305 5313.

118. Gajek, Z. and Mulak, J., A new method of calculation of interionic Coulomb exchange contribution to the crystal field parameters. *International Journal of Quantum Chemistry*, 1985, **28**, 889 894.

119. Garcia D. and Faucher, M. Crystal field parameters in rare earth compounds: extended charge contribution. *Physical Review B*, 1984, **30**(4), 1703 1707.

120. Raychaudhuri, A. K. and Ray, D. K., Effect of the ligand charge distribution on the crystalline electric field of rare earth ions. *Proceedings of Physical Society*, 1967, **90**, 839 846.

121. Ellis, M. M. and Newman, D. J., Crystal field in rare earth trichlorides. III.Charge penetration and covalency contributions in $PrCl_3$. *Journal of Chemical Physics*, 1968, **49**(9), 4037 4043.

122. Barnett, M. P. and Coulson, C. A., The evaluation of integrals occurring in the theory of molecular structure. Parts I and II. *Phil. Trans. Roy. Soc. A*, 1951, **243**, 221 249.

123. Buehler, R. J. and Hirschfelder, J. O., Bipolar expansion of Coulombic potentials. *Physical Review*, 1951, **83**(3), 628 633.

124. Buehler, R. J. and Hirschfelder, J. O., Bipolar expansion of Coulombic potentials. Addenda. *Physical Review*, 1952, **85**(1), 149 149.

125. Phillips, J. C. and Kleinman, L., New method for calculating wavefunctions in crystal and molecules. *Physical Review*, 1959, **116**(2), 287 294.

126. Mc Clure, D. S., *Phonons in Perfect Lattices and Lattices with Point Imperfections.* ed. R. W. H. Stevenson. Oliver and Boyd, 1966, pp. 314 .

127. Harrison, W. A., *Solid State Theory.* Mc Graw-Hill Book Company, New York, 1970.

128. Watson, R. E. and Freeman, A. J., Covalency in crystal field theory: $KNiF_3$. *Physical Review A*, 1964, **134**(6), 1526 1546.

129. Knox, K., Shulman, R.G. and Sugano, S., Crystal field splitting in $KNiF_3$. *Bulletin of American Physical Society (II)*, 1960, **5**, 415 415.

130. Soulié, E., Un commentaire a "D_{3d} distortion in $CsUF_6$" par A. F. Leung. *Journal of Physical Chemistry of Solids*, 1978, **39**(6), 695 698.

131. Schäffer, C. E., A perturbation representation of weak covalent bonding. The symmetry basis for the angular overlap model of the ligand field. In *Structure and Bonding*, Vol. 5, Springer Verlag, Berlin, Heidelberg, New York, 1968, pp. 68 95.

132. Gajek, Z. and Mulak, J., The ab initio computational basis of the angular overlap model and its phenomenological effectiveness in the $5f$ ion compounds. *Journal of Physics: Condensed Matter*, 1992, **4**, 427 444.

133. Bishton, S. S., Ellis M. M., Newman, D. J. and Smith, J., Crystal field in rare earth trichlorides. II.Exchange charge effects in $PrCl_3$, *Journal of Chemical Physics*, 1967, **47**(10), 4133 4136.

134. Curtis, M. M., Newman, D. J. and Stedman, G. E., Crystal field in rare earth trichlorides. IV.Parameter variations. *Journal of Chemical Physics*, 1969, **50**(3), 1077 1085.

135. Curtis, M. M. and Newman, D. J., Crystal field in rare earth trichlorides. V.Estimation of ligand–ligand overlap effects. *Journal of Chemical Physics*, 1970, **52**(3), 1340 1344.

136. Dexter, D. L., Exciton models in the alkali halides. *Physical Review*, 1957, **108**, 707 712.

137. Wybourne, B. G., Energy levels of trivalent gadolinium and ionic contributions to the ground–state splitting. *Physical Review* 1966, **148**(1), 317 327.

138. Moskowitz, J. W., Hollister C., Hornback, C. J. and Basch, H., Self-consistent field study of the model in ionic salts. I. NiF_6^{4-}. *Journal of Chemical Physics*, 1970, **53**(7), 2570 2580.

139. Mc Weeny, R. and Sutcliffe, B. T., *Methods of Molecular Quantum Mechanics*, Academic Press, London, New York 1969, chpts. 3, 7.

140. Mc Weeny, R. and Sutcliffe, B. T., *Methods of Molecular Quantum Mechanics*, Academic Press, London, New York 1969, chpt. 5.

141. Burns, G. and Axe, J. D., Covalent bonding effects in rare earth crystal fields. In *Optical Properties of Ions in Crystals*, ed. H. M. Crosswhite and H. W. Moos. Interscience, 1967, pp. 53 71.

142. Garcia, D. and Faucher, M., Crystal field parameters in rare earth compounds: a mixed "covalo–electrostatic model". *Journal of Chemical Physics*, 1985, **82**(12), 5554 5564.

143. Faucher, M., Garcia, D. and Moune, O. K., Effect of second neighbours on the crystal field in rare earth compounds. *Journal of Alloys and Compounds*, 1992, **180**, 243 249.

144. Faucher, M., Garcia, D. and Moune, O. K., Crystal field effect in rare earth compounds. Influence of second nearest neighbours. *Journal of Luminescence*, 1992, **51**, 341 351.

145. Sternheimer, R. M., Shielding and antishielding effects for various ions and atomic systems. *Physical Review*, 1966, **146**(1) 140 160.

146. Watson R. E. and Freeman A. J., Shielding and distortion of rare earth crystal field spectra, *Physical Review A*, 1964, **133**(6), A1571 A1584.

147. Sternheimer, R. M., Blume, M. and Peierls R. F., Shielding of crystal fields at rare earth ions, *Physical Review*, 1968, **173**(2), 376 389.

148. Ahmad, S., Many body calculation of crystal field shielding in the Pr^{3+} ion. *Journal of Physics C: Solid State Physics*, 1981, **14**, 2759 2769.

149. Watson, R. E. and Freeman A. J., Electronic polarizabilities and Sternheimer shielding factors. *Physical Review*, 1963, **131**(1), 250 255.

150. Sternheimer, R. M., Electronic polarizabilities of ions. *Physical Review*, 1957, **107** 1565 1569.

151. Dalgarno A., Atomic polarizabilities and shielding factors. *Advances in Physics*, 1962, **11** 281 315.

152. Burns, G., Shielding and crystal fields at rare earth ions. *Physical Review*, 1962, **128**(5) 2121 2130.

153. Burns, G., Crystal field at rare earth ions. *Journal of Chemical Physics*, 1965, **42**(1), 377 390.

154. Lenander, C. J. and Wong, E. Y., Crystal field shielding in $PrCl_3$ *Journal of Chemical Physics*, 1963, **38**(11), 2750 2752.

155. Watson R. E. and Freeman A. J., Nonlinear and linear shielding of rare earth crystal field interactions. *Physical Review*, 1965, **139**(5), A1606 1615.

156. Ghatikar, M. N., Raychaudhuri, A. K. and Ray, D. K., Electronic shielding of the crystalline electric field in rare earth ions. *Proceedings of Physical Society*, 1965, **86**, 1235 1238.

157. Sternheimer, R. M., Electronic polarizabilities of the alkali atoms. *Physical Review*, 1962, **127**(4), 1220 1223.

158. Edmonds, A. R., *Angular Momentum in Quantum Mechanics*. Princeton University Press, Princeton, New York, 1957, chpt. 3, p. 47.

159. Edmonds, A. R., *Angular Momentum in Quantum Mechanics*. Princeton University Press, Princeton, New York, 1957, chpt. 6, p. 95.

160. Erdös, P. and Kang, J. H., Electronic shielding of Pr^{3+} and Tm^{3+} ions in crystals. *Physical Review B*, 1972, **6**(9), 3393 3408.

161. Hutchinson, C. A. Jr. and Wong, E. Y., Paramagnetic resonance in rare earth trichlorides. *Journal of Chemical Physics*, 1958, **29**(4), 754 760.

162. Hutchings, M. T. and Ray, D. K., Investigation into the origin of crystalline electric field effects on rare earth ions. I.Contribution from neighbouring induced moments. *Proceedings of Physical Society*, 1963, **81**, 663 676.

163. Morrison, C. A., Dipolar contributions to the crystal fields in ionic solids. *Solid State Communications*, 1976, **18**, 153 154.

164. Edmonds, A. R., *Angular Momentum in Quantum Mechanics*. Princeton University Press, Princeton, New York, 1957, chpt.5, p.72.

165. Edmonds, A. R., *Angular Momentum in Quantum Mechanics*. Princeton University Press, Princeton, New York, 1957, chpt.4, p.61.

166. Slater, J. C, Ferromagnetism and the band theory. *Review of Modern Physics*, 1953, **25**, 199 210.

167. Wannier, G. H., The structure of electronic excitation levels in insulating crystals. *Physical Review*, 1937, **52**(3), 191 197.

168. Ziman, J. M., *Principles of the Theory of Solids*. Cambridge University Press, Cambridge, 1972.

169. Stevens, K. W. H., Many electron theory of distorted lattices. *Journal of Physics C: Solid State Physics*, 1973, **6**, 2191 2202.

170. Kohn, W. and Onffroy, J. R., Wannier functions in a simple nonperiodic system. *Physical Review B*, 1973, **8**(6), 2485 2495.

171. Gay, J. G. and Smith, J. R., Theory of localized defects in solids. *Physical Review B*, 1974, **9**(10), 4151 4164.

172. Bates, C. A., Dixon, J. M., Fletcher, J. R. and Stevens K. W. H., The E-type orbitals of octahedrally co-ordinated ions. *Journal of Physics C: Solid State Physics*, 1968, **1**, 859 871.

173. Stevens, K. W. H., Equivalent angular momentum operators for second quantised operators. *Physics Letters A*, 1974, **47**, 401 403.

174. Elliot, R. J. and Thorpe, M. F., Orbitals effects on exchange interactions. *Journal of Applied Physics*, 1968, **39**, 802 807.

175. Rotenberg, M., Bivins R., Metropolis N., and Wooten, J. K. Jr, *The 3-j and 6-j Symbols*. MIT Press, Cambridge Massachusetts, 1959, p. 4.

176. Dixon, J. M., Wardlaw, R. S. and Stevens, K. W. H., Second Quantisation and Equivalent Two-Body Operators. *Physica B*, 1984, **124**, 149 155.

177. Bishton, S. S. and Newman D. J., Parameterization of the correlation crystal field. *Journal of Physics C: Solid State Physics*, 1970, **3**, 1753 1761.

178. Touborg, P., Crystal fields in Sc, Y, and the heavy rare earth metals Tb, Dy, Ho, Er, Tm and Lu. *Physical Review B*, 1977, **16**(3), 1201 1211.

179. Schmitt, D., Morin, P. and Pierre, J., Crystal field and magnetic properties in some cubic holmium compounds. *Physical Review B*, 1977, **15**(4), 1698 1705.

180. Barbara, B., Berthier, Y., Devine, R. A. B. and Rossignol, M. F., Neutron spectroscopy of rare earths in $LaAl_2$. *Journal of Physics F.: Metal Physics*, 1982, **12**, 2625 2632.

181. Barnes, S. E., Baberschke, K. and Hardimen, M., Anomalous crystal field splittings of lanthanides s-states in metals. *Physical Review B*, 1978, **18**(6), 2409 2421.

182. Buzukin, S. V. and Khaliullin, G. G., Crystal field theory of the rare earth ions in metals (Russ.) *Fiz. Met. Metalloved.*, 1988, **66**(2), 213 221.

183. Hummler, K. and Fähnle, M., Full potential linear-muffin-tin-orbital calculations of the magnetic properties of rare earth transistor metal intermetallics. Description of the formalism and application to the series RCo$_5$ (R = rare earth atom). *Physical Review B*, 1996, **53**(6), 3272 3289.

184. Diviś, M. and Kuriplach, J., Electronic structure and crystal field in REMg and RERh (RE=rare earth) intermetallics. *Physica B*, 1993, **183**, 25 32.

185. Newman, D. J., Models of lanthanide crystal fields in metals. *Journal of Physics F.: Metal Physics*, 1983, **13**, 1511 1518.

186. Williams, G. and Hirst, L. L., Crystal field effects in solid solutions of rare earths in noble metals. *Physical Review*, 1969, **185**(2), 407 415.

187. Gunnarsson, O., Jonson, M. and Lundqvist, B. I., Description of exchange and correlation effects in inhomogeneous electron sysytems. *Physical Review B*, 1979, **20**(8), 3136 3164.

188. Newman, D. J. and Ng, B., Ab initio crystal field calculation for small internuclear separations. *Journal of Physics C: Solid State Physics*, 1986, **19**, 389 394.

189. Dixon, J. M. and Dupree, R., The effect of d electrons on crystal field potentials in rare metals and dilute alloys. *Journal of Physics F.: Metal Physics*, 1973, **3**, 118 124.

190. Duthie, J. C. and Heine, V., Theory of screening effects on crystalline electric fields in metallic rare earth materials. *Journal of Physics F.: Metal Physics*, 1979, **9**(7), 1349 1355.

191. Turberfield, K. C., Passel, L., Birgeneau, R. J. and Bucher, E., Crystal fields in rare earth metallic compounds. *Physical Review Letters*, 1970, **25**(11), 752 755.

192. Murani, A. P., Magnetic susceptibility and electrical resistivity of some gold-rare-earth alloys. *Journal of Physics C: Solid State Physics, Metal Physics Suppl.*, 1970, 2, S153 S158.

193. Ashcroft, N. W., Mermin, N. D., *Solid State Physics*, Holt, Rinehart and Winston, New York, 1976.

194. Raimes, S., *The Wave Mechanics of Electrons in Metals*, North-Holland, Amsterdam, 1961.

195. Sachs, M., *Solid State Theory*, McGraw-Hill, New York, 1963.

196. Steinbeck, L., Richter, M. and Eschrig, H., Calculated crystal field parameters for rare earth impurities in noble metals. *Physical Review B*, 1944, **49**(23), 16289 16293.

197. Bross, H. and Ehrnsperger, M., Calculation of the dielectric function of Li metal. *Zeitschrift für Physik B*, 1995 **97**(1), 17 24.

198. Walker, A. B. and Taylor, R., Density-dependent potentials for simple metals. *Journal of Physics: Condensed Matter*, 1990, **2**, 9481 9499.

199. Richter, M., Oppeneer, P. M., Eschrig, H. and Johansson, B., Calculated crystal field parameters of $SmCo_5$. *Physical Review B*, 1992, **46**(21) 13913 13927.

200. Cohen, M. L. and Heine, V., The fitting of pseudopotentials to experimental data and their subsequent application. *Solid State Physics*, 1970, **24**, 37 248.

201. *International Tables for X-Ray Crystallography*. The Kynoch Press, Birmingham, England, 1952.

202. Harrison, W. A., *Electronic Structure and the Properties of Solids*. W. H. Freeman and Company, San Francisco, 1980.

203. Pines, D., *Solid State Physics Vol.1*, Academic Press, New York, 1955, p. 367.

204. Raimes S., *Many Electron Theory*, North-Holland, Amsterdam, 1972.

205. Debye, P. P. and Hückel, E., Zur Theorie der Electrolyte. I. Gefrierpunktserniedrigung und Verwandte Erscheinungen. *Physikalische Zeitschrift*, 1923, **24**(15), 185 206.

206. Dixon, J. M. and Dupree, R., Crystalline electric fields of rare earth ions in metals. I.Theory. *Journal of Physics F: Metal Physics*, 1971, **1**(5), 539 553.

207. Christodoulos, F. and Dixon, J. M., A reassesment of the role played by a $5d$ virtual bound state in the determination of crystal fields for heavy rare earth ions in gold and silver. *Physics Letters A*, 1987, **124**(8), 437 441.

208. Anderson, P. W, Localized magnetic states in metals. *Physical Review*, 1961, **124**(1), 41 53.

209. Eagles, D. M., Models for the heavy rare earth metals and $(RE)Fe_2$ compounds involving $5d$ and $6s$ electrons. *Phys. Kondens. Materie*, 1973, **16**, 181 199.

210. Dixon, J. M., $5d$ and $4f$ mixing coefficients for rare earth ions in the pure metals and in silver and gold. *Solid State Communications*, 1973, **12**, 789 794.

211. Kikkert, P. J. W., Crystal field of Dy in non-magnetic metals. Ph.D. thesis, University of Croningen, 1980.

212. Devine, R., d resonance effects associated with rare earths and dilute rare earth alloys. *Journal of Physics F.: Metal Physics*, 1974, **4**, 1447 1453.

213. Eagles, D. M., The contribution of $5d$ electrons to the saturation magnetic moments and magnetic hyperfine fields in the heavy rare earth metals. *Solid State Communications*, 1973, **12**, 291 294.

214. Devine, R. and Dixon., J. M., Conduction-electron contribution to electric field gradients in rare earth metals. *Physical Review B*, 1973, **7**, 4902 4907.

215. Lacueva, G., Levy, P. M. and Fert, A., Unified approach to some transport and EPR properties of noble metals with rare earth impurities. *Physical Review B*, 1982, **26**(3), 1099 1124.

216. Chow, H. C., Exchange corrections to the crystalline field parameter C_4 for rare earths in the noble metals. *Physical Review B*, 1973, **7**, 3404 3405.

217. Raimes, S., *Many Electron Theory*, North-Holland, Amsterdam, 1972, chpt. 4.

218. Löwdin, P. O., On the non-orthogonality problem connected with the use of atomic wave functions in the theory of molecules and crystals. *Journal of Chemical Physics*, 1950, **18**(3), 365 375.

219. Nellis, W. J. and Brodsky, M. B., Virtual-bound-state formalism in Pd-U alloys, *Physical Review B*, 1970, **2**(11), 4590 4596.

220. Harrison, W. A., *Solid State Theory*. Mc Graw-Hill Book Company, New York, 1970, chpt.2.

221. Harrison, W. A., *Electronic Structure and the Properties of Solids*. W. H. Freeman and Company, San Francisco, 1980, part II.

222. Harrison, W. A., *Pseudopotential in the Theory of Metals*. Benjamin, New York, 1966.

223. Schrieffer, J. R. and Wolf, P. A., Relation between the Anderson and Kondo Hamiltonians. *Physical Review*, 1966, **149**(2), 491 492.

224. Schrieffer, J.R., The Kondo effect – the link between magnetic and non-magnetic impurities in metals. *Journal of Applied Physics*, 1967, **38**, 1143 1150

225. Harrison, W. A., Electron structure of f-shell metals. *Physical Review B*, 1983, **28**(2), 550 559.

226. Harrison, W. A. and Straub, G. K., Electronic structure and properties of d- and f-shell metal compounds. *Physical Review B*, 1987, **36**(5), 2695 2706.

227. Andersen, O. K., Klose, W. and Nohl, H., Electronic structure of Chevrel phase high-critical field superconductors. *Physical Review B*, **17**(3), 1209 1237.

278

228. Siebert, J. F., Dodds, S. A. and Silsbee, R. H., Transition electron spin resonance in dilute rare earth aluminium alloys. *Physical Review B*, 1976, **14**(11), 4813 4825.

229. Garifullin, I. A., Farzan, T. O., Khaliullin, G. G. and Kukovitsky, E. F., Study of orbit lattice coupling of Er impurities in *Cu* and *Ag* hosts. *Journal of Physics F.: Metal Physics*, 1985, **15**(4), 979 990.

230. Takahashi, H. and Kasuya, T., Anisotropic *p-f* mixing mechanism explaining anomalous magnetic properties in Ce monopnictides: I. Effective 4*f* level. *Journal of Physics C.: Solid State Physics*, 1985, **18**, 2697 2707.

231. Takahashi, H. and Kasuya, T., Anisotropic *p-f* mixing mechanism explaining anomalous magnetic properties in Ce monopnictides: II.Crystal field splitting in rare earth pnictides. *Journal of Physics C.: Solid State Physics*, 1985, **18**, 2709 2719.

232. Coqblin, B. and Schrieffer, J. R., Exchange interaction in alloys with cerium impurities. *Physical Review*, 1969, **185**(2), 847 853.

233. Siemann, R. and Cooper, B. R., Planar coupling mechanism explaining anomalous magnetic structures in cerium and actinide intermetallics. *Physical Review Letters*, 1980, **44**, 1015 1019.

234. Dixon, J. M. and Wardlaw, R. S., The interaction of a single *s* band with a substitutional potential of arbitrary range. *Journal of Physics C.: Solid State Physics*, 1986, **19**, 943 961.

235. March, N. H. and Deb , B. M., *The Single-Particle Density in Physics and Chemistry*. Academic Press, 1987.

236. Parr R. G. and Yang W., *Density Functional Theory of Atoms and Molecules*. Oxford University Press, 1989.

237. Eschrig, H., *The Fundamentals of Density Functional Theory*. ed. B. G. Teubner Verlagsgesellschaft. Stuttgart, 1996.

238. Richter, M., Electronic structure and magnetism of lanthanide-, actinide- and transition metal systems.
Habilitationsschrift zur Erlangung des akademischen Grades Dr. rer. nat. habil., Technischen Universität Dresden, 1997.

239. Trygg, J., First principles studies of magnetic and structural properties of metallic systems. Ph.D. thesis, Acta Universitatis Upsaliensis, Uppsala, 1995.

240. Hohenberg P. and Kohn W., Inhomogenous electron gas. *Physical Review B*, 1964, **136**(3), 864 870.

241. Almbladh, C. O. and Pedroza, A. C., Density functional exchange-correlation potentials and orbital eigenvalues for light atoms. *Physical Review A*, 1984, **29**, 2322 2330.

242. Ceperley D. M. and Alder B. J., Ground state of the electron gas by a stochastic method. *Physical Review Letters*, 1980, **45**, 566 569.

243. Kenny S. D., Rajagopal G., Needs R. J., Leung W. K., Godfrey M. J., Williamson A. J. and Foulkes W. M. C., Quantum Monte Carlo calculations of the energy of the relativistic homogeneous electron gas. *Physical Review Letters*, 1996, **77**, 1099 1102.

244. Dirac P. A. M., Note on exchange phenomena in the Thomas atom. *Proc. Cambridge Phil. Soc.*, 1930, **26**, 376 385.

245. Levy, M., Electron densities in search of Hamiltonians. *Physical Review A*, 1982, **26**(3), 1200 1207.

246. Lieb, E. H., Density functionals for Coulomb systems. *International Journal of Quantum Chemistry*, 1983, **24**, 243 277.

247. Slater, J. C. The self-consistent field for molecules and solids. In *Quantum Theory of Molecules and Solids*, Vol. 4. McGraw-Hill, New York, 1974.

248. Janak, J. F., Proof that $\partial E / \partial n_i = \epsilon_i$ in density-functional theory. *Physical Review B*, 1978, **18**, 7165 7168.

249. Theophilou, A. K., The energy density functional formalism for excited states. *Journal of Physics C*, 1979, **12**, 5419 5430.

250. Gunnarsson O. and Lundqvist B. I., Exchange and correlation in atoms, molecules, and solids by the spin-density-functional formalism. *Physical Review B*, 1976, **13**(10), 4274 4298.

251. Von Barth, U., Local-density theory of multiplet structure. *Physical Review A*, 1979, **20**, 1693 1703.

252. Theophilou, A. K., Denity Functional theory for excited states. In *The Single–Particle Density in Physics and Chemistry*. ed. N. H. March and B. M. Deb. Academic Press, 1987.

253. Gross, E. K. U., Oliveira, L. N., and Kohn, W., Density-functional theory for ensembles of fractionally occupied states. I.Basic formalism. *Physical Review A*, 1988, **37**(8), 2809 2820.

254. Brooks, M. S. S., Eriksson, O., Wills, J. M. and Johansson B., Density functional theory of crystal field quasiparticle excitations and the ab initio calculation of spin Hamiltonian parameters. *Physical Review Letters*, 1997, **79**(13), 2546 2549.

255. Diviš, M., Richter, M., Forstreuter, J., Koepernik, K. and Eschrig, H., Crystal-field and magnetism of RGa_2 (R=Ce, Er) compounds derived from density-functional calculations. *Journal of Magnetism and Magnetic Materials*, 1997, **176**, L81 L88.

256. Kuz'min, M. D., Richter, M. and Eschrig, H., Ab initio calculation of the crystal field in RFe_{12} and RFe_8M_4 (M=Ti,V,Mo). *Journal of Magnetism and Magnetic Materials*, 1999, **196-197**, 762 764.

257. Capitani, J. F., Nalewajski, R. F. and Parr, R. G., Non Born-Oppenheimer density functional theory of molecular systems. *Journal of Chemical Physics*, 1982, **76**, 568 573.

258. English, H. and English, R., Exact density functionals for ground-state energies. I.General results. *physica status solidi (b)*, 1984, **123**, 711 721.

259. Boroński, E. and Nieminen, R. M., Electron-positron density-functional theory. *Physical Review B*, 1986, **34**(6), 3820 3831.

260. Von Barth, U. and Hedin, L., A local exchange-correlation potential for the spin polarized case. *Journal of Physics C*, 1972, **5**, 1629 1642.

261. Rajagopal, A. K. and Callaway, J., Inhomogeneous electron gas. *Physical Review B*, 1973, **7**(5), 1912 1919.

262. Cortona, P., Doniach, S. and Sommers, C., Relativistic extension of the spin-polarized local-density-functional theory: study of the electronic and magnetic properties of the rare-earth ions. *Physical Review A*, 1985, **31**(5), 2842 2850.

263. Eschrig, H., Seifert, G. and Ziesche, P., Current density functional theory of quantum electrodynamics. *Solid State Communications*, 1985, **56**(9), 777 780.

264. Jansen, H. J. F., Magnetic anisotropy in density-functional theory. *Physical Review B*, 1988, **38**(12), 8022 8029.

265. Richter, M. and Eschrig, H., Electronic structure of ferromagnetic rare-earth-transition-metal compounds. *Physica B*, 1991, **172**, 85 93.

266. Białynicki-Birula, I. and Białynicka-Birula, Z., *Quantum electrodynamics.* Pergamon Press, Oxford, New York, 1975.

267. Steinbeck, L., Richter, M., Nitzsche, U. and Eschrig, H., Ab initio calculation of electronic structure, crystal field, and intrinsic magnetic properties of Sm_2Fe_{17}, $Sm_2Fe_{17}N_3$, $Sm_2Fe_{17}C_3$, and Sm_2Co_{17}. *Physical Review B*, 1996, **53**, 7111 7127.

268. Eriksson, O., Brooks, M. and Johansson, B., Orbital polarization in narrow–band system. Application to volume collapses in light lanthanides. *Physical Review B*, 1990, **41**, 7311 7314.

269. Severin, L., Brooks, M. S. S. and Johansson, B., Relationship between the Coulomb integral U and the Stoner parameter I. *Physical Review Letters*, 1993, **71**, 3214 3217.

270. MacDonald, A. H., Daams, J. M., Vosko, S. H. and Koelling, D. D., Non-muffin-tin and relativistic effects on the electronic structure of noble metals. *Physical Review B*, 1982, **25**, 713 725.

271. Daalderop, G. H. O., Kelly, P. J. and Schuurmans, M. F. H., *Journal of Magnetism and Magnetic Materials*, 1992, **104–107**, 737.

272. Perdew, J. P. and Zunger, A., Self-interaction correction to density-functional approximatioms for many-electron systems. *Physical Review B*, 1981, **23**(10), 5048 5079.

273. Svane, A., Electronic structure of cerium in the self-interaction-corrected local-spin-density approximation. *Physical Review B*, 1996, **53**(8), 4275 4286.

274. Perdew, J. P., Yue, W., Accurate and simple density functional for the electronic exchange energy: Generalized gradient approximation. *Physical Review B*, 1986, **33**(12), 8800 8802.

275. Perdew, J. P., Chevary, J. A., Vosko, S. H., Jackson, K. A., Pederson, M. R., Singh, D. J. and Fiolhais, C., Atoms, molecules, solids, and surfaces: Applications of the generalized gradient approximation for exchange and correlation. *Physical Review B*, 1992, **46**(11), 6671 6687.

276. Kohn, W., Becke, A. D. and Parr, R. G., Density Functional Theory of electronic structure. *Journal of Physical Chemistry*, 1996, **100**, 12974 12980.

277. Fähnle, M., Mapping of the density functional theory on the crystal field theory of rare earth systems. *Journal of Magnetism and Magnetic Materials*, 1995, 151, L5 L7.

278. Diviš, M., Nekvasil, V. and Kuriplach, J., Density functional calculation of the crystal field interaction in rare-earth cuprates. *Physica C*, 1998, **301**, 23 28.

279. Liu, Z., Richter, M., Diviš, M. and Eschrig, H., Calculation of paramagnetic susceptibilities and specific heats by density-functional-crystal-field theory: $PrPd_2X_3$ and $PrPd_2X_3$ (X=Al, Ga). *Physical Review B*, 1999, **60**(11), 7981 7992.

280. Goldschmidt, Z. B., Atomic properties (free atoms). In *Handbook on the Physics and Chemistry of Rare Earths*, ed. by K. A. Gschneider Jr and L. Eyring. North-Holland, Amsterdam, 1978.

281. Carnall, W. T. and Crosswhite H. M., *Optical spectra and electronic structure of actinide ions in compounds and in solution.* Argonne National Laboratory Report, ANL-84/90, 1985.

282. Chang, N., Gruber, J., Leavitt, R. and Morrison, C. A., *Journal of Chemical Physics*, 1982, **76**, 3877 .

283. Auzel, F. and Malta, O. L., *Journal de Physique (Paris)*, 1983, **44**, 201

284. Yeung, Y. Y. and Newman, D. J., Crystal field invariants and parameters for low-symmetry sites. *Journal of Chemical Physics*, 1984, **82**, 3747 3752.

285. Faucher, M., Garcia, D. and Jørgensen, C. K., Crystal field strength: angular overlap analysis of ligand field effects in some rare-earth compounds. *Chemical Physics Letters*, 1986, **129**(4), 387 391.

286. Gajek, Z., Mulak, J. and Krupa, J. C., Actual and idealized crystal field parameterisations for uranium ions in UF_4. *Journal of Solid State Chemistry*, 1993, **107**, 413 427.

287. Jørgensen, C. K., Faucher, M. and Garcia, D., The 'ligand field' energy differences between 4f orbitals are mainly provided by the kinetic energy operator. *Chemical Physics Letters*, 1986, **128**(3), 250 254.

288. Malta, O. L., A simple overlap model in lanthanide crystal-field theory. *Chemical Physics Letters*, 1982, **87**(1), 27 29.

289. Schäffer, C. E. and Jørgensen, C. K., *Molecular Physics*, 1965, **9**, 401 .

290. Gajek, Z., Reliability of conventional crystal field models in f-electron systems. *Journal of Alloys and Compounds*, 1995, **219**, 238 243.

291. Gajek, Z., First-principle estimation of electronic structure of uranium oxychalcogenides UOY, Y = S, Se, Te. Application to the INS spectra of UOS. *Journal of Physics: Condensed Matter*, 2000, **12**, 415 437.

292. Urland, W., The assessment of the crystal-field parameters for f^n-electron systems by the angular overlap model. Rare-earth ions M^{3+} in $LiMF_4$. *Chemical Physics Letters*, 1981, **77**(1), 58 62.

293. Urland, W., The assessment of the crystal-field parameters for f^n-electron systems by the angular overlap model. Rare-earth ions M^{3+} in Cs_2NaMCl_6. *Chemical Physics Letters*, 1981, **83**(1), 116 118.

294. Gajek, Z. and Krupa, J. C., Strong crystal field effect in Np^{4+}:$ThCl_4$ – optical absorption study. *Journal of Physics: Condensed Matter*, 1998, **10**, 11841 11855.

295. Yeung, Y. Y. and Newman, D. J., Angular overlap and superposition models of the orbit-lattice interaction. *Journal of Chemical Physics*, 1985, **122**, 415 417.

296. Urland, W., Electronic structure and co-ordination geometry of Pr^{3+}-doped rare earth trichlorides. *Journal of Less-Common Metals*, 1989, **148**, 151 157.

297. Mulak, J. and Gajek, Z., Crystal field of complex parameterization and metal-ligand distance dependence of the parameters. *Polish Journal of Chemistry*, 1993, **67**, 1489 1504.

298. Born, M. and Oppenheimer, J. R., Zur Quantentheorie der Molekeln. *Annalen der Physik (Leipzig)*, 1927, **84**(4), 457 484.

299. Bersuker, I. B. and Polinger, V. Z., *Vibronic Interactions in Molecules and Crystals*. Springer Series in Chemical Physics, Vol. 49. Springer Verlag, Berlin, Heidelberg, 1989.

300. Bersuker, I. B., *The Jahn-Teller Effect and Vibronic Interactions in Modern Chemistry*. Plenum Press, New York, London, 1984.

301. Landau, L. D. and Liphsitz, E. M., *Kwantovaya Mekhanika. Nerelativistskaya Teoriya*. Nauka, Moscow, 1974.

302. Bradley, C. J. and Cracnell, A. P., *The Mathematical Theory of Symmetry in Solids: Representation Theory for Point Group and Space Groups*. Claredon, Oxford 1972.

303. Sturge, M. D., The Jahn-Teller Effect in Solids. *Solids State Physics*, 1967, **20**, 91 211.

304. Jahn, H. A. and Teller, E., Stabililty of polyatomic molecules in degenerate electronic states. I.Orbital degeneracy. *Proceedings of Royal Society, London, Ser A*, 1937, **161**, 220 235.

305. Feynman, R. P., Forces in Molecules. *Physical Review*, 1939, **56**, 340 343.

306. McLachlan, A. D., The wave functions of electronically degenerate states. *Molecular Physics*, 1961, **4**, 417 423.

307. Köppel, M., Damcke, W. and Cederbaum, L. S., Theory of vibronic coupling in linear molecules. *Journal of Chemical Physics*, 1981, **74**(5), 2945 2968.

308. Griffith, J. S., *The Irreducible Tensor Method for Molecular Symmetry Group*. Prentice-Hall, Englewood Clifs, New Jersay, 1962.

309. Bleaney, B. and Bowers, K. D., The cupric ion in a trigonal crystalline electric field. *Proceedings of Physical Society, London Ser A*, 1952, **65**, 667

310. Low, W., *Paramagnetic Resonance in Solids*. Academic Press, New York, 1960.

311. Jahn, H. A., *Proceedings of Physical Society A*, 1938, **164**, 117.

312. Avvacumov, V. I., Deformation of $Cr^{3+}(H_2O)_6$) and $Ni^{2+}(H_2O)_6$) complexes and the Jahn-Teller splitting of spin levels of Cr^{3+} and Ni^{2+} (Russ.). *Optika i Spectroscopiya*, 1962, **13**, 588 .

313. Dunitz, J. D. and Orgel, L. E., Electronic properties of transition metal oxides – I.Distortions from cubic symmetry. *Physical Chemistry of Solids*, 1957, **3**, 20 29.

314. Elliot, R. J., Harley, R. T., Hayes, W. and Smith, S. R. P, Raman scattering and theoretical studies of Jahn-Teller induced phase transitions in some rare earth compounds. *Proceedings of Physical Sociecty, London, Ser A*, 1972, bf 328, 217 266.

284

315. Öpik, U. and Pryce, M. H. L., *Proceedings of Physical Society A*, 1957, **238**, 425 .

316. Lier, A. D. and Ballhausen, C. J., Inherent configurational instability of octahedral inorganic complexes in E_g electronic states. *Annals of Physics (N.Y.)*, 1958, **3** 304 319.

317. Herzberg, G., Twelfth Spiers Memorial Lecture. Deterination of the structure of simple polyatomic molecules and radicals in electronically excited states. *Discussions of Faraday Society*, 1963, **35**, 7 29.

318. Bersuker, I. B. and Vekhter, B. G., The vibronic theory of ferroelectricity. *Ferroelectrics*, 1978, **19** 137 150.

319. Van Vleck, J. H., *Theory of Electric and Magnetic Susceptibilities*. Oxford University Press, Oxford, 1992.

320. Bader, R. F. W. and Bandrauk, A. D., Relaxation of the molecular charge distribution and vibrational force constant. *Journal of Chemical Physics*, 1968, **49**(4), 1666 1675.

321. Bersuker, I. B., Mutual vibronic influence of weakly coordinated molecular systems in chemical reactions and catalysis. *Chemical Physics*, 1978, **31**, 85 93.

322. Bersuker, I. B., *Teor. Eksp. Khim. [Theor. Exp. Chem. (Eng. Transl.)]*, 1978, **14**, 3 12.

323. Rajnak, K. and Krupke, W. F., Energy levels of Ho^{3+} in $LaCl_3$. *Journal of Chemical Physics*, 1967, **46**, 3532 3542.

324. Newman, D. J., Parameterization of crystal induced correlation between f-electrons. *Journal of Physics C: Solid State Physics*, 1977, **10**, 4753 4764.

325. Rajnak, K. and Wybourne, B. G., Configuration interaction in crystal field theory. *Journal of Chemical Physics*, 1964, **41**, 565 569.

326. Copland, G. M., Balasubramanian, G. and Newman, D. J., Configuration interaction contributions to the correlation crystal field in Pr^{3+}. *Journal of Physics C: Solid State Physics*, 1978, **11**, 2029 2037.

327. Newman, D. J., Bishton, S. S., Curties, M. M. and Taylor C. D., Configuration interaction and lanthanide crystal field. *Journal of Physics C: Solid State Physics*, 1971, **4**, 3234 3248.

328. Judd, B. R., Ligand field theory for actinides. *Journal of Chemical Physics*, 1977, **66**(7), 3163 3170.

329. Judd, B. R., Modifications of Coulombic interactions by polarizable atoms. *Math. Proc. Camb. Phil. Soc.*, 1976, **80**, 535 539.

330. Newman, D. J., Siu, G. G. and Fung, W. Y. P., Effect of spin-polarization on the crystal field of lanthanide ions. *Journal of Physics C: Solid State Physics*, 1982, **15**, 3113 3125.

331. Crosswhite, H. and Newman, D. J., Spin correlated crystal field parameters for lanthanide ions substituted into $LaCl_3$. *Journal of Chemical Physics*, 1984, **81**(11), 4959 4962.

332. Becker, P. J., Crystal field parameters of Ho^{3+} in YPO_4. *physica status solidi*, 1970, **38**, 379 384.

333. Eisenstein, J. C., Spectrum of Er^{3+} in $LaCl_3$. *Journal of Chemical Physics*, 1963, **39**(9), 2128 2140.

334. Siu, G. G. and Newman, D. J., Spin-correlation effects in lanthanide-ion spectroscopy. *Journal of Physics C.: Solid State Physics*, 1983, **16**, 7019 7025.

335. Newman, D. J. and Ng, B., Two-electron tensor operator expressions in crystal field theory. *Journal of Physics C.: Solid State Physics*, 1985, **18**, L803 L806.

336. Judd, B. R., Correlation crystal fields for lanthanide ions. *Physical Review Letters*, 1977, **39**(4), 242 244.

337. Newman, D. J. and Ng, B, Spin-correlated crystal field parameters for trivalent actinides. *Journal of Physics C.: Solid State Physics*, 1988, **21**, 3273 3276.

338. Yeung, Y. Y. and Newman, D. J., A new approach to the determination of lanthanide spin correlated crystal field parameters. *Journal of Physics C.: Solid State Physics*, 1986, **19**, 3877 3884.

339. Yeung, Y. Y., Newman, D. J., Orbitally correlated crystal field parameterization for lanthanide ions. *Journal of Chemical Physics*, 1987, **86** (12), 6717 6721.

340. Ng, B. and Newman, D. J., Many body crystal field calculations. I.Methods of computation and perturbation expansion. *Journal of Chemical Physics*, 1987, **87**(12), 7096 7109.

341. Ng, B. and Newman, D. J., Many body crystal field calculations. II. Results for the system $Pr^{3+}-Cl^-$. *Journal of Chemical Physics*, 1987, **87**(12), 7110 7117.

342. Rajnak, K. and Wybourne, B. G., Configuration interaction effect in l^N configurations. *Physical Review*, 1963, **132**(1), 280 290.

343. Rajnak, K. and Wybourne, B. G., Configuration interaction in crystal field theory. *Journal of Chemical Physics*, 1964, **41**(2), 565 569.

344. Garcia, D. and Faucher, M., Configuration interaction and crystal field calculation for Pr^{3+} and Nd^{3+}. *Journal de Chimie Physic*, 1989, **86**(5), 961 965.

345. Faucher, M. and Garcia, D., Crystal field simulation by full configuration interaction. *Journal of Luminescence*, 1990, **46**, 375 379.

346. Denning, R. G., Berry, A. J. and Mc Caw, C. S., Ligand dependence of the correlation crystal field. *Physical Review B*, 1998, **57**(4), R2021 R2024.

347. Faucher, M. D., Moune, O. K., Garcia, D. and Tanner, P. Evidence for strong interaction between the $5f^2$ and $5f^{17}p^1$ configurations of U^{4+} in the octahedral sites of Cs_2UBr_6 and Cs_2ZrBr_6. *Physical Review B*, 1996, **53**(15), 9501 9504.

348. Garcia, D. and Faucher, M. D., A full calculation of multiconfiguration interaction effects up to 120000 cm^{-1} (15 eV) on the ground configuration state levels of $PrCl_3$. Zeeman effect interpretation. *Journal of Chemical Physics*, 1989, **91**(12), 7461 7466.

349. Reid, M. F., Correlation crystal field analyses with orthogonal operators. *Journal of Chemical Physics*, 1987, **87**(5), 2875 2884.

Author Index

Abragam, A. 53, (12, 95)
Ahmad, S. 116, (148)
Alder, B. J. 191, (242)
Almbladh, C. O. (241)
Altshuler, S. (42)
Andersen, H. (21)
Andersen, O. K. (227)
Anderson, P. W. 167, 178, 179, (71, 75, 208)
Ashcroft, N. W. (193)
Auzel, F. (283)
Avvacumov, V. I. (312)
Axe, J. D. (141)

Baberschke, K. (181)
Bacquet, G. (36)
Bader, R. F. W. (320)
Baker, J. M. (56)
Balasubramanian, G.(326)
Balcar, E. (17)
Ballhausen, C. J. (38, 316)
Bandrauk, A. D. (320)
Barbara, B. 143, (180)
Barnes, S. E. (187)
Barnett, M. P. 74, (122)
Basch, H. (138)
Bates, C. A. 135, (172)
Becke, A. D. (276)
Becker, P. J. (332)
Beem, K. M. (43)
Berry, A. J. (346)
Bersuker, I. B. (299, 300, 318, 321, 322)

Berthier, Y. (115, 180)
Bessel, F. W. 153, 164, 165
Bethe, H. A. 53, (68)
Białynicka-Birula, Z. (266)
Białynicki-Birula, I. (266)
Biegański, Z. (19)
Birgeneau, R. J. (14, 106, 191)
Bishton, S. S. (133, 177, 327)
Bivins, R. (31, 175)
Bleaney, B. (12, 97, 309)
Bloch, F. 132, 133, 135, 203, 209
Blume, M. 109, (147)
Bohm, D. 159
Bohr, N. 160, 197, 198
Boltzmann, L. E. 168
Born, M. 231, 232, (298)
Boroński, E. (259)
Bowers, K. D. (309)
Bradley, C. J. (302)
Bramley, R. 28, (54)
Bravais, A. 150, 151
Bray, K. L. (117)
Brillouin, L. 132
Brodsky, M. B. (219)
Brooks, M. S. S. (254, 268)
Bross, H. (197)
Bucher, E. (106, 191)
Buckmaster, H. A. 12, 17, 22, (32, 47)
Buehler, R. J. 74, (123, 124)
Burns, G. (141, 152, 153)
Buzukin, S. V. 157, 158, 176, (182)
Bührer, W. (15)

Callaway, J. 199, (261)
Candela, G. A. (9)
Capitani, J. F. (257)
Carnall, W. T. (5, 281)

Cederbaum, L. S. (307)
Ceperley, D. M. 191, (242)
Chang, N. C. (282)
Chatterjee, R. (47)
Chevary, J. A. (275)
Chow, H. C. (216)
Christodoulos, F. 73, (70, 207)
Clark, M. G. (66)
Clebsch, R. F. A. 234, 236, 237, 242, 259
Cohen, M. L. 166, (200)
Coles, B. R. 166
Condon, E. 14, 108, 111
Cooper, B. R. (10, 233)
Copland, G. M. (326)
Coqblin, B. 183, (232)
Cortona, P. (262)
Coulomb, C. A. 5, 8, 46, 50, 52, 70, 73, 74, 81, 82, 83, 84, 87, 113, 133, 136, 140, 150, 154, 159, 167, 169, 173, 174, 179, 184, 186, 203, 213, 214, 247, 253, 254
Coulson, C. A. 74, (122)
Cracnell, A. P. (302)
Cromer, D.T. (89)
Crosswhite, H. (331)
Crosswhite, H. M. (281)
Curtis, M. M. (134, 135)
Czopnik, A. (59)

Daalderop, G. H. O. (271)
Daams, J. M. (270)
Dalgarno, A. (151)
Damcke, W. (307)
Da Mommio, A. (49)
Deb, B. M. (235)
Debye, P. P. 157, 161, (205)
Denning, R. G. (346)
Desclaux, J. P. (88, 104, 114)
Devine, R. A. B. (115, 180, 212, 214)
Dexter, D. L. 94, (136)
Dieke, G. H. (1)
Dirac, P. A. M. 52, 198, 199, (244)
Diviš, M. (184, 255, 278, 279)
Dixon, J. M. 37, 176, (77, 172, 176, 189, 206, 207, 210, 214, 234)
Dodds, S. A. (228)

Doniach, S. (262)
Dugas, J. (36)
Dunitz, J. D. (313)
Dupree, R. (189, 206)
Duthie, J. C. (190)

Eagles, D. M. (209, 213)
Eckart, C. 16, 136, 236, 244, 250, 260
Edelstein, N. (7)
Edmonds, A. R. (28, 102, 158, 159, 164, 165)
Ehrnsperger, M. (197)
Eisenstein, J. C. (333)
Elliot, R. J. (13, 174, 314)
Ellis, M. M. 86, 100, (107, 121, 133)
English, H. (258)
English, R. (258)
Erdös, P. (160)
Eriksson, O. (254, 268)
Eschrig, H. (196, 199, 237, 255, 256, 263, 265, 267, 279)
Escribe, C. (36)
Euler, L. 11, 20, 187, 257, 258

Faucher, M. 72, 74, (80, 105, 119, 142, 143, 144, 285, 287, 344, 345, 347, 348)
Farzan, T. O. (229)
Fähnle, M. 205, (183, 277)
Fermi, E. 147, 154, 155, 156, 157, 158, 160, 162, 165, 166, 179, 183
Fert, A. (215)
Feynman, R. P. 235, (305)
Fiolhais, C. (275)
Fischer, C. F. (87)
Fletcher, J. R. (172)
Fock, V. A. 11, 47, 52, 85, 93, 100, 109, 110, 191, 201
Forstreuter, J. (255)
Foulkes, W. M. C. (243)
Fourier, J. 132, 147, 148, 149, 150, 152, 153, 154, 156
Freeman, A. J. 86, 97, (48, 88, 104, 114, 128, 146, 149, 155)
Friedel, J. 6, 156, 165
Fung, W. Y. P. (330)
Furrer, A. (15, 16, 25)

Gaber, J. Jr (104)
Gaite, J. M. (36)
Gajek, Z. 86, 125, (80, 113, 132, 286, 290, 291, 294, 297)
Garcia, D. 72, 74, (105, 119, 142, 144, 285, 287, 344, 345, 348)
Garifullin, I. A. (229)
Gaunt, J. A. 108, 111
Gay, J. G. (171)
Gerloch, M. 216, 217, (26)
Ghatikar, M. N. (156)
Godfrey, M. J. (243)
Goldschmidt, Z. B. 213, (280)
Goodman, G. L. (5)
Gordan, P. 234, 236, 237, 242, 259
Green, H. S. 167
Gregers-Hansen, P. E. (21)
Griffith, J. S. (100)
Gross, E. K. U. 193, 194, (253)
Grossgut, P. (49)
Gruber, J. B. (282)
Gunnarson, O. 193, (187, 250)

Hardimen, M. (181)
Harding, J. H. (26)
Harley, R. T. (314)
Harrison, W. A. 181, (127, 202, 220, 221, 222, 225, 226)
Hartree, D. R. 11, 47, 52, 85, 188, 190, 191, 201
Hayes, W. (314)
Hedin, L. 197, (260)
Heer, H. (15)
Heine, V. (190, 200)
Heitler, W. 93
Hellmann, H. 235
Hellwege, K. H. (96)
Herman, F. 92, (85)
Hermite, C. 232
Herzberg, G. (317)
            item Hilbert, D. 193, 212
Hirschfelder, J. O. 74, (123, 124)
Hirst, L. L. 166, (186)
Hohenberg, P. 186, 187, 188, 189, 192, 193, 194, 195, 196, 200, 201, 202, 203, 204, (240)
Holland-Moritz, E. (18)
Hollister, C. (138)

Holm, E. (21)
Hopgood, F. R. A. (81)
Hopkins, T. A. (6)
Hornback, C. J. (138)
Hornyak, E. J. (43)
Hubbard, J. 86, 93, 97, (81)
Hummler, K. (183)
Hund, F. 8, 167, 200
Hutchings, M. T. (39, 162)
Hutchison, C. A. Jr (9)
Hutton, D. R. (52)
Hückel, E. 161, (205)
Hüfner, S. (2)

Jackson, K. A. (275)
Jahn, H. A. 227, (304, 311)
Janak, J. F. 193, (248)
Jansen, H. J. F. (264)
Johansson, B. (199, 254, 268, 269)
Johnston, D. R. (109)
Jones, D. A. 20, 21, (56)
Jonson, M. (187)
Jørgensen, C. K. 85, 220, (26, 78, 285, 287, 289)
Judd, B. R. (44, 67, 328, 329, 336)

Kang, J. H. (160)
Kaplan, I. G. (30)
Kasuya, T. 183, (23, 112, 230, 231)
Kelly, P. J. (271)
Kenny, S. D. (243)
Khaliullin, G. G. 157, 158, 176, (182, 229)
Kibler, M. (40)
Kikkert, P. J. W. (211)
Klein, M. P. (53)
Kleiner, W. H. 69, 70, (73)
Kleinman, L. (125)
Klose, W. (227)
Knight, W. D. 167
Knox, K. 86, (129)
Koepernik, K. (255)
Koelling, D. D. (270)
Kohn, W. 157, 186, 187, 188, 189, 190, 191, 192, 193, 194, 195, 196, 197, 198, 199, 200, 201, 202, 203, 204, 205, 209 (90, 170, 240, 253, 276)

290

Koster, G. F. 16, (45)
Kozyrev, B. M. (42)
Köppel, M. (307)
Kramers, H. A. 8, 26, 53, 236, 238, (69)
Kronecker, L. 4, 82
Krupa, J. C. (8, 286, 294)
Krupke, W. F. (323)
Kukovitsky, E. F. (229)
Kuriplach, J. (184, 278)
Kuzmin, M. D. (256)

Lacueva, G. (215)
Lagrange, J. 187, 203
Landau, L. D. (301)
Lander, G. H. (18, 104)
Laplace, P. S. 12, 17, 54
Lea, K. R. 26, (46)
Leask, M. J. M. 26, (46)
Leavit, R. P. (4, 282)
Legendre, A. M. 17, 54, 56, 120, 153, 174
Lenander, C. J. (103, 154)
Leung, W. K. (243)
Levy, M. (245)
Levy, P. M. 193, 196, (24, 215)
Liberman, D. A. (89)
Lieb, E. H. 193, (246)
Lier, A. D. (316)
Lindgard, P. A. (62)
Lindhard, J. 147, 155, 156
Liphsitz, E. M. (301)
Liu, Z. (279)
London, F. 93
Lovesey, S. W. (17)
Low, W. (61, 98, 310)
Löwdin, P.O. 175, 184, 261, (218)
Lundqvist, B. I. 193, (187, 250)

MacDonald, A. H. (270)
Maclaurine, C. 17, 54
Madelung, O. 85, 93, 100, 140, 150, 220, 226
Majta, J. P. (106)
Malta, O. 218, (283)
March, N. H. (235)
Margenau, H. (99)
Margolis, J. S. 86, (108)

Mc Caw, C. S. (346)
Mc Clure, D. S. (126)
Mc Lachlan, A. D. (306)
Mc Weeny, R. (79, 82, 139, 140)
Mermin, N. D. (193)
Mesot, J. (25)
Metcalf, D. H. (6)
Metropolis, N. (31, 175)
Michoulier, J. (36)
Miller, W. H. 152
Mistra, S. K. (57)
Morin, P. (179)
Morrison, C. A. (4, 163, 282)
Moskovitz, J. W. 97, (138)
Moune, O. K. (143, 347)
Mulak, J. (11, 59, 80, 118, 132, 286, 297)
Mulliken, R. 70
Murani, A. P. (192)
Murhpy, G. H. (99)

Nalewajski, R. F. (257)
Needs, R. J. (243)
Nekvasil, V. (278)
Nellis, W. J. (219)
Neumann, J. 164, 165
Newmann, D. J. 86, 88, 89, 216, 217, (27, 50, 65, 107, 121, 133, 134, 135, 284, 295, 324, 327, 330, 331, 335, 337, 338, 339, 340, 341)
Newton, I. 20
Ng, B. (65, 188, 335, 337, 340, 341)
Nielson, C. W. 16, (45)
Nieminen, R. M. (259)
Nitzsche, U. (267)
Nohl, H. (227)

Oliveira, L. N. 193, 194, (253)
Onffroy, J. R. (170)
Oppeneer, P. M. (199)
Oppenheimer, J. R. 231, 232, (298)
Orbach, R. 166
Orgel, L. E. (313)
Osborn, R. (17)
Oyamada, A. (23)
Öpik, U. (315)

Pappalardo, R. 85, 220, (64, 78)
Parr, R. G. 187, (236, 257, 276)
Passel, L. (106, 191)
Pauli, W. 33, 75, 77, 167
Pederson, M. R. (275)
Pedroza, A. C. (241)
Peierls, R. F. 109, (147)
Penney, W. G. 53, (72, 93, 94)
Perdew, J. P. 203, (272, 274, 275)
Phillips, J. C. (125)
Pierre, J. (179)
Pines, D. 159, (203)
Planck, M. K. 181
Polinger, W. Z. (299)
Pope, D. F. D. (56)
Prather, J. L. (101)
Pryce, M. H. L. 53, (95, 315)
Purvins, H-.G. (15)

Racah, G. 1, (29)
Raimes, S. (194, 204, 217)
Rajagopal, A. K. 199, (261)
Rajagopal, G. (243)
Rajnak, K. (5, 323, 325, 342, 343)
Rana, R. S. (5)
Rao, V. U. S. (20)
Rasmussen, P. G. (43)
Ray, D. K. (120, 156, 162)
Raychaudhuri, A. K. (120, 156)
Reid, M. F. (349)
Renner, R. 236
Richardson, F. S. (6)
Richter, M. 185, 206, 207, (196, 199,
        238, 255, 256, 265, 279)
Rimmer, D. E. (81)
Ritz, W. 94
Rossignol, M. F. (180)
Rotenberg, M. 16, (31, 175)
Rudowicz, C. 22, 28, (34, 37, 41, 54,
        55, 60)
Russell, H. N. 15, 252
Rydberg, J. R. 160

Sachs, M. (195)
Satten, R. A. (109)
Saunders, F. A. 15, 252
Schäffer, C. E. 220, 221, (131, 289)
Schlapp, R. 53, (72, 93, 94)

Schmidt, E. H. W. 261
Schmidtke, H. H. 85, 220, (78)
Schmitt, D. 143, (179)
Schoenflies, J. R. 24
Schreiber, C. L. (109)
Schrieffer, J. R. 179, 183, (223, 224,
        232)
Schrödinger, E. 2, 95, 107, 108, 109,
        132, 133, 163, 165, 187, 230,
        236
Schuurmans, M. F. H. (271)
Segal, E. 26, (58)
Seifert, G. (263)
Severin, L. (269)
Sham, L. J. 187, 188, 189, 190, 191,
        192, 193, 194, 195, 196, 197,
        198, 199, 200, 201, 202, 203,
        204, 205, 209, (90)
Sharma, R. R. 67, 74, (116)
Shen, Y. (117)
Shibuya, M. (86)
Shing, Y. H. (47)
Shirley, P. A. (53)
Shortley, G. 14, 108, 111
Shulman, R. G. (74, 129)
Siebert, J. F. (228)
Siemann, R. (233)
Silsbee, R. H. (228)
Silver, B. L. (63)
Singh, D. J. (275)
Siu, G. G. (330, 334)
Skillman, S. 52, (85)
Slater, J. C. 31, 43, 52, 82, 131, 133,
        134, 135, 137, 142, 170, 172,
        173, 175, 189, 193, 201, 204,
        (83, 166, 247)
Smith, D. 12, 136, 138, 169, 180, (23)
Smith, H. (21)
Smith, J. (133)
Smith, J. R. (171)
Smith, S. R. D. (314)
Sommers, C. (262)
Soulié, E. 86, (130)
Stark, J. 31
Stedman, G. E. (134)
Steinbeck, L. 205, (196, 267)
Sternheimer, R. M. 107, 108, 109,

292

111, 112, (145, 147, 150, 157)

Stevens, K. W. H. 8, 13, 14, 15, 16, 17, 18, 21, 22, 25, 49, 54, 131, 133, 145, 169, 171, 180, 207, (13, 35, 76, 97, 169, 173, 176)

Straub, G. K. (226)

Sturge, M. D. (303)

Sugano, S. (74, 129)

Sutcliffe, B. T. (79, 139, 140)

Svane, A. (273)

Synek, M. (49)

Takagi, S. (23)

Takahashi, H. 183, (230, 231)

Tanner, P. (347)

Taylor, A. D. (17)

Taylor, C. D. (327)

Taylor, R. (198)

Teller, E. 227, (304)

Theophilou, A. K. 193, (249, 252)

Thomas, L. H. 147, 154, 155, 156, 162

Thornley, J. H. M. 12, 136, 138, 169, 180, (33)

Thorpe, M. F. (174)

Touborg, P. 143, (178)

Troup, G. J. (52)

Trygg, J. 185, (239)

Turberfield, K. C. (106, 191)

Urban, W. (50)

Urland, W. (292, 293)

Van Vleck, J. H. 53, 245, (22, 91, 92, 319)

Vekhter, B. G. (318)

Vogt, O. (21)

Von Barth, U. 193, 197, (251, 260)

Vosko, S. H. (270, 275)

Waber, J. T. (89)

Wakim , F. G. (49)

Walker, A. B. (198)

Walker, E. (15)

Wallace, W. E. 26, (20, 58)

Wang, Y. L. (10)

Wannier, G. H. 132, 133, 135, 140, 169, 178, 184, (167)

Wardlaw, R. S. 37, 176, (77, 176, 234)

Watson, R. E. 86, 97, (48, 128, 146, 149, 155)

White, J. A. (22)

Wickman, H. H. (53)

Wigner, E. P. 13, 16, 136, 236, 244, 250, 259, 260

Williams, G. 166, (186)

Williamson, A. J. (243)

Wills, J. M. (254)

Wolf, P. A. (223)

Wolf, W. P. 26, 179, (46)

Wong, E. Y. (109, 154)

Wooley, R. G. (26, 51)

Wooten, J. K. Jr (31, 175)

Wybourne, B. G. 13, 14, 15, 18, 32, 49, 137, 180, 213, (3, 137, 325, 342, 343)

Yang, W. 187, (236)

Yeung, Y. Y. (284, 295, 338, 339)

Yue, W. (274)

Zener, C. 177, 183

Zhang, S. (24)

Ziesche, P. (263)

Ziman, J. M. (168)

Zunger, A. 203, (272)

Żołnierek, Z. 61, 218, 228, (110, 111)

Subject Index

actinide ions 105, 129, 163, 192, 208, 211, 212, 221, 238, 247, 253, 254

additivity of crystal field potential 7, 37, 53, 54, 56, 250, 258

adiabatic approximation (model, assumption) 231, 232, 241
potential 235-240, 242-245

admixture amplitude 5

Ag 161, 168, 176

Al 154, 156-158, 176, 182, 183

alpha expansion technique of Sharma 67, 74

analytical model of crystal field 5, 6

Anderson model (Hamiltonian) 167, 178, 179

angular excitation 108, 117
hybridization coefficient η 181
momentum 11-13, 16, 19, 23, 107, 179-181, 193, 247, 254, 259
overlap model (AOM) 3, 7, 20, 37, 85, 102, 174, 218, 219, 223-227

anharmonicity of adiabatic potential 231, 239, 244

anhydrous chlorides 253

anisotropic charge distribution 144
correlation crystal field 253

annihilation (destruction) operator 32, 136, 137, 140, 169, 178, 234

antiscreening (antishielding) 105, 106, 116, 117, 147, 158, 159, 255

antisymmetrizer $\hat{A}$ 42

Archimedean antiprism 26

asphericity of potential 152

asymptotic Bessel function 164
Neumann function 164

atomic core 11, 163, 165, 166, 174, 175
radii 181
units 1, 160

Au 157, 158, 168, 170-172, 175, 176, 182, 183

averaged central potential 110

axial field 26, 54
symmetry, $C_{\infty v}$ 3, 37, 54, 60, 217, 218, 221, 228, 239, 240, 257, 258
vector 19

azimuthal angle 20, 22, 66

Bloch function 132, 133, 135, 203, 209

Bohm-Pines unitary transformation 159

Born-Oppenheimer adiabatic approximation 185, 231
function (product) 232

bound ion 1, 2

Bravais lattice 150, 151

breathing mode 242

Brillouin zone 132

Buckmaster operator 17, 22

Buckmaster-Smith-Thornley (BST) operator 12

Buehler-Hirschfelder expansion of $1/r_{12}$ 74

C_{3h} point symmetry 129

C_{3v} point symmetry 58

C_{4v} point symmetry 56, 57

$C_{\infty v}$ point symmetry 3, 37, 54, 59, 60, 101, 217, 218, 221, 228, 239, 240, 257, 258

Ce monopnictides 183

cellular partitioning 217
central (global, main) coordinate sys-
    tem 4, 123, 128, 153, 182,
    218-220, 257
    ion 2-4, 6-8, 11, 23, 24, 91, 92,
    97, 100, 103
    potential 1, 110, 163
centrifugal potential 111
characteristic equation 19, 28
charge aureole 156
    density distribution 39, 122, 147-
    150, 177
    penetration 6, 65, 70-72, 74, 75,
    100, 103, 122
    redistribution effect 103
    transfer 5, 33, 35, 39, 91-94, 97,
    100, 102, 103, 196, 255
chemical potential 187
Cl^- ion 94
Clebsch-Gordan coefficient 234, 236,
    237, 242, 259
cluster (ionic complex) 2, 41, 49, 50,
    52, 91, 94, 98, 99, 218, 223,
    224, 257
CN^- cyanide anion 8
collective normal coordinates 233
    oscillations of electron plasma
    148, 159, 160
condensation of conduction electrons
    163, 170
Condon-Shortley coefficient 108, 111
    convention 14
conduction band states 177, 179, 183,
    184
    electrons 6, 147, 150, 158, 161,
    163, 166, 170, 174, 175
configuration interaction (CI) 31, 93,
    94, 102, 108, 109, 144, 213,
    254
contact covalency 5, 74, 84-86, 102,
    221, 248
    or exclusion contributions 84
    polarization 5, 84, 88, 89, 223
    screening 5, 84, 87
convergence factor 152
cooperative Jahn-Teller effect 241
coordination factor 4, 21, 22, 124,
    182
Coqblin-Schrieffer interaction 183
core electrons 196
correlation crystal field 96, 247, 250,
    252, 253
    effect 31, 93, 109, 116, 143, 144,
    191, 197, 247, 251, 252
    exchange contribution 207
Coulomb interelectronic interaction
    2, 8, 81-83, 133, 140, 159,
    166, 167, 169, 173, 174
    direct (J) and exchange (K) op-
    erators 70, 73, 74
    exchange energy 136, 140, 253
    interconfiguration interaction 254
Coulson-Barnett zeta function 74
covalency 4, 6, 36, 39, 85, 86, 91-94,
    96, 97, 99, 101-103, 121,
    177, 183, 221, 248, 253
covalo-electrostatic model 103
creation operator 32, 136, 137, 140,
    169, 178, 234
cross-section of inelastic neutron scat-
    tering 39
cross-matrix element 13, 94, 96-98,
    100
critical value of k in screening func-
    tion 159, 175
crystal field effect 3, 6-8, 41, 103,
    143, 147, 156, 192, 219, 241,
    245, 254, 255
    mode 117
    parameters 7, 14, 17-19, 23, 26,
    27, 33, 37, 39, 153, 157,
    158, 204-208, 211-216, 222,
    241, 247, 248, 251, 252
    potential (Hamiltonian) 8, 11,
    12, 19, 23, 26-39, 142, 143,
    148, 149, 178, 182-184, 216,
    217, 233, 247, 252, 254-258
    splitting 26, 59, 227, 233, 244,
    245
    strength parameter 215
crystal lattice potential 147, 149, 152,
    156
$Cs_2NaPrCl_6$ 226
$Cs_2NaTbCl_6$ 226

Cs_2UCl_6 129
$CsUF_6$ 222, 226
Cu 176
cubic symmetry 23, 59, 60, 168, 171,
      175, 176
cubooctahedron 25

$^1D_2$ term of f^2 configuration 252
D_{3d} point symmetry 129
d-band 183
d-electron metal 167, 177
d-electron shell (ion) 32, 163, 166,
      168, 169, 171, 173, 175, 177
d-type hybridization admixture 175
Debye frequency 157
    -Hückel theory 161
    length 161
    sphere radius 161
degenerate ground state 192, 194
degree of electron localization in vbs
      165, 166, 168, 175
density function (matrix) 43-45, 48,
      186
    functional theory (DFT) 7, 143,
      144, 148, 149, 185-187, 190,
      192, 194, 198, 199, 207, 208,
      228
    of states 149, 154, 157, 182
determinantal function 92, 98
diagonal unit tensor 127
diamagnetism 106
dielectric function $\varepsilon(\mathbf{q})$ 147, 149, 151,
      154-158
    polarization 106
differential tensor operator 123
dilute Er-Au alloy 171
dipole moment 2, 62, 63, 119, 121,
      123, 124, 126, 127, 129
    polarization contribution 129
Dirac-Fock method 52
    -Slater method 52
Dirac matrices 199
direct Coulomb interaction 73, 74,
      113, 115-117, 136, 140, 169-
      175
    operator 82
discontinuity of Fourier spectrum of
      periodic potential 151, 154

dispersion dependence of plasma os-
      cillations 160
displacement of atomic positions 216,
      229
divalent rare earths 238
Dixon-Wardlaw perturbational approach
      37, 176
double bar (reduced) matrix element
      12, 16, 236, 237, 251
    tensor operator 248-250
dynamic instability (Renner effect)
      236
    Jahn-Teller effect 235-243
    mechanism of screening in met-
      als 159, 175
e_g orbital doublet 168, 171
e_μ ($\mu = \sigma, \pi, \delta, \ldots$) parameters 3, 4,
      102
effective charge, $Z_{eff}e$ 61
    crystal field potential 6, 31, 33,
      38, 77, 79, 87, 89, 254, 255
    distance, R_{eff} 61
    Hamiltonian 3, 8, 38, 69, 84, 96,
      97, 101, 121, 204, 208, 212,
      213, 228
    potential 189, 190, 199, 201, 202,
      207, 251
eight order (or higher) crystal field
      terms 143
elastic force 236
electrical field gradient 125, 126, 167
    field intensity 126
    resistivity 2, 167
electron gas in metal 175, 188, 189,
      198, 201, 204
    -lattice coupling constant 236
    -lattice states 238
    spectroscopy 2, 39
    transfer 92, 93
    vibrational eigenstates 235, 238
electronic excitation 1, 4, 35
    spectra 212, 247, 252, 254
    term 14, 26, 31, 33, 235, 244
electrostatic contribution of atomic
      core 163, 174
    correlated crystal field 255
    crystal field Hamiltonian 121

empty-core approximation 158
energy denominator 93, 96, 102, 243, 244, 253
ensemble density 193, 194
    energy 194
EPR (ESR) 2, 20, 29, 31, 53, 102, 167, 241
equiensemble 194, 195
Er^{3+} ion 170, 171
Er^{3+}: Au system 172, 175
Er^{2+} ion 31
Euler angles 11, 20, 257, 258
Euler-Lagrange equation 187
excess magnetic moment in metals 167
exchange charge cloud 85, 100
    -correlation energy 189, 191, 192, 195, 198, 200, 201, 204
        -potential 190, 202, 205
    effect 5, 65, 68, 71, 74
    interaction 33, 36, 89, 98-100, 108, 113, 115, 117, 136, 140, 141, 144, 167, 168, 171, 177, 179, 184
excitation mode 107-109, 111, 116
exclusion contribution 84
    model 35, 77, 100
    principle 33, 75, 77, 117, 159, 167
expansion of plane wave into spherical waves 153
expectation value 44, 46, 50, 124
extended charge 103
    parameterization 250
external electronic states 166

F^k Slater parameter (integral) 1, 31, 142, 170, 172, 173, 201, 211, 248, 249, 255
f-electron ion 32, 166, 171, 177, 198, 200, 205, 207-209, 211-214, 219, 225
    -electron metal 177
    -type hybridization admixture 175
face centered cubic (fcc) lattice (matrix) 151-153, 156, 182
Fermi energy (level) 157, 158, 160, 165, 166, 179, 183

surface 160
    vector k_F 158
ferroelectricity 241
fitting procedure 17, 26, 27, 206, 213, 214, 216, 217, 225, 227, 244
force constant 239, 242-245
Fourier transform of crystal field potential 149
    lattice potential 147, 149-153, 156
fourth-order crystal field parameters 107, 144, 170, 173, 176, 183
four-vector current 199
    potential 199
fractional parentage coefficient method 250
free electron gas 154, 160
    energy 245
    ion Hartree-Fock ground state energy 93
    parameters (data) 249
free ion polarizability 103
Friedel aureole 156
    condensation of conduction electrons 6
    sum rule 165
further neighbours 33, 36, 60, 103, 217, 221-223, 227

Gaunt coefficient 108-111
Gd^{3+} ion 31, 94
generalized electrostatic model 121, 124
    gradient approximation (GGA) 204, 206, 208
    product function formalism 2, 4, 5, 41-44, 47, 48, 68, 79, 85, 92, 97, 98, 109
    two-electron direct Coulomb integral 46
    two-electron exchange Coulomb integral 46
    Wannier function 132, 133
geometric factor 18, 110
global crystal field parameter 4, 55, 217, 218, 221
    potential 7, 53, 55, 56, 103
gradient 125, 126

Green function 167
ground electron configuration 34, 39
    term 7
group product function 38, 91, 92,
    98, 100
        theoretical selection rules 242,
        244

halogenide ions 8
harmonic approximation 230, 231
    oscillator 232
harmonicity of plasma oscillations 159,
    160
Hartree energy 188
    potential 190
Hartree-Fock Hamiltonian (energy)
    85, 100, 110, 201
        method (approach) 47, 52, 93,
        109, 191
Hartree-Lindhard approximation 147,
    155, 156
heat capacity 2
Heitler-London model 93
Hellmann-Feynman theorem 235
Herman-Skillman procedure 52
Hermite polynomial 232
hermiticity of Hamiltonian (poten-
    tial) 14, 114, 249
hexagonal system 26
highly degenerate periodic system 8
    ionized actinides 238
Hilbert space 193, 212
Hohenberg-Kohn theory 186-189, 193-
    196, 200, 201, 204
Hund rule 8, 167, 200, 207
hybridization 6, 37, 144, 163, 173-
    175
hyperfine structure (field) 105, 167

icosahedron 26
identity representation 233
imaginary crystal field parameters 15,
    25, 28
indistinguishability of electrons 43,
    45, 132
induced multipole moments 119, 121,
    126, 127

inelastic neutron scattering (INS) 2,
    39, 206
inhomogeneous Schrödinger equation
    108
intensity of transition 39
interchange of electrons 249, 250
interconfiguration interaction (mix-
    ing) 254, 255
inter-electron direct Coulomb inter-
    action 1, 46, 47, 50, 148,
    159, 178
        exchange Coulomb interaction
        1, 5, 33, 46, 47, 50, 68, 70,
        167-171, 173
        interaction 1, 2, 166, 167, 174,
        188, 201, 202, 213, 214, 247,
        254
interference order 152
inter-ionic charge transfer excitation
    5, 33, 35, 39, 102
    excitations 5, 35, 102
    overlap integral 77, 78, 85, 88,
        100
intermediate crystal field case 8
internuclear repulsion 229
intra-atomic ($4f$-$5d$) crystal field ef-
    fect 168, 174, 175
    direct Coulomb integral 179
    exchange Coulomb integral 179
    excitations 33, 35
intra-configuration interaction 23
intrinsic crystal field parameters 4,
    21, 37, 55, 86, 144, 220-
    224, 227, 250, 257
    potential 3, 65
inversion 236, 255
ionic complex 2, 4, 5, 41, 91, 255,
    257
ionization energy 103
irreducible representation 168, 212,
    219, 233, 238, 242, 243

J quantum number 8, 212, 239, 252
Jahn-Teller active ion 241
    active mode 242
    distortion 239, 241
    effect (JTE) 235-243
    energy 227, 237, 238

instability 237
stabilization energy 239
structural phase transition 241
theorem 235, 236
Jones transformation 20, 21
Jørgensen-Pappalardo-Schmidtke approach 85

Kleiner correction 69, 70
$KNiF_3$ 93, 102, 103
Knight shift 167
Kohn effect 157
Kohn-Sham-Dirac equations 199
Kohn-Sham equations 190, 192, 197, 202, 203
    orbitals 189, 205
Kramers degeneracy 8, 236, 238
    doublet 8, 26
    system 238
Kronecker delta 4, 82

L quantum number 212, 252
l^2 configuration 250, 251
l^N configuration 248, 249, 252, 254
Lagrange multiplier 187, 203
lanthanide metallic materials 143
lanthanium ethyl sulphate 94
lanthanide ions 92, 96, 105, 116, 117, 129, 144, 163, 183, 212, 221, 247, 253, 254
Laplace equation 12, 17, 54
lattice contribution 207, 208
    dynamics 8, 241, 245
    potential 119, 144
    sum 148-150
    vibrations 148, 232, 233
Legendre polynomial 17, 54, 56, 120, 153, 174
Levy-Lieb functional 193
ligand 33-37, 53-63, 215-228, 248, 250, 255, 257
Lindhard method 147, 156
linear combination of atomic orbitals (LCAO) 101, 149
    E-e problem (JTE) 239, 240
linear ligator 218, 223-225
linearity of contributions 6

linear screening 106-109, 114, 116, 117
    vibronic constant 242, 243
$LiTbF_4$ 226
$l^2 LS$ term 252
$|LM>$ eigenfunctions 250
local charge density 122
    coordinate system 3, 4, 34, 127, 128, 181, 217, 219, 250, 255
    density approximation (LDA) 7, 144, 149, 181, 191, 192, 198, 202-206
    mode 235
    spin density approximation (LSDA) 198, 202, 203, 207
localization of free electrons 165-168, 175
localized electron system 8, 177, 179
long-wave approximation 147, 151, 155
LS term 254, 255

Maclaurine expansion of crystal field potential 17
Madelung constant 150
    potential 85, 93, 140
magnetic measurements 2
    moment 214, 245
    properties 198, 227, 254
    quantum number 107, 169, 178
main quantum number 7
many-configurational approximation 35, 36
    -electron crystal field 254
mapping DFT on effective Hamiltonian 192, 204, 208
matrix elements of rotation 11, 128, 257
mean radii powers $< r^k >$ 17
metallic system 6, 37, 143
Miller indices 152
Mn^{2+} ion 238
modified point charge model (MPCM) 61, 218
molecular orbital (MO) method 42, 93, 101, 102, 220, 224, 245
monoclinic system 25-29
monopole 121, 123

Monte-Carlo method 191
Mulliken approximation 70
multi-determinantal function 92
multiplicity of irreducible representation 238
multipolarity index 107, 108, 115
multipole correlated fitting procedure
28
expansion 54, 61, 63, 105, 107,
119, 120, 128
interactions 119
moments 4, 62, 63, 119, 121,
124, 126, 127
polarizability 121
mutual charge penetration 33, 144

n-electron correlations 247
NaCl 94
Nd,Er:RBa$_2$Cu$_3$O$_7$ (R=Y,La) 206
Nd:R$_2$CuO$_4$ (R=Y,La) 206
negative intrinsic parameters 144
neutron spectroscopy 2, 39, 102, 206
noble metal (Ag, Au) matrix 168
non-additive mechanisms (contributions) 37
-adiabatic mixing of states 231
-diagonal linear vibronic constants
243
-linear screening 117
-local approach 38
-orthogonality of states 1, 5, 91,
175, 184, 261
$< a|\chi >$ 34, 35, 77, 82, 101
$< \xi|\chi >$ 34, 35, 77, 83, 84, 87
$< \chi_{t_1}|\chi_{t_2} >$ 34, 35, 77, 83,
84, 88
of virtual bound state 174
-scalar coupling of tensors 248
-spherical part of crystal lattice
potential 148
-totally symmetric modes 237,
241
normal components of multipole moments 125
coordinates 232-234, 242
modes 234
normalized Stevens operator equivalents 13, 14

$U_q^{(k)}$ operator 16
norm of crystal field parameters 14
Np^{4+}:ThCl$_4$ system 226
NpO$_2$ 226
nuclear magnetic resonance (NMR)
102, 241
multipole moments 105
number of independent parameters
19, 23-28, 215, 225, 248,
250-253

O_h point symmetry 60, 61, 129
$O(3)$ symmetry group 233
occupation number representation 32,
131
octahedron 25, 59
offset character of vbs mechanism 173
one-configuration model 34, 49, 52,
65, 77, 91, 95, 135, 212,
255
-determinantal function 79, 92,
109
-electron approach 91, 102, 189
crystal field potential 3, 31,
32, 53, 143, 144
density function 44, 45
excitation 35, 91
on-site contribution 207, 208
open-core-shell treatment 149
operator equivalent method 16, 17,
21, 49
optical spectra 2, 53
orbital angular momentum operator
169, 200, 259
vibronic constants 245
orbitally correlated crystal field (LCCF)
254
orthogonality 94, 98, 99, 101, 159,
166
of Legendre polynomials 174
of 3-j symbols 113, 128
orthogonalization 8, 38, 77, 97, 261
orthogonalized plane wave (OPW) 177
orthogonalizing potential (field) 77,
83, 84, 144
procedure by Löwdin 184, 261
orthovanadates 238
oscillator force constant 239

mode 244
quantum number 232
overlap 5, 33, 37, 65, 74, 77, 78, 87,
        108, 121, 166, 174, 177, 182,
        184
    matrix 261
oxygen anion 94

p-type hybridization admixture 175
paramagnetic susceptibility 207, 245
parameterization of crystal field po-
        tential 11, 28, 31, 32, 247,
        248, 250, 254, 255
parity 18, 23, 172, 255
partial crystal field parameters 169,
        170
    PCM potential 56
partition function 8
Pauli exclusion principle 33, 75, 77,
        117, 159, 167
penetration effect 5, 33, 65, 68, 70-
        73, 168
perturbational scheme by Stevens 144
p-f mixing mechanism 183
phenomenological parameterization of
        crystal field 1, 7, 216, 219,
        250
    SCCF model 252
    screening factor 107
phase shift 165, 166
phonon spectrum 157
Planck constant 181
plasma oscillations 159, 160
plasmon 148, 159, 161, 175
point charge model (PCM) 4, 33, 53-
        55, 58-65, 68, 71, 75, 106,
        110, 121, 158, 216, 224
    contribution 124, 127, 144
point symmetry of central ion 33, 234
    group 233, 238, 242
polar angle 20, 22, 66
    vector 19
polarization 4, 6, 33, 35, 37, 91, 93,
        102, 103, 119, 121, 127-129,
        144, 200, 205-208, 241, 248,
        252, 253
polarizability 33, 103, 117, 121, 126,
        128

$PrCl_3$ 129
Pr^{3+} ion 116, 129
Pr^{3+}:$LiYF_4$ system 226
primary force constant 243, 244
projection operator technique 38, 131-
        135, 144, 201, 234
proper anharmonicity 244
pseudo-degenerate levels 235, 238
    -Jahn-Teller effect 238
    -vector 19
    -point charge model 72
pseudopotential 75, 165, 177
pseudopotential parameter 157, 158
PuO_2 226

q-space 149, 151
quadratic Jahn-Teller effect 231, 239,
        240
    vibronic constant 243
quadrupole moment 2, 62, 63, 119,
        126
    antishielding factor 105
quantization axis 23, 25, 27
quantum-mechanical tunneling 239
quasi-molecular model 233, 234

Racah coefficients 1
radial coefficient $\beta_{kq}(r)$ 32
    distribution of virtual bound state
        173
    of wave function $P_{nl}(r)$ 1, 18,
        32, 55, 62, 63, 97, 99, 108,
        111-113, 117, 153, 157, 171,
        205, 242, 244, 252, 253
radial excitation 107, 109, 117
    (breathing) mode 242
random phase approximation (RPA)
        159, 161, 175
rare earth compounds 103
    metals 143, 144, 168
reciprocal lattice vector 148-151, 156,
        157
    space 149
recoupling coefficient 260
reduced matrix element 12, 16, 236,
        237, 260
reference function (state) 42, 92, 100,
        109

system of standardization 19, 27, 28

relativistic contraction of core states 71

relativistic density functional theory 199, 207

reliability factor 214, 215

Renner effect 236

renormalization process 3, 77-87, 92, 94, 95, 101

renormalized Slater integrals 255

states 33

resonance mode 235

regime 178

state 6, 166, 178

rhombic system 25, 27, 28

Ritz variational principle 94

rotation group $SO(3)$ 12, 20

matrix 11, 20, 21

operator 11

rotational invariance 135

Russell-Saunders coupling 15, 252

Rydberg 160

s-band 183

s-d Zener interaction 183

s-type hybridization admixture 175

S-total spin moment 238, 253

S-term crystal field splitting 31

scalar convolution of irreducible tensors 234, 243

scale factor 214

scaling relation 71

scheelite matrix 29

Schmidt orthogonalization process 261

Schoenflies symbol 24

Schrieffer-Wolf transformation 179

Schrödinger equation 2, 95, 107-109, 163, 165, 187, 229, 230, 236

screened electrostatic contribution 151, 154, 157, 162, 182

screening 4, 6, 36, 37, 71, 72, 91, 102, 106, 108, 112, 114, 116, 117, 149, 152, 156-159, 208, 213, 256

constant 162

factors 107, 109, 121

function 161, 175

in metals 6, 37, 144, 147, 148, 150, 158, 170

of virtual bound state 173, 174

radius 154

second neighbour effect 103

order crystal field parameters 106, 107, 117, 144

phase transition 241

second quantization formalism 32, 131, 132, 135-137, 169, 178

secular equation 95, 97, 109, 235, 236

selection rules 101, 107

self-consistent field problem 42, 48, 148

Hartree field 147

exchange 201

self-interaction correction 149

self-consitent correction equations 203-207

semilocalization 163, 167

Sharma expansion method 67, 74

simple electrostatic model 4

overlap model 218

product 233

simulataneous diagonalization 212, 227

single particle density 149

singularity of $\varepsilon(\mathbf{q})$ 155, 156

sixth order crystal field parameters 106, 107, 144, 174-176, 183

size parameters of spin orbital, α 157, 158

Slater determinant function 43, 133-135, 137

integrals 1, 31, 82, 142, 170-173, 201, 211, 248, 249, 255

type orbital (STO) 52, 175

X_α method 189, 193, 204

Smith-Thornley operators 136, 138, 169, 180

soft mode 241

space distribution of electron density 33, 36

specific intrinsic parameters $b_l(\mathbf{q})$ 153

spectrochemical series 8, 174

spectrum of energy levels in crystal field 255

spherical angles 120

Bessel function 164, 165
components 125
harmonics (operator) 11–13, 20–22, 113, 114, 122, 157, 164
harmonic addition theorem 55, 120
Neumann function 164
tensor (operator) 2, 11-13, 23, 170, 172, 254
wave 181
spherically symmetric problem 1, 3, 163, 247
spin correlation 167
correlated crystal field (SCCF) 252-254
density 102
Hamiltonian 13, 14, 29
momentum operator 169
-orbit coupling 1, 8, 31, 159, 168, 179, 200, 211, 238
-orbital 11
polarization 200, 206-208, 252, 253
of virtual bound state 167, 168
spin polarized electrons 196
spin quantum number 178, 212, 252
split-off state from conduction band 184
stabilization Jahn-Teller energy 239
standardization of crystal field Hamiltonian 20, 27, 28
static crystal field 229, 244, 245
dielectric function 147, 153
Jahn-Teller effect 239, 240
Stark splitting 31
stationary conditions 47, 48, 50, 95, 101
Sternheimer equation 108-112
Stevens multiplication factor 17
operator equivalents 13, 14, 17, 21, 180
parameterization 171, 207
polynomial 54
strong crystal field case 8, 168
orthogonality requirement 44, 47, 52
structure factor 151

structural phase transition 220, 241
superconducting materials 157
superposition model 3, 7, 19, 21, 37, 55, 88, 127, 143, 144, 166, 174, 216, 218, 222-224, 227, 250, 253-255, 258
symbol 3-j 4, 12, 16, 62, 87, 113, 115, 116, 170-172, 181, 259
6-j 115, 116, 259
9-j 259
3n-j 8, 117, 259
symmetrical product 242
symmetry adapted system 19, 24, 27, 28
symmetry descent technique 215
symmetry properties of 3-j and 6-j symbols 116, 128

t_{2g} orbital triplet 168, 171
tensor formalism 14, 16, 49, 117
operators 31
term dependent crystal field (TDCF) 96, 248, 250-252
tesseral harmonics (operators) 12, 13, 15, 21
tetragonal symmetry 25, 28
Thomas-Fermi model 147, 154-156, 162
vector 154
three electron interaction 255
tight binding approach 181
time even operator 119
odd operator 119
reversal symmetry (inversion) 8, 18, 19
total angular momentum 11-13, 16, 254
wave function (electronic + nuclear) 230, 231
transition density function (matrix) 43-45
metal ions 238
metals 158, 159, 177
translational invariance of crystal lattice 131, 132, 149
Trees parameters, T^k 248, 249
triangle rule 23, 107, 170
triclinic system 19, 23, 27, 28

trivalent actinide ions 253
    rare earths 238, 253
two-centre expansion of $1/r_{ij}$ 73
    integrals 73
two-electron correlation crystal field 86, 213, 247-251
    Coulomb interaction 73, 98
    density function 43

U^{4+} ion 129
UBr_3 226
UCl_3 226
UCl_4 226
UF_5 226
UI_3 226
uncertainty principle 154
unimodular orthogonal group $SO(3)$ 12, 20
unitary transformation 259
unity matrix 242, 261 universal functional of $n(\mathbf{r})$, $F[n(\mathbf{r})]$ 187, 192, 194, 195
unrestricted Hartree-Fock (UHF) function 109, 117
UO_2 226
UOS 226
UOSe 226
UOTe 226
uranium compounds 86, 87
usual spherical Bessel function $j_l(kr)$ 153

valence band 183
    electrons 196
    of central ion 8
Van Vleck formula for paramagnetic susceptibility 245
variational-perturbational method 108
vector coupling coefficient 126, 259
    theory 14, 259
vibration frequency 244
vibrational level 240
    mode 232
vibronic anharmonicity 245
    constants 243, 244
    degeneracy 238, 240
    Hamiltonian 238
    interaction 230-232, 235, 241, 242

structure 245
virtual bound state 6, 37, 52, 144, 163, 166-176, 182-184
virtual ligands 217, 218, 221
von Barth-Heiden formulation od DFT 197

Wannier function 132-135, 140, 178, 184
water molecule (as ligand) 8
wave vectors $|\tau LSJM\rangle$ 212
weak crystal field case 8
Wigner coefficient 259
    -Eckart theorem 16, 136, 236, 243, 250, 260
    rotation matrix 13
Wybourne parameterization 13-15, 18, 137, 180

Yb^{3+} ion 117

Zener theory 177
zero-field splitting 14, 29, 31, 94
zero-order Slater integral 82
zero-point plasmon 148, 159

2^l-pole moment 121, 125, 128
2^k-pole crystal field component 108
$2k_F$-oscillations 156
$3d$, $4d$, $5d$, $4f$, $5f$ shells 7
3-j symbol 4, 62, 87, 113, 115, 116, 259
3n-j symbol 8, 117, 259
$4f$ open shell electrons 167-175
$4f$-$5d$ interaction (direct and exchange) 171, 174, 175
$4f$-$5d$ penetration effect 168
$5d$-electron density 169
$5d$-virtual bound state 163, 167, 168, 173
$5f$-$6d$ system 163
6-j symbol 115, 116, 259
9-j symbol 259
α, β, γ parameters of f electron system 255
$\varepsilon(\mathbf{q})$ plots 154-156
σ, π, δ, ϕ, bonds 181
v-representable 187

Printed and bound by CPI Group (UK) Ltd, Croydon, CR0 4YY

03/10/2024

01040410-0005